知味郫都饮食文化

汉英对照

A TASTE OF PIDU—DISCOVERING CULTURE THROUGH FOOD

主编◇杜 莉 陈祖明

四川科学技术出版社

图书在版编目（CIP）数据

知味·郫都饮食文化 = A TASTE OF PIDU — DISCOVERING CULTURE THROUGH FOOD : 汉英对照 / 杜莉，陈祖明主编. -- 成都 : 四川科学技术出版社, 2021.9
ISBN 978-7-5727-0264-8

Ⅰ. ①知… Ⅱ. ①杜… ②陈… Ⅲ. ①饮食—文化—成都—汉、英 Ⅳ. ①TS971.202.711

中国版本图书馆CIP数据核字(2021)第175834号

汉英对照

知味·郫都饮食文化

ZHIWEI PIDU YINSHI WENHUA

主编：杜 莉 陈祖明

出 品 人 程佳月
责任编辑 程蓉伟 张 琪
出版发行 四川科学技术出版社
装帧设计 程蓉伟
封面设计 程蓉伟
责任出版 欧晓春
制 作 成都华桐美术设计有限公司
印 刷 成都市金雅迪彩色印刷有限公司
成品尺寸 210mm × 285mm
印 张 19.5
字 数 300千
版 次 2021年9月第1版
印 次 2021年9月第1次印刷
书 号 ISBN 978-7-5727-0264-8
定 价 238.00元

地接膏腴　江流清白　东企锦城　西瞻玉垒

岷山南峙　沱水北流　天府奥区　神皋沃野

——《郫县志》

《知味·郫都饮食文化》编纂委员会

A Taste of Pidu —
Discovering Culture through Food Compilation Committee

Director
Lu Yi

Associate Directors
Xu Liang, Zhong Dazuo, Du Li, Li Xiang, Chen Zuming, Zhang Yuan

Members
Wei Zhizhong, Wang Shengqi, Wang Qian, Wang Shengpeng, Yin Chuan, Feng Fei, Liu Xiaoyi, Sun Huaiju, Li Yang, Li Hejin,
Zhang Qian, Yang Ting, Chen Long, Chen Xiao, Chen Lilan, Chen Hongguang, Chen Jing, Zhao Tianqi, Zheng Wei,
Guan Liao, Yao Shili, Yuan Antai, Cheng Rongwei, Tong Xun, Lei Yong, Zhan Songcheng, Xiong Xiaoli, Pan Wei

Editors-in-Chief
Du Li, Chen Zuming

Deputy Editors-in-Chief
Wang Shengpeng, Zhang Yuan, Zheng Wei, Chen Lilan, Yuan Antai

Chinese Writing and Proofreading
Du Li, Chen Zuming, Wang Shengpeng, Zheng Wei, Chen Lilan, Yuan Antai,
Zhang Qian, Liu Junli, He Shu, Tong Xun, Guan Liao, Chen Hongguang

English Translation and Proofreading
Zhang Yuan, Yin Chuan, Zhang Qi, Li Xiaoxiao, Hu Yin, Li Yang

Dish Design and Production
Chen Zuming, Zhan Songcheng, Chen Lilan, Ran Wei, Tong Xun, Guan Liao,
Chen Hongguang, Wang Xilin, Su Rui, Lan Li, Zhou Yuanfu, Xu Guiqiang

Photography and Dish Modeling
Cheng Rongwei, Chen Xiao, Zhan Songcheng, Feng Fei, Li Kai

Assist in Photographing
Pidu District Catering Trade Association, Yang's Chicken(Moshangrenjia),
Pidu District Photographers Association, Pidu District Fusion Media Center

A Taste of Pidu — Discovering Culture through Food Compilation Units
Chengdu Shudu Sichuan Cuisine Industry Investment and Development Co., Ltd.
Sichuan Tourism University

郫都区三道堰水乡　（杨健/摄）

郫都区位于成都平原的腹心地带，古称“郫邑”，是古蜀国望帝杜宇和丛帝鳖灵建都立国之地，因“杜宇化鹃”传说而别称“鹃城”。公元前314年秦灭蜀后，以郫邑为郡县，称“郫县”。2016年，郫县撤县而改设郫都区，属成都市管辖。郫都历史悠久，汉代思想家严君平和汉代辞赋家、思想家扬雄都是郫都人，可以说，郫都文脉传承有源、发展有根。纵观郫都两千余年来的发展历程，可谓文化昌隆，特别是饮食文化，亮点尤为突出。

说到郫都的饮食文化，首先要说郫县豆瓣。郫县豆瓣是川菜最为重要的调味品之一，也是川菜有别于其他地方风味流派的核心特征之一，被誉为“川菜之魂”。清代自福建移民而来的陈氏族人，率先用辣椒、蚕豆瓣和食盐等原料制作出“辣子豆瓣”，渐渐远近闻名，因其产自郫县，人们便称之为“郫县豆瓣”。郫县豆瓣中的辣椒经过长时间发酵，从而演变为一种辣而不燥的复合味，与单纯的辣椒相比，它可使烹制出的菜品味道更为醇厚，事实上，许多著名川菜都离不开郫县豆瓣的参与，如豆瓣鱼、回锅肉和川味麻辣火锅等。

第二要说郫筒酒。它是历史悠久的名酒，相传为西晋名士山涛出任郫县县令时所创。宋代赵抃《成都古今集记》中记载道：“成都府西五十里，曰郫县，以竹筒盛美酒，曰郫筒。”诗圣杜甫云：“鱼知丙穴由来美，酒忆郫筒不用酤。”大文豪苏轼言：“所恨蜀山君未见，他年携手醉郫筒。”陆游更赞道：“未死旧游如可继，典衣犹拟醉郫筒。”清代大美食家袁枚在其所著的《随园食单》中，也有郫筒酒的记载。由于郫都区所在地域水质优良，湿度、温度、土壤及微生物种群都有利于食品发酵，从而为郫县豆瓣和郫筒酒的诞生提供了良好的自然条件，如今，四川全兴酒厂新址落户郫都区也是有科学道理的。

第三要说农家乐。20世纪80年代，在成都周边区域出现了一种全新的餐饮形式，不仅解决了外来游客的就餐问题，也促进了当地农民的就业增收。郫都区农科村本是园艺、苗圃之乡，前来购买树苗和盆景的顾客，往往都有在当地就近解决日常饮食的愿望。在这种情况下，以种植树苗和做园艺为主业的徐家大院，率先以提供简餐方式来解决客人的吃饭难题。四川省政协老领导廖伯康、冯元蔚等也曾到此一游，廖老有感于当地的田园风光、乡土气息、吃农家菜、享农家乐，说这类餐饮店可称之为“农家乐”，于是，冯老便挥毫题写了“农家乐”三个大字，农家乐从此名扬巴蜀内外。

第四要说犀浦鲶鱼。四川方言常将鲶鱼说成鲢鱼、鲇鱼，其实，从制作原料来看，鲶鱼准确无误。犀浦鲶鱼是一道四川传统经典名菜，起源于清末，是犀浦镇三合居包席饭店的特色菜。此菜以犀浦当地的子鲶为主料，采用郫县豆瓣、大蒜等软烧而成，以肉嫩、味美、形整等著称。

郫都饮食文化底蕴深厚、内涵丰富。适逢“第四届世界川菜大会”举办之际，由四川旅游学院杜莉教授、陈祖明教授率领项目团队主编的《知味·郫都饮食文化》一书面向国内外出版、发行，是对郫都饮食文化系统而全面地梳理、传承和弘扬，将有利于促进川菜及中国饮食文化的交流与发展。期许以此为契机，进一步扛起传承文明、弘扬优秀传统文化的重任，牢固树立中华民族的文化自信。

是为序。

卢一

2021年7月29日　晨

PREFACE

Located at the heart of Chengdu Plain, Pidu District was called "Piyi (namely County of Pi)" in the ancient time. It's the land where King Wang (Du Yu) and King Cong (Bie Ling) established the ancient Shu Kingdom. It was called City of Cuckoo because of the legend that Du Yu turned into Cuckoo finally. After the Qin Empire conquered Shu in 314 B. C., it changed "Piyi" into "Pixian County" of Shu Prefecture. After more than 2,300 years of development, Pixian County merged into Chengdu City and designated as today's Pidu District in 2016. Therefore, Pidu District has a long history. The ancient thinker of the Han Dynasty Yan Junping, the essayist and thinker of the Han Dynasty Yang Xiong were both Pidu people. It can be said that there are sources and roots of the inheritance and development of Pidu culture. Throughout the history of Pidu, the culture, especially the food culture, has been developed very well.

When we talk about the food culture of Pidu, we must mention about Pixian chili bean paste firstly. Pixian chili bean paste is one of the most important seasonings in Sichuan cuisine and the core characteristics which differs Sichuan cuisine from others. It is the soul of Sichuan cuisine. In the Qing Dynasty, Chen family immigrated from Fujian Province make bean paste with chili, broad bean, water and salt, etc. Because it was famous far and near and produced in Pixian County, people called it Pixian chili bean paste. Comparing to ordinary chili, chili in Pixian chili bean paste is mellow, not pungent and has compound flavor after long time fermentation. It creates thick taste to dishes. Lots of famous Sichuan dishes are inseparable from Pixian chili bean paste, such as Chili Bean Paste Flavored Fish, Twice-Cooked Pork and Sichuan spicy hotpot.

The second is Pitong Wine. It is a famous wine with a long history. It is said that Pitong Wine was created when Shan Tao was the magistrate of Pixian County in the Jin Dynasty. *Ancient and Modern Collections of Chengdu* written by Zhao Bian in the Song Dynasty recorded, "There is a place about 25 kilometers away from Chengdu City called Pixian. People held good wine with bamboo tubes, then called the wine Pitong Wine". The Poet Sage Du Fu said, "Ya Fish in Qionglai County is delicious and Pitong Wine is excellent." The literature giant Su Shi said, "It is so regretful that you could not see the Shu Mountain. Let us drink Pitong wine together next time." Lu You praised, "I hope I could visit the old place again before death. I would like to hock my clothes for Pitong Wine." Yuan Mei, The great gourmet in the Qing Dynasty, recorded Pitong Wine in his *Food List*. Pidu District is suitable for food fermentation

due to its excellent water quality, humidity, temperature, soil and microbial population, which provide good conditions for making chili bean paste and wine. Currently, it is reasonable for Sichuan Quanxing Distillery to settle in Pidu District.

The third is happy farmhouse. In the 1980s, new catering forms appeared around Chengdu, which not only solved the dining problem of travelers, but also promoted farmers' local employment and income. Nongke Village in Pidu District is originally the hometown of horticulture and nursery. Customers who come to buy saplings and bonsai want to facilitate their meals. Hence, places doing business on horticulture and tree nursery like Courtyard of Xu' started to sell easy dishes to customers. Liao Bokang, Feng Yuanwei, leaders from the Chinese People's Political Consultative Conference of Sichuan, have once visited this place. Mr. Liao was impressed by the pastoral scenery, local flavor, farmhouse dishes and fun. He said that such restaurants could be called happy farmhouse. Mr. Feng wrote three big words "Nong Jia Le" (means happy farmhouse). Since then, happy farmhouse has become very popular in Sichuan.

The fourth is Xipu Catfish. It is a famous Sichuan dish invented at the end of the Qing Dynasty, and the special dish of Sanheju Restaurant in Xipu Town. This dish is stewed with local baby catfish, Pixian chili bean paste, garlic, etc. It is famous for its tender meat, delicious taste and complete shape.

Pidu has profound food culture. On the occasion of "The 4th Sichuan Cuisine Conference", *A Taste of Pidu — Discovering Culture through Food (Chinese and English Version)* edited by the project team led by Professor Du Li and Professor Chen Zuming of Sichuan Tourism University, was published and distributed both at home and abroad. It is a systematic and comprehensive sorting, inheritance and promotion of Pidu food culture. This book will also promote the interaction and development of Sichuan cuisine and Chinese food culture. We hope to take this opportunity to further shoulder the responsibility of inheriting and carrying forward our traditional culture, and firmly establish the cultural confidence of the Chinese nation.

That's all.

Lu Yi

Morning, July 29th, 2021

前言

天府成都，安逸郫都。郫都区位于成都平原西北部，辖区面积438平方千米，经济综合实力连续20余年进入四川省“十强”，已成为成都建设全面体现新发展理念城市的中心城区。郫都区地处都江堰灌区核心区，气候宜人、八河并流、土地肥沃，勤劳智慧的郫都人依托优势资源，不断进取、开拓创新，创造出辉煌灿烂的饮食文化，铸就了繁荣兴旺的特色美食产业，不仅满足了民众对美好饮食生活的需要，也有力地促进了第一、第二、第三产业的有机融合，乃至成都“世界美食之都”的建设和乡村振兴战略的实施。

郫都饮食文化有着丰富的内涵和深厚的底蕴，涉及历史、食材、菜点、制作技艺、美食集聚区、餐饮店、名人与饮食典故等。早在古蜀时期，望帝杜宇教民农桑，留下了杜鹃催耕的动人传说；丛帝鳖灵治理水患，开启了治水兴蜀的千秋伟业。郫都饮食文化自此开始，经过漫长岁月，不断创新发展，在清代创制出郫县豆瓣，在一定程度上促进了川菜的发展。到20世纪80年代，伴随着改革开放的东风，郫都区农科村诞生了中国第一家农家乐，成为中国农家乐旅游发源地。进入21世纪以来，郫都区率先建设了川菜产业园区，建立了与川菜文化息息相关的“三馆”，即成都川菜博物馆、中国·川菜文化体验馆和中国川菜博览馆，使郫都饮食文化更加丰富多彩。食材是饮食文化发展的物质基础，郫都出产了许多名优食材，如云桥圆根萝卜、唐元韭黄、德源大蒜、新民场生菜、郫县豆瓣等，有的已成为国家地理标志保护产品。其中，郫县豆瓣的制作技艺被列入国家级非物质文化遗产名录，还研发出系列新型复合调味品并运用于川菜创新之中。这里还建有袁隆平杂交水稻科学园，已成为国家优质水稻种植试验基地。郫都人善于兼收并蓄、包容创新，他们以本土特色食材和从外地、外国引入的食材为基础，匠心独运，烹制出一道道颇具特色的佳肴美馔，囊括冷菜、热菜、面点小吃等类别，名品有犀浦鲶鱼、郫县豆瓣鱼、苕菜狮子头、太和牛肉、唐昌板鸭等，不胜枚举。郫都人将美食与旅游有机融合，形式多样，有升级换代的乡村美食休闲游，特色农家乐遍地开花，乡村酒店如雨后春笋般兴起；有风情各异的街镇美食品鉴游，太清路美食街、唐昌大椿巷特色小吃街区等传统特色彰显，郫都区BLOCK街区、石犀里特色商业街区等现代气息浓厚，“川菜世纪”市集、犀浦夜市街区等夜市美食经济异军突起；还有魅力无穷的川菜文化深度游，川菜文化“三馆”、川菜小镇、郫县豆瓣工业旅游基地等，可充分满足广大游客观赏、体验、品鉴、研学等多样化的美食需求。此外，郫都区饮食名人辈出、群贤毕集、薪火相传，古代的扬雄、山涛等，近现代的张大千等，他们都与郫都饮食结下了深深的不解之缘，为传承、弘扬郫都饮食文化做出了重要贡献。

2021年10月，“第四届世界川菜大会”将在郫都区召开，这是郫都区饮食文化和产业发展进程中的一个新里程碑。为了系统梳理、传承和弘扬郫都饮食文化，加强郫都饮食文化的国际交流，成都蜀都川菜产业投资发展有限公司与四川旅游学院合作，开展了《知味·郫都饮食文化》一书的编撰工作。本书由四川旅游学院川菜发展研究中心牵头，组建了包括饮食文化、烹饪技术与艺术、英语翻译等方面的专家和学者，以及烹饪大师在内的项目团队，于2021年4月开始工作。团队多次赴郫都实地调研，并与相关部门、行业协会、餐饮企业家、烹饪大师及名师等座谈，搜集、整理相关资料，首先设计全书框架大纲，并在征求多方意见和建议的基础上，针对本书的名特食材、菜点、农家乐与乡村酒店、美食街、名人与典故等内容，制定了明确的入选标准和原则，并据此编撰出全书条目及初稿，再次提交学院领导、专家及各方征求意见和建议，通过反复修改、完善，最终形成定稿，并进行英文翻译。

《知味·郫都饮食文化》一书，全面、系统地梳理和总结了郫都饮食文化的整体情况，内容丰富、图文并茂、中英文对照，集学术性、实用性、趣味性于一体。全书由杜莉设计框架大纲并负责总体统筹，共分为六

章，其中，第一章、第六章主要由王胜鹏执笔；第二章、第五章主要由郑伟执笔；第三章主要由陈丽兰、陈祖明执笔；第四章主要由陈祖明负责并执笔，袁安泰、童逊、官燎、陈洪光等人也参与其中，最后，由杜莉统稿，负责审核中文稿，张茜、刘军丽、贺树校正，由张媛、尹川、张琪、李潇潇、胡音、李扬等翻译成英文并审校。本书共选录菜点100道，其中，第三章第三节以郫县豆瓣为特色调料的40道菜点，主要由陈祖明、詹淞丞、陈丽兰、徐向波、李晓、卢黎等制作；第四章的郫都60道名特菜点，主要由郫都区相关餐饮企业的厨师精心制作；图片摄影及菜点造型设计主要由程蓉伟负责并予以实施，陈筱、詹淞丞、冯飞、李凯等参与。此外，成都蜀都川菜产业投资发展有限公司为本书提供了相关资料和图片，并做了许多组织协调工作。

在本书内容的研讨和编撰过程中，不仅得到四川旅游学院、郫都区相关部门和四川科学技术出版社等相关单位领导的大力支持，而且还得到郫都区餐饮同业公会、郫都区美食推广大使团队、杨鸡肉（陌上人家）、郫都区摄影家协会、郫都区融媒体中心，以及餐饮界同仁、出版社编辑的积极配合与协作。在本书即将出版之际，对给予关心、指导、支持与帮助的各位领导、专家学者、业界同仁表示衷心感谢！同时，也对所有参与此项工作，并为此付出大量心血与宝贵时间，尤其是被占用了周末及暑假的朋友们表示衷心感谢！

郫都饮食文化内容丰富、底蕴深厚，必须进行深入、持久地研究，但因开展相关工作并编撰成书、出版的时间短暂，加之能力所限，书中难免存在不足甚至错漏之处，恳请领导、专家学者、业界同仁和广大读者不吝赐教，以便今后修订完善。

编　者

2021年8月于成都

FORWORD

Chengdu, the kingdom of abundance; Pidu, the place of comfortable. Pidu District is located in the northwest of Chengdu Plain. It covers an area of 438 square kilometers. Its comprehensive economic strength has been the top 10 of Sichuan Province in 20 consecutive years. It has already become the central urban area of Chengdu, which is comprehensively realizing the new concept of development. Located in the core of Dujiangyan gravity irrigation area, Pidu District has pleasant climate, fertile soil, and 8 parallel rivers. On the base of endowed resources, hard-working and talented Pidu people keep on moving forward and take innovative steps to create brilliant food culture and to establish prosperous food industrial, which not only meet people's requests for better food and life, but also promote the organic integration of the primary, secondary and service industries, the construction of Chengdu as the "World Gourmet Capital", as well as the realization of rural revitalization strategy.

Food culture of Pidu District has abundant context and profound basis. It is composed of history, ingredients, various dishes, cooking skill, gourmet blocks, restaurants, celebrities and food allusions. In ancient Shu, King Wang (Du Yu) taught people farming and sericulture and left the legend of cuckoo expediting tillage; King Cong (Bie Ling) started the great cause of controlling flood and prospering Shu for thousands of years. Since then, Pidu food culture has been innovating and developing for a long time. Pixian chili bean paste was created in the Qing Dynasty, which promoted the development of Sichuan cuisine. After reform and opening up in 1980s, Nongke Village in Pidu District opened the first happy farmhouse in China and become the birthplace of Chinese happy farmhouse tourism. In 21st century, Pidu District has established the Industrial Zone of Sichuan Cuisine, Sichuan Cuisine Museum of Chengdu, China · Sichuan Cuisine Cultural Experience Museum, Sichuan Cuisine Exhibition Hall of China, which make Pidu food culture more colorful. Ingredients are the material base of food culture. Lots of quality ingredients have been produced in Pidu, like Yunqiao spring radish, Tangyuan Chinese Chive, Deyuan garlic, Xinmingchang lettuce, Pixian chili bean paste, etc. All of them are National Geographical Indication Products. Among them, the producing process of Pixian chili bean paste is in the

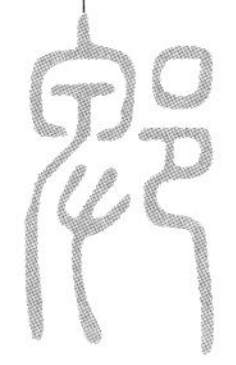

National Intangible Culture Heritage List. New compound seasonings have been invented for the innovation of Shichuan dishes too. There also exists Yuan Longping Hybrid Rice Science Park, which has become a national high-quality rice planting experimental base. Pidu people are good at incorporating things of diverse nature, and being open-minded. They created lots of special dishes including cold dishes, hot dishes, and pasty and snacks by using local characteristic ingredients, ecdemic ingredients and foreign ingredients. There are numerous famous dishes like Xipu Catfish, Chili Bean Paste Flavored Fish, Lion's Head Meatballs with Shaocai, Taihe Beef, Tangchang Flat Duck. Pidu people have combined fine food with tourism. They have updated the rural gourmet leisure tour, opened many characteristic happy farmhouses and village hotels. There are colorful gourmet streets like traditional Taiqing Road Gourmet Street, Dachun Alley Snack Block at Tangchang Town, modern Pidu BLOCK, Shixili Business and Gourmet Block, and night markets like "Chuancaishiji" Market, Xipu Night Market Block. There are charming Sichuan cuisine cultural tourism, like Sichuan Cuisine Museum of Chengdu, China • Sichuan Cuisine Cultural Experience Museum, Sichuan Cuisine Exhibition Hall of China, Lvcheng Sichuan Cuisine Town and Pixian Chili Bean Paste Industrial Tourism Base, which can meet the demands of tourists on visiting, experiencing, researching and studying. Moreover, there are many celebrities related with delicacies from generation to generation. There are Yang Xiong and Shan Tao in ancient time, Zhang Daqian in modern time. They have an indissoluble bond with Pidu delicacy and have made important contributions to inheriting and carrying forward Pidu food culture.

In October, 2021, "The 4th Sichuan Cuisine Conference" will be held in Pidu District. It is a new milestone for the development of Pidu food culture and its industry. Chengdu Shudu Sichuan Cuisine Industry Investment and Development Co., Ltd. cooperates with Sichuan Tourism University in order to systematically sort out, inherit and carry forward Pidu food culture and strengthen international exchanges, and have complied *A Taste of Pidu — Discovering Culture through Food (Chinese and English Version)* cooperatively. Sichuan Cuisine Development and Research Center of Sichuan Tourism University has built a team including experts in food culture, cooking skill and art, English translation and famous chefs. Since April, 2021, the team has carried out field study in Pidu District, interviewed related departments, guilds, entrepreneurs of catering industry and famous chefs, which have collected and sorted out rich data. First of all, the framework of the book was designed; and on the basis of soliciting various opinions and suggestions, a more clear selection criteria and principle foe the famous and special food ingredients, dishes, happy farmhouses and rural hotels, food streets, celebrities and allusions were developed. Accordingly, the entries and first draft of the book were compiled and submitted to the college leaders, experts and all parties for comments and suggestions, revised and improved to form a final draft and English translation.

A Taste of Pidu — Discovering Culture through Food (Chinese and English Version) has comprehensively and systematically sorted out and summarized the overall situation of Pidu food culture. The book is rich in content, illustrated and bilingual, which integrates academic nature, practicality and interests. Du Li designed the framework and outline of the book and was responsible for overall planning. It is divided into six chapters. The 1st and 6th chapters are mainly written by Wang Shengpeng. The 2nd and 5th chapters are mainly written by Zheng Wei. The 3rd chapter is mainly written by Chen Lilan and Chen Zuming. The 4th chapter is mainly written by Chen Zuming, while Yuan Antai, Tong Xun, Guan Liao, Chen Hongguang, etc. are also participating in writing. Du Li edited and verified the Chinese version finally. Zhang Qian, Liu Junli and He Shu were responsible for the proofreading. Zhang Yuan, Yin Chuan, Zhang Qi, Li Xiaoxiao, Hu Yin， Li Yang translated it into English and proofread the English version. 100 dishes have been chosen into this book. 40 dishes with the seasoning of Pixian chili bean paste in Section 3 of Chapter Three have been made by Chen Zumin, Zhan Songcheng, Chen Lilan, Xu Xiangbo, Li Xiao, Lu Li, etc. 60 special Pidu dishes in Chapter Four have been made by chefs of relevant enterprises of Pidu. Photography and modeling design were carried out by Chen Rongwei, with the participation of Chen Xiao, Zhan Songcheng, Feng Fei, Li Kai, etc. Moreover, Chengdu Shudu Sichuan Cuisine Industry Investment and Development Co., Ltd. has provided related materials and photos, and took the responsibility of organization and coordination.

In the process of research and compiling, this book has not only received the strong support of leaders of Sichuan Tourism University, leaders of relevant departments in Pidu District and leaders of Sichuan Science and Technology Publishing House, but also received the active cooperation of Pidu District Catering Trade Association, Pidu District Food Promotion Ambassador Team, Yang's Chicken (Moshangrenjia), Pidu District Photographers Association, and colleagues in the catering industry and editors of the publishing house. Here, we would like to dedicate gratitude to all leaders, experts, scholars and colleagues in this industry for their care, guidance, support and help. Meanwhile, we would like to express thanks to all friends who have devoted their energy and time, especially in weekends and summer vacations.

Food culture of Pidu District has abundant context and profound basis, which needs deeper and lasting research. However, due to the short time on research, compilation and publication, as well as limit of ability, there are inevitably deficiencies or even mistakes in this book. We sincerely request leaders, experts and scholars, colleagues in the industry and readers to give us advice for future revision and improvement.

Editor

Chengdu, August, 2021

目录 CONTENTS

郫都区菁蓉湖生态公园　（杨　健/摄）

第一章 CHAPTER ONE

郫都饮食 源远流长

The Distant Origin and Long Development of Pidu Food Culture

郫都区古称“郫邑”，地处成都平原腹心，位于四川省成都市的西大门，历史悠久、物产丰富，建县至今已有2 300多年。在距今2 800年前，以杜宇为首领的蜀族部落逐渐强大，进而建立起古蜀国，并把都城定于郫邑，于是，郫邑成了古蜀国政治、经济、文化交流中心。公元前314年，秦灭蜀以后，实行郡县制，始改“郫邑”为“郫县”，隶属于蜀郡。唐代仪凤二年（公元677年）于今唐昌镇置唐昌县，属益州，北宋开宝四年（公元971年）改唐昌为永昌，崇宁元年（公元1102年）改永昌为崇宁。元代时，郫县属成都路。郫、崇两县在明清时同属成都府，1914年属西川道，1935年属第一行政督察区。1958年，崇宁县撤销，大部分地区并入郫县。1968年，郫县属温江地区。1983年，郫县改属成都市，并于2016年撤县改为郫都区。

In ancient times, Pidu District was called “Piyi (namely, County of Pi)”. Located at the heart of Chengdu Plain and the western gateway of Chengdu, Pidu boasts abundant natural resources as well as a long history of some 2,300 years since being established as a county. 2,800 years ago, the Shu Tribe led by Du Yu gradually evolved into the ancient Shu Kingdom. Designated as the capital, Piyi became the political, economic and cultural center of the ancient Shu Kingdom. After the Qin Empire conquered Shu in 314 B. C., it changed “Piyi” into “Pixian County” of Shu Prefecture according to its system of prefectures and counties. In the second year of Yifeng Era during the Tang Dynasty (667 A. D.), Tangchang County of Yi Prefecture was established in today’s Tangchang Town; the name Tangchang was changed into Yongchang and Yongchang into Chongning separately in the fourth year of Kaibao Era (971 A. D.) and the first year of Chongning Era (1102 A. D.) of the Northern Song Dynasty. When it entered Yuan Dynasty, Pixian County belonged to Chengdu Lu (the administrative level roughly equal to today’s city). Pixian County and Chongning County belonged to Chengdu Fu (former Chengdu Lu) during the Ming and Qing Dynasties. In 1914, Pixian County belonged to Xichuan Dao (the administrative level roughly equal to today’s province). In 1935, it belonged to the First Administrative Supervision District (an administrative level lower than province and higher than county) of the Republic of China. In 1958, Chongning County was removed and most of it merged into Pixian County. In 1968, Pixian County belonged to Wenjiang Prefecture. In 1983, Pixian County merged into Chengdu City and designated as Pidu District in 2016.

在郫都漫长的发展历程中，得益于优越的自然条件和丰富的物产资源，辛勤的郫都人民创造了源远流长、丰富多彩的饮食文化，总体上可以分为古代、近代、现代三个阶段。

Thanks to its excellent natural conditions and abundant resources, the diligent people of Pidu created an enduring and colorful food culture in the long course of development, which can be roughly divided into three periods, namely, the ancient, modern and contemporary times.

第一节 古代郫都饮食文化

Section One Pidu Food Culture in Ancient Times

一、先秦时期：郫都饮食文化初步萌芽

根据考古成果发现，成都平原上拥有众多早于三星堆遗址和金沙遗址的史前古城遗址，其中就包括郫都的三道堰古城。该古城遗址位于郫都区三道堰，处于成都平原的腹心地带，其年代大约在距今4 500～3 700年前，总体上属宝墩文化时期。1996年、1997年和1998年，考古学家对该遗址进行了大范围发掘，发现房屋基址12座，其中有祭祀功能的大殿1座，墓葬1座，陶器有花边口绳纹罐、敞口圈足尊、盘口圈足尊等。城址及一系列石器、陶器文物都展现了新石器时代郫都先民的生活细节。可以想见，当时郫都大地上遍布森林和湖泊，而这些房屋组成了许多城邦、聚落。郫都先民们一方面使用石器、棍棒和弓箭等工具打猎、捕鱼，另一方面也在土地上开始耕种粮食，甚至还养起了猪、牛、羊、鸡、狗等家畜和家禽。人们劳作之后，使用陶器来做饭、饮水，炊烟袅袅。可以说，此时郫都饮食文化已经开始萌芽。

I. Pre-Qin Period: Pidu Food Culture Sprouting

According to archaeological findings, there are many prehistoric ancient city sites on the Chengdu Plain earlier than Sanxingdui Site and Jinsha Site, including the ancient city of Sandaoyan in Pidu. The ruins situate at today's Sandaoyan village of Pidu District in the heart of Chengdu Plain. It boasts a history of about 3,700 to 4,500 years, belonging to the Baodun Culture. In 1996, 1997 and 1998, the archaeologists conducted large-scale excavation in Sandaoyan ruins and discovered 12 home bases, among which there was 1 hall with a sacrificial function, 1 tomb and numerous potteries including rope-textured jars with lace mouth, round-bottom zun (a kind of pottery) with flat mouth, round-bottom zun with plate-shaped mouth, etc. The ruins and a series of stone artifacts and porcelains demonstrate the details of ancestors' life during the Neolithic Age. It can be inferred that Pidu was covered with forests and lakes at that time, and that the houses of ancestors constituted many city-states and settlements. On the one hand, the ancestors used stone artifacts, sticks, bows and arrows and other tools to hunt or fish; on the other hand, they already began planting grains and even feeding pigs, cows, goats, chickens, dogs and other livestock or poultry. After a day's work, they would cook and drink with porcelains, demonstrating such a tranquil picture. It can be said that Pidu food culture has begun to sprout then.

至3 000多年前，在成都平原上存在着一个可与中原夏商文明相媲美的古蜀王国。该王国初创于夏商之际，灭于战国晚期，共经历了古蜀开国之蚕丛、柏灌、鱼凫、杜宇、开明五代王朝，持续发展达1 500～1 600年之久。古蜀国第四代首领望帝杜宇将都城迁至郫邑，从山上迁至平原定居，结束了以游猎为主的生存状态，开始从事农耕渔牧的生产活动。晋代常璩《华阳国志·蜀志》载：“后有王曰杜宇，教民务农。”杜宇率民开创了按季耕种、春播秋收、圈养禽畜、养蚕织衣的农耕文明，进一步推动了古蜀农耕文化和饮食文化的发展。

其山林澤漁園囿果瓜四節代熟靡不有焉　有周
之世限以秦巴雖奉王職不得與春秋盟會君長莫同
書軌　周失綱紀蜀先稱王有蜀侯蠶叢其目縱始稱
王死作石棺石槨國人從之故俗以石棺槨為縱目人
冢也　次王曰柏灌次王曰魚鳧魚鳧王田於湔山忽
得仙道蜀人思之為立祠　後有王曰杜宇教民務農
一號杜主時朱提有梁氏女利遊江源宇悅之納以為
妃移治郫邑或治瞿上　七國稱王杜宇稱帝號曰望
欽定四庫全書　華陽國志
帝更名蒲卑自以功德高諸王乃以褒斜為前門熊耳
靈關為後戶玉壘峨眉為城郭江潛綿洛為池澤以汶
山為畜牧南中為園苑會有水災其相開明決玉壘山
以除水害帝遂委以政事法堯舜禪授之義遂禪位於
開明帝升西山隱焉時適二月子鵑鳥鳴故蜀人悲子
鵑鳥鳴也巴亦化其教而力農務迄今巴蜀民農時先
祀杜主君　開明位號曰叢帝叢帝生盧帝盧帝攻秦
至雍生保子帝　帝攻青衣雄張獠僰九世有開明帝

《华阳国志·蜀志》书影

About 3,000 years ago, there was an ancient Shu Kingdom on the Chengdu Plain comparable to the Xia and Shang Civilizations in the Central Plains. Founded in the Xia and Shang Dynasties and destroyed in the late Warring States Period, the kingdom lasted for about 1,500 to 1,600 years and experienced five dynasties, namely, Cancong, Baiguan, Yufu,

并矣惠王曰善　周愼王五年秋秦大夫張儀司馬錯
都尉墨等從石牛道伐蜀蜀王自於葭萌拒之敗績王
遯走至武陽為秦軍所害其傅相及太子退至逢鄉死
於白鹿山開明氏遂亡凡王蜀十二世　冬十月蜀平
司馬錯等因取苴與巴焉　周赧王元年秦惠王封子
通國為蜀侯以陳壯為相置巴郡以張若為蜀國守戎
伯尚強乃移秦民萬家實之　三年分巴蜀置漢中郡
六年陳壯反殺蜀侯通國秦遣庶長甘茂張儀司馬
欽定四庫全書　華陽國志
錯復伐蜀誅陳壯　七年封子惲為蜀侯司馬錯率巴
蜀衆十萬大船船萬艘米六百萬斛浮江伐楚取商於
之地為黔中郡　五年惠王二十七年儀與若城成都
周迴十二里高七丈郫城周迴七里高六丈臨邛城周
迴六里高五丈造作下倉上皆有屋而置觀樓射圃
成都縣本治赤里街若徙置少城内城營廣府舍置鹽
鐵市官并長丞脩整里闤市張列肆與咸陽同制其築
城取土去城十里因以養魚今萬歲池是也惠王二十

江別支流雙過郡下以行舟船岷山多梓栢大竹頹隨
水流坐致材木功省用饒又溉灌三郡開稻田於是蜀
沃野千里號為陸海旱則引水浸潤雨則杜塞水門故
記曰水旱從人不知饑饉時無荒年天下謂之天府也
外作石犀五頭以厭水精穿石犀溪於江南命曰犀牛
里後轉為耕牛二頭一在府市市橋門今所謂石牛門
是也一在淵中　乃自湔堰上分穿羊摩江灌江西於
玉女房下自涉郵作三石人立三水中與江神要水竭
欽定四庫全書　華陽國志
不至足盛不没肩時青衣有沫水出蒙山下伏行地中
會江南安觸山脇溷崖水脈漂疾破害舟船歷代患之
氷發卒鑿平溷崖通正水道或曰氷鑿崖時水神怒氷
乃操刀入水中與神鬭迄今蒙福　僰道有故蜀王兵
闌亦有神作大灘江中其崖嶄峻不可鑿乃積薪燒之
故其處懸崖有赤白五色　氷又通笮通汶井江徑臨
邛與蒙溪分水白木江會武陽天社山下合江又導洛
通山洛水或出瀑口經什邡郫別江會新都大渡　又

《华阳国志 · 蜀志》书影

Duyu and Kaiming. Du Yu, also called King Wang, the fourth-generation leader of ancient Shu, moved the capital to Piyi. Living at the plain changed the lifestyle of hunting to farming, fishing and herding. According to *The Records of Huayang • Chronicles of Shu Kingdom* written by Chang Qu of the Jin Dynasty, "there was a king called Du Yu, who taught his people to farm." Under the leadership of Du Yu, the kingdom entered an agricultural era. People learned how to plant grains according to seasons, feed livestock and poultry, as well as keep silkworms to make clothes, which further developed the farming culture and food culture of the ancient Shu Kingdom.

战国时期，秦惠王开始移民巴蜀。《华阳国志・蜀志》载："秦惠王封子通国为蜀侯，以陈壮为相。置巴郡。以张若为蜀国守。戎伯尚强，乃移秦民万家实之。"秦昭王时，李冰为蜀郡守，兴修都江堰水利工程，使蜀地农业生产得到迅速发展，极大地丰富了人们的食物资源。《华阳国志・蜀志》言："溉灌三郡，开稻田。于是蜀沃野千里，号为陆海。旱则引水浸润，雨则杜塞水门。故记曰：水旱从人，不知饥馑，时无荒年，天下谓之天府也。"而郫都正处于都江堰灌区上游，得益于水利工程的修建，农业生产得到了较快发展。

During the Warring States Period, King Hui of Qin began emigrating people to Bashu. According to *The Records of Huayang • Chronicles of Shu Kingdom*, "King Hui of Qin conferred leader of Zitong State as vassal king of Shu, and Chen Zhuang as the grand councilor. Then the Ba Prefecture was established, with Zhang Ruo in charge. As the King of Xirong (barbarians in the west) was still powerful, King Hui moved ten thousand households to Shu in order to consolidate the place." When King Zhao of Qin ruled, he sent Li Bing as the chief of Shu Prefecture, who organized the building of Dujiangyan irrigation system, rapidly developing the agricultural production here and greatly enriching the food resources of local people. According to *The Records of Huayang • Chronicles of Shu Kingdom*, "Dujiangyan Dam can provide irrigation water for three prefectures, where rice can be planted. Therefore, thousands of miles of rice paddies stretch in Shu, which people call 'sea on the land'. When it is dry, the dam could provide enough water for the grains; when there is too much rain, the dam could hold the water within. People all across the country call Shu Prefecture 'Tianfu (Land of Abundance)' because the irrigation is controllable thanks to Dujiangyan Dam, thus the people live here never know how it feels to be hungry." Located at the upstream of Dujiangyan irrigation area, the agriculture of Pidu developed rapidly.

二、秦汉魏晋至唐宋时期：筵宴与名酒交相辉映

（一）秦汉时期

至汉代，郫都的农业已较为发达。1966年，在郫都犀浦一座东汉砖室残墓中出土了一块残碑，其中记载了当时田舍、奴婢、耕畜等生产资料的占有情况，具有极高的考古研究价值。通过专家研究发现，到东汉时，郫都家产三十万钱到五十万钱的人已较多，犀浦残碑上的几户人资产都在三十万钱至六十万钱之间。此外，犀浦残碑上还记载了当时牛、奴婢、田产等的价值、数量。这些信息反映出汉代时郫都的农业生产已经处于较高水平，物质生活丰富，而农业的发展也为当时饮食文化的发展奠定了较好基础。20世纪

1972年郫都竹瓦铺东汉砖石墓一号石棺——宴客·乐舞·杂技石刻

70年代初，在郫都竹瓦铺汉代砖石墓中出土了多块与宴饮内容相关的画像石，从中可以一窥当时的饮食生活状况。在1972年郫都竹瓦铺东汉砖石墓一号石棺——“宴客·乐舞·杂技石刻”中，右上部为一间硬山式的厨房，内有一灶，灶上置有耳釜和甑，灶前一人正匍匐加柴，另有一厨师在案上做菜，房外有送食的侍者。画面正中是一正厅，楼上有栏，栏左设双扇门；楼下大厅内两侧席前分别置有食具，右三人正在饮酒，左二人用所执物和另一人用手指向助兴的乐舞，他们正在一边饮食，一边欣赏乐舞杂技。在1974年郫都竹瓦铺东汉砖石墓五号石棺——“饮宴·乐舞石刻”中，同样展现了一幅一边宴饮一边欣赏乐舞的场景。画面分上下两格，上格是宴饮，在长的席案上，宾主席前有碗、案等，在长席的左、右、中各有一人侍立。下格右侧有二人端坐，前有三案，其前有一鼎，似为主人正在观赏杂技。其左有耍盘、跳舞、七盘舞，左侧有乐队。同样，在1974年郫都竹瓦铺东汉砖石墓六号石棺——“饮宴”（摹本）中，除了可见宾主正在宴饮、赏乐外，还可见右下角的庖厨正在烧釜烹饪，以及厨房内悬挂着的肉类食材等。这些石刻画面所展现的宴饮情景可与郫都人扬雄《蜀都赋》中的记载相印证：“若其吉日嘉会……置酒乎荥川之闲宅，设座乎华都之高堂。延帷扬幕，接帐连冈。众器雕琢，早刻将皇。”且尚有“厥女作歌”“舞曲转节”等描述。从以上内容可以看出，汉代时的郫都，不仅筵宴的奢华程度已初具规模，而且在宴饮过程中还伴有乐舞、杂技等现场助兴活动，甚至拥有了较为专业的厨师群体，其烹饪、饮食器具种类也是丰富多样。

II. From Qin and Han, Wei and Jin Dynasties to Tang and Song Dynasties: Feasts and Wines all the Time

1. Period of Qin and Han Dynasties

Entering Han Dynasty, the agriculture in Pidu had been relatively well-developed. In 1966, a broken stone tablet was unearthed in a brick and stone tomb of the Eastern Han Dynasty in Xipu of Pidu, which records the possession situation of land, houses, servants, livestock and other production materials at that time, and is of high archaeological value. Through experts' research, it is found that in the Eastern Han Dynasty many households were able to earn 300 to 500 thousand coins in Pidu. Several households possessed assets of 300 to 600 thousand coins according to the stone tablet in Xipu. Moreover, the tablet recorded the value and number of the cows, servants, land, etc. Such information reflects the development of agriculture and material life in Pidu. A relatively-high level of agriculture laid good foundation for the development of the then food culture. In the early 1970s, several stone paintings about feasts were found at the Han Dynasty brick and stone tomb in Zhuwapu of Pidu, from which the food and lifestyle of that time could be inferred. On the stone carving of the No.1 Sarcophagus of the 1972 Pidu Zhuwapu Eastern Han Brick and Stone Tomb — "Treating the Guests, Music Dance and Acrobatics", there is a kitchen with hard hill roofs on the upper right of the carving, in which there is an stove. Above the stove there is an ear kettle and steamer. A man is adding firewood into the stove. In the meantime, a chef is cooking while a servant is waiting outside the kitchen. In the middle of the carving is the main hall. The second floor of the hall is surrounded by handrails, to the left side of which there is a double door. In the first floor, three people at the right side are drinking, while three others are pointing at the performers, in front of whom there are tableware and food for each person. These people are enjoying music, dance, and acrobatics while drinking and eating. The stone carving of the No.5 Sarcophagus of the 1974 Pidu Zhuwapu Eastern Han Brick and Stone Tomb —“Carnival • Music and Dance” shows a similar picture of drinking and appreciating the music and dance during a feast. The carving is vertically divided into two parts. The upper part describes a drinking feast, on which there are bowls and catering boards on the long banquet table and one servant at the left, center and right side respectively. On the right side of the lower part, there are two people, maybe the hosts, sitting straightly and appreciating the acrobatics. In front of them there are three catering boards and a vessel. On the left, there is a band and people performing plate-spinning and the seven-plate dance. Similarly, the stone carving of No.6 Sarcophagus of the 1974 Pidu Zhuwapu Eastern Han Brick and Stone Tomb — “Drinking Feast” (facsimiled version) demonstrates a picture where guests and hosts appreciate music and drink. At the right bottom of the carving, there is a kitchen, in which the meat and other food materials are hung and the cooks are cooking with the kettles. These occasions reflected on the stone carvings perfectly match with that recorded in the *Story of Chengdu* by Yang Xiong, a poet born in Pidu. It is depicted in the book “when people gather together in a good day, they would prepare good wine in a leisure house near the river or a proper hall in the city. The guests are so many that the host has to keep adding tables, some of which even reach to the mountains nearby. And the carved decorations are beautiful.” In the meantime, “there are girls singing” and “accompaniment along with the beat of dancing”. Through the pictures described above, the feasts in Pidu have been of certain scale in the Han Dynasty. Music, dances and acrobatics have been part of the feast. Moreover, there have been professional cooks, servants and various kinds of utensils.

（二）魏晋至唐宋时期

至魏晋南北朝时期，尽管与郫都饮食相关的文献记载和出土文物相对较少，但依然可以从三国曹操的《四时食制》和晋代左思《蜀都赋》所描述的内容中一窥当时的饮食情况。曹操在《四时食制》中记载了当时郫都出产的一种“子鱼”及制法：“郫县子鱼，黄鳞，赤尾，出稻田，可以为酱。”曹操一生南征北战，从未到过蜀国，却对郫都盛产的一种小鱼及其制作的鱼子酱十分清楚，可见当时这一美食已名声在外。对于蜀中宴饮盛况，左思《蜀都赋》言：“若其旧俗，终冬始春，吉日良辰，置酒高堂，以御嘉宾。金罍中坐，肴烟四陈。觞以清醥，鲜以紫鳞。羽爵执竞，丝竹乃发。巴姬弹弦，汉女击节。……合樽促席，引满相罚。乐饮今夕，一醉累月。”这一场景与扬雄在其《蜀都赋》中所描述的内容极其相似，无不反映出当时郫都富贵人家的豪宴之盛。

2. Period from Wei and Jin to Tang and Song Dynasties

Although there are relatively few written records and unearthed relics about Pidu diet till Wei, Jin, Northern and Southern Dynasties, we can still get a glimpse of the diet at that time from the contents described in *Arrangement of Food Throughout the Year* by Cao Cao, head of Wei State during the Three Kingdoms (namely Wei, Shu and Wu) Period, and *Story of Chengdu* by Zuo Si of the Jin Dynasty. In *Arrangement of Food throughout the Year*, Cao recorded a type of baby fish from the then Pidu and the cooking method. “Growing in paddy field, the baby fish of carp with yellow scale and red tail can be used to make sauce.” Cao, who fought on many fronts, had never been to Shu State. However, he knew clearly about the fish popular in Pidu and the sauce made with such baby fish. It is obvious that this dish was well-known at that time. In terms of the bustling scene of feasts in Shu State, there are words in *Story of Chengdu*. “Just following its old tradition, the wine is prepared in the hall on a chosen lucky day during late winter and early spring to entertain the guests. A golden and urn-shaped wine-vessel is placed in the central hall with plates full of delectable foods around the guests. The cups are filled with pure wine and the fresh seasonal vegetables are offered. The guests toast to each other with bird-shaped wine cups in hand, while the sound of music instruments recalls. Girls from Ba Prefecture play the string instrument; others from Chengdu beat the rhythm… The invited are drinking wine with each cup filled to the brim, while sitting closer to each other and enjoying themselves. People who have fun on this one night will feel hangover in the next days or even months." Such a scene is very similar to that described in *Story of Chengdu* by Yang Xiong from Western Han Dynasty, which shows the feast of the rich in Pidu.

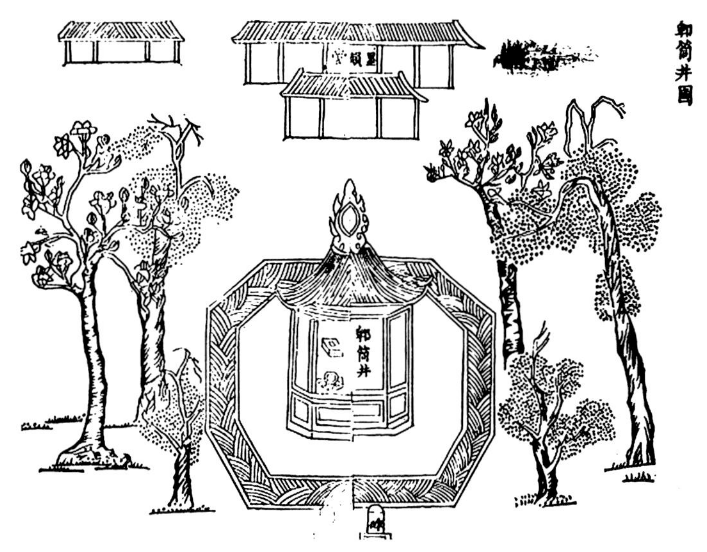

清同治年间《郫县志》所载郫都郫筒井

自秦汉时期开始，酒一直在宴饮活动中扮演着至关重要的角色，而在郫都，最为著名的则非郫筒酒莫属了。据唐代张周封《华阳风俗录》云："郫署有池，池旁有大竹，郡人刳其节，倾酿于筒，苞以藕丝，蔽以蕉叶，信宿，香达于外。然后断之，俗号郫筒酒。"唐宋时期，众多文人墨客也对郫筒酒进行了称颂。诗圣杜甫客居成都时，曾在其《将赴成都草堂途中有先寄严郑公五首》中吟到："得归茅屋赴成都，直为文翁再剖符。……鱼知丙穴由来美，酒忆郫筒不用酤。"他旅居成都时，最为称道的佳肴美酒是丙穴鱼、郫筒酒，只要忆及郫筒酒，就不再想买其他酒来喝了。宋代文豪苏轼也有诗云："所恨蜀山君未见，他年携手醉郫筒。"由此可见，当时的郫筒酒已受到众多名人雅士的广泛喜爱。

Since the Qin and Han Dynasties, wine has always been important in each feast, among which the most famous is the Pitong Wine. According to the *Records of Huayang Customs* by Zhang Zhoufeng of the Tang Dynasty, "there is a pool in Pi Prefecture, and there are big bamboos next to the pool. The local people cut a piece of bamboo tube, pour wine into it, wrap the tube with lotus stem and cover it with banana leaves. After two nights waiting, the wine smells good. Then cut the bamboo tube off and offer. It is called Pitong Wine." During the Tang and Song Dynasties, many men of letters appreciated the Pitong wine. While Du Fu, the poet-sage, lived in Chengdu, he recited in his *Five Poems Specially Written for Yan Wu on My Way to Chengdu Thatched Cottage*. "I will return to my cottage in Chengdu only because my friend Yan Wu (the former governor of Langzhou) resumed his post. The Bingxue (now Qianglai City) fish has always been delicious, and once thinking of the nice Pitong wine, I don't want to buy other wines anymore." As he traveled through Chengdu, what impressed him most was the Bingxue fish and Pitong wine. Once he thought of Pitong Wine, he never bought other wines. Su Shi, the eminent writer of the Song Dynasty, once wrote in his poem that "it is a pity that you have never been to Sichuan. If there is a chance someday in the future, we can drink the Pitong Wine till all's blue." We can see that the Pitong Wine has already been popular among the famous people at that time.

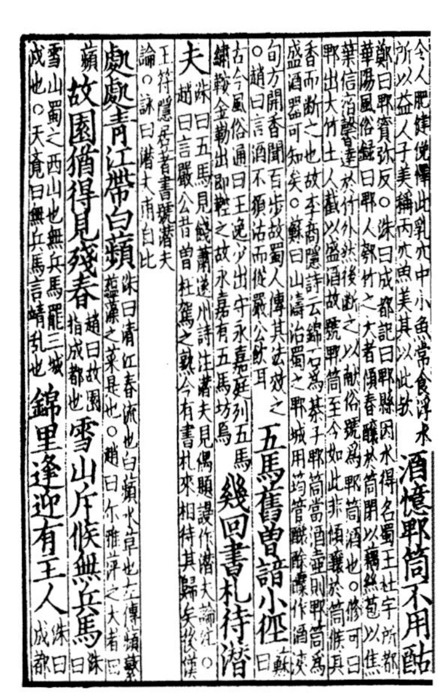
酒憶郫筒不用酤
五馬舊曾諳小徑
幾回書札待潛夫
處處青江帶白蘋
故園猶得見殘春
雪山斥候無兵馬
錦里逢迎有主人

杜甫《分门集注杜工部诗》

唐宋时期，对于郫都的饮食，杜甫、苏轼和陆游等诗文大家除了赞颂郫筒酒之外，还喜爱并记录了犀浦芋头。杜甫《赠别贺兰铦》诗言："我恋岷下芋，君思千里莼。"宋代的苏东坡和陆游也对犀浦芋头心向往之，作为一位热衷烹饪美食的著名文学家，苏轼在诗中写道“朝行犀浦催收芋，夜度绳桥看伏龙”。陆游在《晚过保福》中也说：“茶试赵坡如泼乳，芋来犀浦可专车。放翁一饱真无事，拟伴园头日把锄。”

Besides the Pitong Wine, men of letters like Du Fu, Su Shi and Lu You showed an interest in and recorded the taro in Xipu during the Tang and Song Dynasties. Du Fu once wrote in his poem, “I am fond of the taro at the foot of Minshan Mountain while you miss the water shield from Qianlihu Lkae.” Till the Song Dynasty, Su Shi and Lu You yearned for the taro in Xipu too. As a famous litterateur who is keen to cook delicious food, Su Shi wrote the poems, “I go to Xipu urging people to dig taro in the morning and go across the Flood Pacifying Cable Bridge to visit the Dragon-Taming Temple in Dujiangyan City at night”. Lu You also recited in the poem *Passing by Baofu Temple at Night*, “I am eager to taste new Zhaopo Tea (popular since the Tang Dynasty) and to bring myself the endless taro in Xipu. I am full and bored, so I want to live a farmer’s life in my garden”.

三、明清时期：郫县豆瓣助推川菜快速发展

明清时期，四川饮食在前代奠定的基础上博采众长、兼收并蓄，逐渐形成了一个特色突出，体系较为完善的地方风味菜系。而郫都生产的郫县豆瓣和当地出产的丰富食材，则成为助推川菜快速发展至关重要的物质保障。

III. Period of Ming and Qing Dynasties: Pixian Chili Bean Paste Boosting the Rapid Development of Sichuan Cuisine

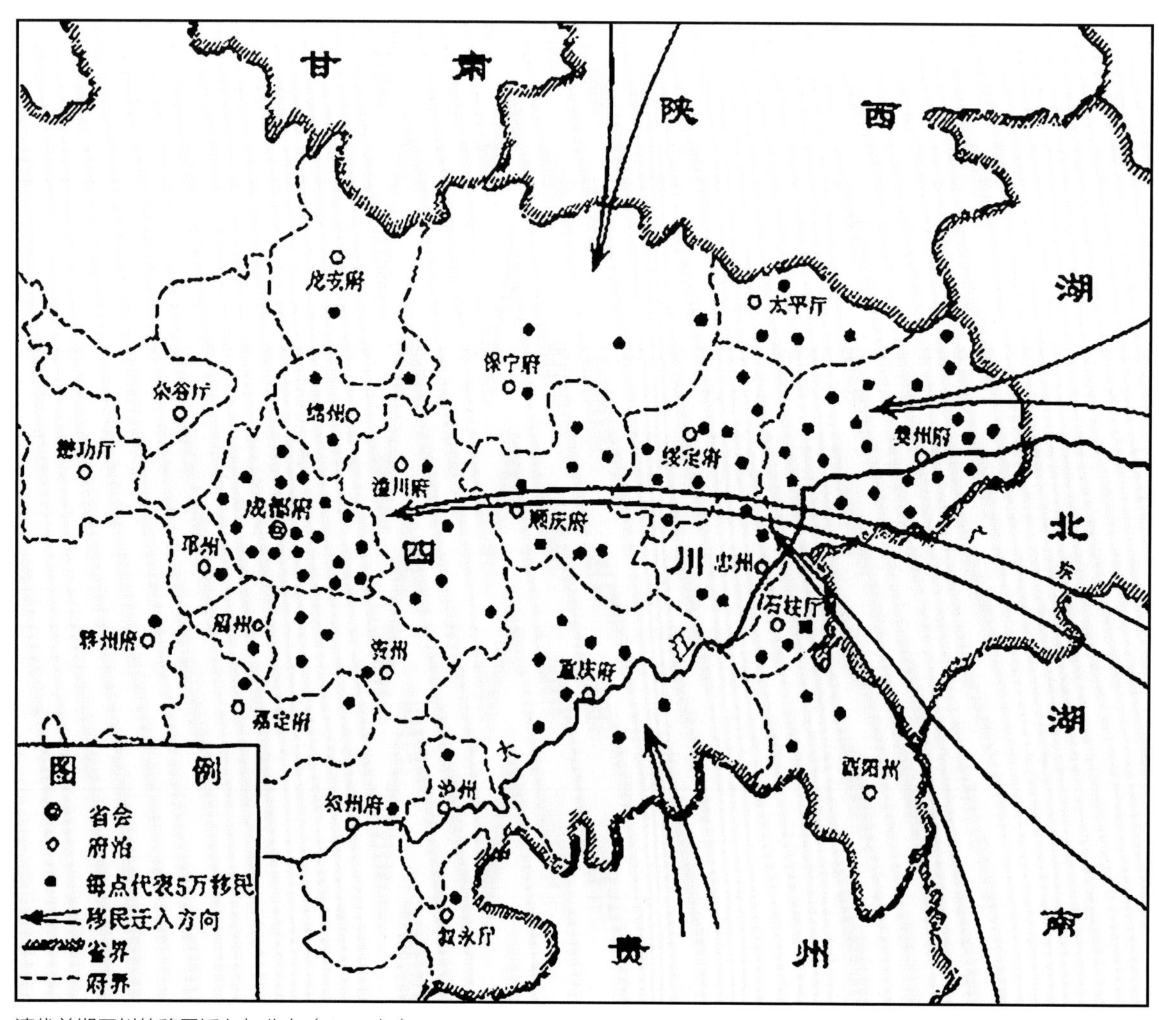

清代前期四川的移民迁入与分布（1776年）

During the Ming and Qing Dynasties, Sichuan cuisine drew on others' successful experience and developed further to form a comprehensive and typical system of local delicacy. The Pixian chili bean paste produced in Pidu as well as other ingredients became the important material guarantee to boost the development of Sichuan cuisine.

明末清初，四川饱受战乱影响，人口骤减，社会经济遭受重创。从康熙年间开始，在清朝政府移民政策的推动下，湖北、湖南等十余个省的大量移民陆续入川，由此出现了持续百年，规模空前的人口迁徙浪潮，史称“湖广填四川”。高达170多万人的大量移民，不仅带来了先进的生产技术和番薯、辣椒等新的农作物，大大丰富了四川人民的生活资料，而且还使四川地区凋敝的经济状况得以迅速恢复。其中，一部分移民来到郫都定居，在此生产、生活，从而对郫都当地的饮食方式产生了深远影响。当今著名的郫县豆瓣就是由清初一位从福建迁徙到郫都的陈姓移民所创制，并且在郫都得以不断改进和推广，最终成为川菜烹饪极为重要的一种核心调料，并成就了川菜菜系中的一些独特味型和著名菜品，如豆瓣鱼、回锅肉、麻婆豆腐等。

In the period of late Ming and early Qing Dynasties, Sichuan suffered from the wars, resulting in a steep decrease of population and heavy economic loss. Since the beginning of Kangxi Period, the immigrants from some ten provinces including Hubei and Hunan moved to Sichuan driven by the immigration policy at that time. Therefore, there occurred the trend of "migration from Hubei, Hunan and Guangdong to Sichuan" lasting for over one hundred years and involving more than 1.7 million people. The massive migration brought the advanced production technology and new crop breed, like sweet potatoes, chilies, etc., facilitating the rapid recovery of Sichuan's economy, and enriching the ingredients here. Some of the immigrants moved to and lived in Pidu and exerted a great influence on Pidu's food. The famous Pixian chili bean paste was originally created by an immigrant surnamed Chen who migrated to Pidu from Fujian in the early Qing Dynasty, and was improved later in Pidu, eventually becoming a key condiment used while cooking Sichuan cuisine and bringing some unique flavor and famous dishes, such as Chili Bean Paste Flavored Fish, Twice-Cooked Pork and Mapo Tofu.

清代郫都的人口变化情况

Population Change of Pidu during the Qing Dynasty

项目 Item / 时间 Year	户数 Number of Household	人数 Population
乾隆十年（1745年） The Tenth Year of Qianlong Period (1745)	7 174	55 998
乾隆四十六年（1781年） The Forty-Sixth Year of Qianlong Period (1781)	11 074	89 919
嘉庆十五年（1810年） The Fifteenth Year of Jiaqing Period (1810)	44 130	134 488
同治八年（1869年） The Eighth Year of Tongzhi Period (1869)	50 198	144 750

数据来源：1989年版《郫县志》

Data source: *Chronicles of Pixian* published in 1989

清代时，郫都的农业生产水平十分发达，既为民众的日常生活提供了丰富的烹饪食材，也为当地的饮食文化发展奠定了重要的物质基础。清朝四川绵州（今绵阳市）人何人鹤，在其《郫县竹枝词》中描绘当时农忙时的情景写道："纤手盘成麦草帽，硫黄熏出白如霜。五月薅秧天气热，付与郎君遮太阳"，"雾鬓风鬟水上梳，江为园圃艇为庐。一双赤足白如水，摇傍依舟卖鳜鱼"。在清嘉庆《郫县志·物产》中，所记录的谷属、麦属、蔬属、蓏属、菌属、果属等就达140余种食材。除了众多果蔬，当时的郫都还是优质稻米的重要产区："川中称为鱼米之地，而郫邑又为西蜀中枢，土性在不燥不湿之间，故其米炊饭则滋润香柔，美过他处。"一些常见食材在当时也有了多种烹饪与食用方法，比如黄豆，"郫邑种此者最多，其豆可食、可酱、可豉、可油、可腐"；再如红豆，"郫邑种此者不少，惟饼饵与馒头用之做馅饵"；另如豌豆，"可煮食，老可炒食，可磨面作粉"。在筵宴方面，据清嘉庆《郫县志》记载："郫邑绅士朴俭好礼者居多，偶逢宴会，在不丰不俭间。客至，主人迎之，子弟捧茶以献。设席，子弟侍立于侧，命之坐则坐，执爵捧壶恂恂执，弟子礼甚恭，风颇近古。"据此可见，当时郫都的民间筵宴保持着良好的礼仪、俭朴之风。

During the Qing Dynasty, Pidu's agriculture is developed. It not only provides rich cooking materials for people's daily life, but also lays an important material foundation for the development of local food culture. *Zhuzhici Poem of Pixian* by He Renhe described the scene when people were busy with farm work. "A woman is making a straw hat with her tender hands. Once fumed by alight sulfur, the hat will get white just as frost. It is in May, and her husband is eradicating the weed in the field. Then the woman takes the hat to him to block out the sun". "Woman's beautiful hair sweeps the water. The river is like her garden, and the boat is the dwelling. With her bare feet being so white, the woman rows the light boat to sell mandarin fish". There were more than 140 kinds of ingredients including grains, wheat, vegetables, various melons, edible mushrooms and fruits recorded in *Chronicles of Pixian • Products* during Jiaqing Period of the Qing Dynasty. Besides, Pidu was also the important growing area of quality rice. "The central Sichuan is called the land of fish and rice. Piyi is the main center of west Sichuan with the land not too dry or too wet. Therefore, the cooked rice produced there is softer and sweeter." Some common ingredients were cooked and eaten in a variety of ways. Soybeans "were planted the most in Piyi, which could be eaten directly, or made into sauce, fermented soy beans, soybean oil and bean curd". Red beans "were planted by many people in Piyi, which were used to make the pancake and steamed bun". And pea "could be boiled, fried and ground into flour". In terms of feast, *Chronicles of Pixian* during Jiaqing Period of the Qing Dynasty recorded. "There are many thrifty and polite gentlemen in Piyi. When they gather together to hold a feast, it would be not too sumptuous or simple. The master welcomes the guests and the servants offer the tea. The servants stand beside where the master sits and could be seated only if told, while taking the wine cup and pot in hand carefully. They are as courteous as they follow the ancient etiquette." It can be seen that the feast among the folks in Pidu maintains a style of good manners and frugality.

第二节 近代郫都的饮食文化

Section Two Pidu Food Culture in Modern Times

至近代，中国社会时局动荡不定，但四川作为大后方，却保持了相对的安定，使得郫都饮食在这一时期也得到了一定程度的发展。

Though Chinese society was in turmoil in modern times, Sichuan maintained peace and tranquility as the home front. Therefore, the catering in Pidu continued to develop during this period.

一、郫都饮食民俗丰富多彩

借助郫都近代多个时期的方志资料，可大致了解到当时郫都民间的一些日常饮食习俗。

首先，主食多为米饭，辅食多为面食。米饭以籼米为主，糯米为辅，主要做法有“捞米甑蒸”和“锅焖”两种。焖锅饭喷香可口，甑蒸饭可于甑脚下同时煮菜，各有所长。糯米在平时大多做甜味或咸味糯米饭，在一些节日则用于打糍粑、包粽子、做汤圆、蒸醪糟等。面粉作为主要辅食的食材，农村多赖以补接秋收前大米之不足，可以做馒头、夹面疙瘩、煎锅摊、烙饼子，也可擀面条、做包子、包抄手、做饺子及糕点等。玉米、高粱等杂粮在郫都的农村种植较少，多用作饲料。春荒期内，缺粮的农家多吃苕菜稀饭，当时农村长期流传的民谣形象地描述道：“三根苕菜抬颗饭，还有两根在拉纤。”

I. Rich and Colorful Food Customs of Pidu

Documents including local chronicles of Pidu in modern times throw light upon food customs among folks.

Firstly, the staple food is rice while the complementary mainly is food made of flour. The rice type primarily is long-shaped rice accompanying with sticky rice. The cooking methods include steaming rice by bamboo steamer tray and stewing rice by pot. The two methods have their own advantages. Rice stewed by pot is delicious, while other dishes can be cooked below the tray when steaming rice by bamboo steamer tray. The sticky rice is made into sweet or salty glutinous rice at ordinary times and glutinous rice cake, Zongzi, sweet soup balls and fermented glutinous rice on festival days. Flour, as the ingredient of main complementary food, fills the stomach of countrymen when rice is insufficient before the autumn harvest. It can be made into steamed bun, doughball, frying pancake and baked pancake as well as noodles, steamed stuffed buns, wontons, dumplings and cakes. Other grains such as corns and sorghum are rarely grown in rural areas, mostly used as fodders. During the spring famine, the rural households which lack grains mainly eat porridge with shaocai. The then folk songs spreading around the rural areas described it picturesquely. “Three shaocai leaves floating inside the porridge and other two are hanging on the bowl”, which means people could make porridge with only few rice grains and more shaocai.

其次，民间肉食多以猪肉为主，市场间或有牛羊肉出售。包括郫都在内的成都地区，当时的城乡民众均很少吃肉，家有雇工者，习惯于每月初二、十六各吃一次，俗称“打牙祭”。肉食中的回锅肉及连锅白肉，深受民众喜爱。此外，每逢节日及宴聚时蒸制“夹沙糕”（甜烧白），也成了城乡之地相沿成习的常规菜式。

Secondly, the meat mainly is pork as beef and mutton are sold in the market sometimes. In fact, people in urban and rural areas in Chengdu region including Pidu could hardly eat meat then. The households hiring labors used to eat meat in the second and sixteenth day of every lunar month, commonly known as “Dayaji” (have a rare sumptuous meal). Twice-Cooked Pork as well as Boiled White Radish Slices and Pork Slices in Pork Soup enjoyed tremendous popularity among people. Besides, the Steamed Pork Slices with Red Bean Paste is a regular feature in festivals or at feasts either in cities or country sides.

第三，郫都蔬菜、豆类品种繁多，田中四季常青，城乡不缺鲜菜。民众喜欢于甑脚下以米汤煮白菜，间以红萝卜、白萝卜片，称为“三下锅”；以芹菜切节炒胡萝卜丝，称为“野鸡红”；用菠菜或豌豆尖煮豆腐，称“清白传家”，这些皆为当时素菜烹饪中出场率很高的佳品。此外，将各种豆类用于菜肴烹饪也极为常见，尤以黄豆上新时制作成的豆花最受大众青睐。郫都民众向来喜食辣椒，在烹饪调味和日常佐餐中都必不可少。除新鲜蔬菜外，腌菜、盐菜、泡菜、豆瓣、豆腐乳、豆豉，因风味别具和易于贮存，也为郫都许多家庭所常备。

Thirdly, there are varieties of vegetables and beans in Pidu. Since vegetables are grown throughout the year, fresh vegetables are available both in cities and country sides all the time. People get used to boil napa cabbages, sometimes slices of carrot and radish below the bamboo steamer tray, which is called “three dishes into the pot”; stir-fry celery pieces with shredded carrots, which is called “red pheasant”; and boil spinach or tine peas with tofu, which implicates “honesty can be family heirloom”. All of these are common delicacies of vegetable dishes. Most varieties of beans can be made into dishes, among which tofu pudding made from newly harvested soybean is the most popular vegetable

1909年1月，美国芝加哥大学地质系教授托马斯·张柏林摄于郫都郊区

dishes. People of Pidu are fond of chilies, which are indispensable as daily condiments and accompaniments. Besides fresh vegetables, pickles, bean paste, fermented bean curd and fermented soybeans are regular food in households of Pidu because of their delicate flavors and easy storage characters.

第四，郫都民间盛行饮茶。民国时期，郫都人家都不备细茶，大多仅备苦丁茶、红白茶。除小康、富裕人家常备茶叶外，绝大多数民众都是用时令植物叶代替茶叶，初春时节泡折耳根（即蕺菜）为饮，初夏代之以金银花，冬至后则以桑叶代茶。早晨沏茶于棕包壶内供全天饮用，冬季置壶于“五更鸡”（一种篾丝编的罩子，罩于小油灯上）上，使茶水常温不凉。少数讲究的人家喜用“盖碗”饮茶。由于民间饮茶之习甚浓，从而在郫都各个场镇都能见到很多大大小小的茶馆，1941年前后，郫筒镇有50多家，唐昌镇有30多家，犀浦、花园各20多家。茶馆成了人们品茗、休息、会友、议事的场所，并为赶集者提供了歇脚之地，这一习俗延续至今，仍未见消减之势。

Fourthly, tea is popular among the people. During the Republican Period, most households of Pidu did not have fine teas but only Kuding tea and Hongbai tea. In addition to the well-off and wealthy families who always had tea, the vast majority of people used seasonal plant leaves instead of tea. Fish mint instead of tea leaves was used to make tea in early spring, honeysuckle in early summer and folium mori after the Winter Solstice. Tea leaves were daily supplies for well-off families. In the morning, they would make tea in a bamboo pot for a day's drink. In winter, the bamboo pot will be put in "Wugengji" (a cover made of thin bamboo strip over a small oil lamp) for heat preservation. A few picky households preferred to drink tea with covered tea bowl. There were numerous tea houses in all villages and towns. Around 1941, there were 50 tea houses in Pitong Town, 30 in Tangchang Town and over 20 in Xipu and Huayuan. The tea house has become a place for people to taste tea, rest, meet with friends and discuss official businesses. Those who went to market could stop on the way for a rest there. The custom continues today.

二、郫都调味品名品涌现

近代时，郫都酿造的调味品极具特色，主要品种有郫县豆瓣、酱油、食用醋，另有少量甜酱、豆豉、豆腐乳、酱菜等，其中尤以郫县豆瓣、犀浦酱油、崇宁红酱油最为著名。

II. Well-Known Condiment Brands of Pidu Emerging

In modern times, condiments brewed in Pidu are highly distinctive, whose major varieties include Pixian chili bean paste, soy sauce and table vinegar with a few sweet bean sauce, fermented soybean, fermented bean curd and vegetable pickled in soy sauce. The most famous ones were Pixian Chili Bean Paste, Xipu Soy Sauce and Chongning Dark Soy Sauce.

民国时期，郫都酿制豆瓣酱的酱园主要有益丰和、元丰源两家，他们互相竞争，共同发展。到20世纪40年代，益丰和已拥有近0.67公顷晒场和5000口晒缸，年利润相当于40公顷的地租收入，元丰源的利润与益丰和不相上下。两家各雇工匠近40人，年产豆瓣酱共约20万千克，均发展成为郫都工商界的实力大户。1949年，两家酱园共生产豆瓣酱13万千克。

During the Republican Period, there were two major shops brewing bean paste in Pidu, namely Yifenghe and Yuanfengyuan. They competed with each other and developed together. By 1940s, Yifenghe has owned sunning grounds of nearly 0.67 hectares and 5,000 earthenware vats, whose annual profit amounted to rent of 600 *mu* croplands. The profit of Yuanfengyuan can be compared to that of Yifenghe. Two shops hired nearly 40 people respectively, annually produced bean paste of 200,000kg and developed into important businesses in Pidu's industrial and commercial circle. In 1949, two shops produced 130,000kg bean paste in total.

民国时期，犀浦镇有德丰园、三义公与合浦园三家酱园生产犀浦酱油。德丰园约建于1901年，是创立年头最早的一家酱园，因业主"杨德丰"而得名，该酱园于1933年在成都提督街开设分号，1939年夏迁回犀浦。1943年前后，德丰园年产犀浦酱油约22.5万千克。三义公与合浦园两家酱园先后建立于1915年和1922年，从此，三家酱园相互竞争，生产工艺不断改进，质量不断提高，品种逐渐增多。

During the Republican period, three shops in Xipu including Dengfengyuan, Sanyigong and Hepuyuan, produced Xipu Soy Sauce. Defengyuan was founded in about 1901, as the earliest shop to brew soy sauce. Its name was originated from the owner Yang Defeng. The shop established a branch in Tidu Street of Chengdu in 1933, which was moved back to Xipu in summer of 1939. Around 1943, Defengyuan produced 225,000kg soy sauce per year. Sanyigong and Hepuyuan were founded successively in 1915 and 1922. From then on, three private shops continually improved their manufacturing technique, raised the quality and expanded the varieties during the intense competition with each other.

犀浦酱油老包装

20世纪初，崇宁县道生昌酱园酿制的红酱油也开始销往县外。崇宁县生产红酱油始于清朝道光年间，到20世纪三四十年代，已与郫县豆瓣、温江白酱油齐名，远销重庆、北京、上海等地。

In the early 20th century, dark soy sauce brewed by Daoshengchang Shop of Chongning County was sold outside the county. Since Daoguang era of Qing Dynasty, Chongning began to produce dark soy sauce. During the 1930s to the 1940s, as well-known as Pixian chili bean paste and Wenjiang white soy sauce, it was sold to Chongqing, Beijing and Shanghai.

三、郫都餐饮业的逐渐兴盛

步入近代，郫都的饮食业得到进一步发展，在郫筒镇、唐昌镇等地汇聚了大量的餐饮店铺，同时还出现了许多地方特色极为鲜明的风味菜点。

1909年1月，美国芝加哥大学地质系教授托马斯·张柏林摄于郫都路边饭馆

III. Catering Industries of Pidu Gradually Prospering

The catering industry of Pidu has developed since the modern times. Various food stores gathered in Pitong Town and Tangchang Town as typical local dishes with distinctive characteristics emerged.

（一）饮食店铺

1989年版的《郫县志》，记载了民国时期郫筒镇、唐昌镇与饮食相关的商业情况。

1. Food Stores

Chronicle of Pixian published in 1989 recorded commercial situation of food stores in Pitong and Tangchang during Republican period.

1. 郫筒镇

西街：干菜业8家，酒店10余家，京果糖食业5家，“三星居”“小花园”“敬堂茶园”等茶馆数家；西外街有饮食业7家（内有回民店1家）。南街：干菜业11家，酒店10余家，糖食业1家，面食业1家；南外街有面食业7家。东街：面食业14家，酒店14家，饮食业4家，干菜业4家。北街：面食业70余家，酒店10余家，干菜业6家，京果糖食业3家。此外，还有茶肆业50余家、肉案业50余家、蔬菜业70余家。

1. Pitong Town

West Street: 8 dried vegetable stores, over 10 restaurants, 5 confectionery stores and several tea houses like “Sanxingju”, “Xiaohuayuan” and “Jingtang Tea Garden”;7 restaurants (including 1 store owned by a Hui person) outside the West Street. South Street: 11 dried vegetable stores, over 10 hotels, 1 confectionery store and 1 cooked wheaten food store; 7 cooked wheaten food stores outside the South Street. East Street: 14 cooked wheaten food stores, 14 hotels, 4 restaurants, 4 dried vegetable stores. North Street: 70 cooked wheaten food stores, over 10 hotels, 6 restaurants, 3 confectionery stores. Besides, there were over 50 tea stores, 50 meat stores and 70 vegetable stores.

2. 唐昌镇

北街：糖果糕点业2家，干菜业3家，饮食业15家。东街：干菜业4家，茶肆业7家，专办筵席饮食业2家。南街：多设摊零售黄糖、挂面。西街：饮食业10家，糕点业2家，茶肆业4家，肉案业4家。

2. Tangchang Town

North Street: 2 confectionery stores, 3 dried vegetable stores and 15 restaurants. East Street: 4 dried vegetable stores, 7 tea stores and 2 catering service stores specialized on holding feasts. South Street: mainly brown sugar and fine dried noodles in retail. West Street: 10 restaurants, 2 cake and pastry stores, 4 tea stores and 4 meat stores.

（二）地方特色风味菜点

在当时郫都的饮食店铺中，出现了许多著名的地方风味菜点，“江西馆”的豆花素饭，“耗子洞”的二分白肉，犀浦“三合居”的红烧鲶鱼，“花园场”的罐罐肉、施鸭子等。

2. Local Specialties

There were lots of well-known local specialties in food stores of Pidu, such as Boiling Tofu with the Five Spices of the restaurant “Liuxiangjiu (which means permanent fragrance)”, Tofu Pudding with Rice of Jiangxi Assembly Hall, Pork with Mashed Garlic of “Haozidong (which means the mouse hole)”, Braised Silver Carp with Soy Sauce of “Sanheju” in Xipu as well as Stewed Pork in Pot and Spiced Duck of the Shi’s in Huayuanchang.

江西馆的“豆花素饭”是县城江西会馆山门外一家专卖豆花饭的馆子，其所作豆花质地鲜嫩而又不失劲道，卤酱麻、辣、鲜、香，诸味调和，所售小菜品种丰富，尤以凉拌黄丝、酱味胡豆和泡甜椒最得食客赞誉。耗子洞位于县城西街，因门面窄小、内堂宽大而被戏称为“耗子洞”。该店善做一种经济名菜，叫作“二分白肉”（因清末民初时，仅用二分白银便可购得一份白肉，故得此名）。犀浦三合居则以红烧鲶鱼闻名，此菜选料上乘、烹制讲究，以色鲜、肉嫩、味美、形整而著称。花园场的罐罐肉则由花园场南巷子的李绍南店经营，此菜肥而不腻、味美可口，日销高达两百余罐，尤其是县城内外的行商坐贾均喜尝此品，甚至有专为尝鲜而来者。此外，民国时期县城衙门外的酥肉豆花、豆汤肥肠粉、牛肉荞面，乃至走街串巷的担担面、烧卖、油酥、拉丝饼、米花糖和部分场镇的油烫鸭，也深受大众喜欢。

“Tofu Pudding with Rice” of Jiangxi Assembly Hall actually is the name of a tofu pudding restaurant outside the Jiangxi Assembly Hall’s gate. The tofu pudding is fresh and tender and its dipping sauce is spicy and and flesh. The restaurant also sold a diversity of snacks, among which Dried Tofu Strips Salad, Lima Bean with Sauce Flavor and Pickled Pepper are the most popular among customers. “Haozidong” was located in the West Street of the county. Because the facade is narrow and the cella is broad, it was called this name. The store is good at making a cheap dish, called Two-Fen Pork with Mashed Garlic (a white meat can be bought at two fen taels in the late Qing Dynasty and the early Republic of China). “Sanheju” in Xipu was famous for its Braised Silver Carp with Soy Sauce. The raw material is selected, cooking method is delicate and the whole dish is well-known for good looking, tender meat, great flavor and tidiness in form. Being thick but not too oily, the Stewed Pork in Pot was a specialty dish of Lishaonan’s Store in South Alley, Huayuanchang. More than 200 pots of pork could be sold in one day. Many people, especially the merchants nearby, were fond of it. Some of them even took a detour simply for a bite. Moreover, the Crispy Pork Slice with Tofu Pudding, Rice Noodles with Intestines in Bean Soup, Beef Buckwheat Noodles were sold just in front of the county government office in the Republican period. Dandan Noodles (Sichuan noodles with peppery sauce), Shao-mai (a steamed snack with rice in it, looks like dumplings), fried cake, cake in spun syrup, crunchy rice candy could be found across the streets. In some towns and villages, oil scorched duck was widely praised.

第三节 现代郫都饮食文化

Section Three Pidu Food Culture in Contemporary Times

中华人民共和国成立以后，郫都的社会经济有了极大发展，与之密切相关的郫都饮食业也进入繁荣创新时期。1962年，先后成立了郫县糖业烟酒公司、郫县蔬菜水产饮食服务公司、郫县食品公司等国营商业服务企业，各自开展不同的饮食商品经营活动。其中，郫县糖业烟酒公司主要经营糖食、糕点、酒类等商品。郫县蔬菜水产饮食服务公司最初主要经营干鲜蔬菜及海产品，同时管理郫筒、唐昌两镇的饮食服务行业，1964年改建为郫县饮食服务总店，原干鲜蔬菜和海产品等业务转由郫县糖业烟酒公司经营，1965年内设蔬菜站，开始经营蔬菜。1976年，开始经营水产品，复名郫县蔬菜水产饮食服务公司。1984年分设为郫县蔬菜水产公司和郫县饮食服务公司两家。郫县食品公司主要负责肉食与禽蛋的收购、供应和外调。此外，在个体饮食经营方面，1956年前后，全县共有个体商业户2 263户，其中，饮食服务业785户，约占个体户总数的三分之一。其后，全县个体商业户有所减少。

Since the foundation of the People's Republic of China, the social economy has developed a lot. As a result, catering business closely related to economy were becoming increasingly prospered and innovative and both state-owned and private catering businesses developed well. In 1962, state-owned business service enterprises like Pixian Sugar, Tobacco and Alcohol Company, Pixian Vegetable and Aquatic Food Service Company and Pixian Food Company were founded one after another, operating different kinds of food and beverage businesses. Among them, Pixian Sugar, Tobacco and Alcohol Company majored in confectionery, cakes and pastries as well as liquor and alcoholic beverage. Pixian Vegetable and Aquatic Food Service Company at first majored in dried and fresh vegetables and marine food products, managing catering services industry in Pitong and Tangchang. In 1964, it was converted as Pixian head office of food service and its original vegetable and dry and fresh aquatic products businesses were transferred to the sugar and wine company; in 1965, it set up the vegetable station and operated vegetable businesses; in 1976, it started to operate aquatic product businesses, renamed as Pixian Vegetable and Aquatic Food Service Company; 1984, it established Pixian Vegetable and Aquatic Food Company and Pixian Food Service Company separately. Pixian Food Company was mainly in charge of acquisition, supply and transfer of meat and poultry eggs. Besides, in terms of private catering businesses, there were 2,263 individual sellers in the county around 1956, among which there were 785 catering service companies accounting for one third of the total. Later, individual sellers started to decrease.

从1978年至今，尤其是进入21世纪后，随着改革开放的不断深入，社会、经济、科技和文化的高速发展，以及全社会生活水平的不断提高，人们对饮食的需求也随之日益增长，从而推动了郫都饮食文化的跨越式发展，主要呈现为以下五个特征。

Since 1978, especially in the beginning of the 21st century, as reform and opening up deepened, society, economy, technology and culture rapidly developed at top speed and people's living standards and their demands for food consistently improved, Pidu's food culture has realized a leapfrog development, which mainly manifests 5 characteristics as following:

一、以农家乐为代表的特色餐饮市场快速发展

1978年改革开放以后，个体商业及餐饮服务业逐渐复苏。1980年，全县共发放个体商业营业执照187户，其中饮食服务业53户。1985年，全县个体商业户增至3 936户，饮食服务业达1 027户，占个体商业总

郫都农科村宣传画

户数的26%。犀浦五粮村有10多户农民，利用住房紧靠成灌公路的优势开设“豆花饭店”，取得可观的经济效益。此时，全县有近200余户农民专门经营鸡、鸭、鹅、兔和猪头等腌卤食品，有200余户农民在各场镇或交通要道经营粽子、糍粑、锅盔、凉面、凉粉、黄糕、凉糕、豆浆馍馍、黄水馍馍、玉米馍馍、珍珠馍馍、担担豆花等小吃，填补了饮食服务业的空隙，活跃了集市贸易，既方便了人民生活，也推动了当地的经济和社会发展。

I. The Special Catering Market Represented by Happy Farmhouse Developing Rapidly

Since the reform and opening up in 1978, individual commerce and catering service business has gradually come to life. In 1980, 187 individual business licenses were issued in Pixian, among which 53 were issued to catering service businesses. In 1985, individual sellers increased to 3,936, among which 1,027 were catering service businesses accounting for 26%. In Wuliang Village of Xipu, a dozen peasants opened “Tofu pudding restaurants”, taking the advantage of their house location which is near Chengdu-Dujiangyan Expressway to gain considerable economic benefits. Then, almost 2,000 households of peasants specialized in pickled and marinated chicken, duck, goose, rabbit and pork head. About 200 households of peasants sold snacks like Zongzi, glutinous rice cake, Guokui (a kind of crusty pancake), cold noodles with hot sauce, steamed cake with millet, cold rice cake, steamed bread made from bean flour, steamed bread made from prosomillet flour, steamed bread with corn flour, leaf warped rice and tofu pudding. They filled the void of catering services industry, made fair trade in villages more active, greatly facilitated people’s life and promoted local economy and social development.

20世纪80年代末90年代初，在郫都诞生了引人瞩目的中国第一家“农家乐”，这种将饮食与旅游相结合的全新模式，对郫都餐饮业的后续发展，产生了极为深远的影响。

From the end of 1980s to the beginning of the 1990s, the first “happy farmhouse” in China was born in the catering service industry of Pidu, which is quite remarkable. It is a new type of catering business integrating food with tourism, which has a far-reaching influence.

1987年，地处“天府之国”腹心地带的郫都农科村适时调整种植结构，农户利用川派盆景、苗圃优势大力发展花卉种植，大量的苗木、花卉种植，造就了农科村绿树成荫、群芳争艳的美丽景象，成为“鲜花盛开的村庄，没有围墙的公园”，吸引了众多游客前来游玩、赏花。以徐纪元为代表的村民，逐渐发现了其中潜藏的商机，率先销售农家菜为游客们提供餐食服务，经济效益显而易见，其他村民也开始纷纷效仿。于是，徐家大院成了中国第一家“农家乐”，这一举措，也让农科村成了中国农家乐的发源地。此后，郫都区的农家乐如雨后春笋般不断涌现，影响范围不断扩大，以至于影响到四川各地乃至全国，各地又因地制宜，进而衍生出渔家乐、牧家乐等。这一时期的郫都，还出现了一些知名餐厅，如20世纪80年代后期，望丛祠内的望从餐厅开始引人瞩目，到90年代初期，其示范效应逐渐向周边发散：桃园宾馆、仙客来宾馆、镜湖园宾馆、普莱斯宾馆、欧罗巴酒店、明园饭店、花园别墅酒楼等，先后成为时尚美食之星。

In 1987, Nongke Village in Pidu, located at heartland of “the land of abundance”, readjusted the structure of plants in time. The local peasant households made use of Sichuan-styled bonsai and nursery garden advantages to grow flowers and plants, which embellished the village with luxuriantly green trees and fragrant flowers. It has become “the village of blossoms, a park without any fence”, which attracted many tourists to have a visit and admire the beauty of flowers. Some peasant households with Xu Jiyuan as a representative gradually discovered the potential commercial opportunity and they provided meals for tourists with farm vegetables. Therefore, the courtyard of the Xu’s has become the first “happy farmhouse” in China. As other villagers imitated his pioneering undertaking, Nongke Village became the cradle of China’s happy farmhouses. From then on, happy farmhouses in Pidu emerged like mushrooms after rain and their influence continually expanded to other places in Sichuan even the whole country. Subject to different local realities, happy farmhouses evolved into fishing fun houses or happy ranch houses in some places. Besides, some well-known restaurants emerged in Pidu during this period. For example, Wangcong Restaurant in Wang Cong Shrine was popular in the late 1980s. In the early 1990s, the restaurant was further welcomed in the surrounding areas. Taoyuan Hotel, Xiankelai Hotel, Mirror Lake Park Hotel, Pulaisi Hotel, Ouluoba Hotel, Mingyuan Hotel and Garden Villa Restaurant successively became the star of delicacy.

进入21世纪后，郫都区市场经济的稳步发展，为当地餐饮业的发展提供了良好的外部环境，使郫都餐饮业在第三产业中始终保持着快速增长的势头，形成了数十家餐饮龙头企业与上千家中小餐饮企业、农家旅游点并存发展的格局。特色菜品更是层出不穷，如魔方豆花、杨鸡肉、蒋排骨，以及红星饭店的沱砣鱼、吴记醪糟肉和醪糟鸡，泰和园的印象五谷丰登鸭等，都各具特色，备受好评。至2004年后，郫都农家乐进入黄金发展期，全县农家乐达500余家，其中，评选出的星级农家乐12家。到2005年，星级农家乐已有26家，其中，四星级农家乐有徐家大院、三源农庄、望阳阁、霜梅鱼庄等。此时的郫都农家乐，已成功走出一条以农业资源为载体，集观光、体验、休闲、餐饮、农产品销售等业态于一体，由政府引导、企业主导和“公司+农户”等多种经营模式相结合的乡村旅游发展道路。2007年，郫都餐饮零售额为72 777万元，占当年社会销售品零售总额的26.41%，已成为全区消费需求最旺盛、增长幅度最大的行业，形成了以川菜（包括小吃、火锅）为主，粤菜、鲁菜、淮扬菜等其他菜系和外资快餐“百花齐放、共同发展”的市场格局。2012年，郫都共有餐饮业3 043家。2014年，郫都餐饮市场发展态势良好，旅游餐饮、家宴、婚庆消费成为其中的亮点，经营特色化、市场细分化更加明显，异地扩张和餐饮集团化、连锁化成为发展趋势，全区持有工商营业执照的餐饮企业和个体商户已达4 000家。到2018年，郫都有餐饮服务企业5 000家，其中规模以上餐饮服务企业56家。2020年，全区共有餐饮业市场主体10 287家，较2019年增长65%，餐饮收入168 678万元。餐饮郫都餐饮业的快速发展，为拉动消费需求、构建和谐社会、增加税收、促进就业均发挥了重要作用。

Since the 21st century, as market economy of Pidu steadily developed, the catering service industry was provided with benign external environment and gained momentum in the development of the tertiary industry. Dozens of large-scale catering enterprises as well as thousands of medium and small-sized catering companies, farmhouse tourist spots and restaurants co-existed and developed. The specialties emerged endlessly, including Cubic Tofu Pudding, Yang’s Chicken, Jiang’s Ribs. Others included Fish Chunks, Wu’s Meat with Fermented Glutinous Rice and Chicken with Fermented

Glutinous Rice in Hongxing Restaurant as well as Oiled Stuffed Duck from Taiheyuan Restaurant. They were all welcomed in the market with distinctive features. Besides, the happy farmhouse also entered into the golden period of development. In 2004, there were over 500 happy farmhouses in the county, among which 12 star-rated happy farmhouses were selected. In 2005, there already were 26 star-rated happy farmhouses. The four-star farmhouses included the Courtyard of the Xu's, Sanyuan Farm, Wangyangge Hotel and Shuangmei Fishing Village. At this time, the happy farmhouse has succeeded in the rural tourism development route with agricultural resources as carriers, integrating sight-seeing, experience, relaxation, catering and agricultural products sale into one and combining multiple business models such as government leading mode, enterprise guiding mode and "enterprise plus household of peasants". In 2007, the retail sale of Pidu's catering businesses totaled at 727.77 million yuan, accounting for 26.41% of the total retail sales of consumer goods with the strongest consumer demands and greatest increases. The market layout has been formed with Sichuan cuisine (including snacks and hotpot) as main part, all other cuisines like Cantonese cuisine, Shandong cuisine and Huaiyang cuisine as well as foreign fast food co-existing. In 2012, there were 3,043 catering service enterprises. In 2014, the catering market showed a good momentum of rapid development. Tourist catering, family feasts and wedding consumption have become the new highlights. The specialized operation and subdivided market were more prominent. The expansions in different areas and collectivization and chain orientation of catering companies have become an inevitable trend. There were 4,000 catering enterprises and individual sellers with industrial and commercial business licenses. In 2018, there were 5,000 catering service enterprises, among which 56 were above designated size. Till 2020, there were 10,287 catering market entities in the district, an increase of 65% compared with 2019, with a catering revenue of 1.686,78 billion yuan. The rapid development of Pidu's catering industry plays an important role in boosting consumers' demands, building harmonious society, increasing tax revenue and promoting employment.

二、餐饮企业不断转型升级

随着社会经济的高速发展和人们对饮食生活需求的日益增长，在郫都区政府的大力推动及引导下，郫都餐饮企业采取多种方式，不断转型升级，取得了可喜的良好收益。2009年，全区餐饮企业积极响应政府号召，投入资金3.72亿元，对就餐环境、文化氛围、菜品创新、服务技能等软硬件条件进行提档升级，主动承接成都市区餐饮消费向外转移的趋势，以生态、文化、乡村风情、特色菜品等诸多特色元素吸引越来越多的顾客。其中，西御园、泰和园、三源农庄、红星饭店、了云98号等餐饮企业，对餐厅环境、包间设备、外围环境营造等方方面面实施内部改造，共投入资金6 145万元。2014年，梦桐泉乡村俱乐部和望阳阁对餐厅、住宿、就餐环境、道路进行提档升级改造；荣胜和火锅、蜀滋源、安德芙蓉包等餐饮企业，到成都市区、绵竹、新都等地开设了5家连锁店，为餐饮业实现标准化生产、连锁化经营、创建品牌都起到了积极的推广作用。2015年，郫县餐饮同业公会针对本地餐饮业的发展现状，积极组织市场调研和多层次、多渠道的转型发展研讨，积极推行健康、科学、理性的消费行为，研发节约型、特色化的菜品，简化餐台布置形式，倡导生态环保，制订了《郫县餐饮行业"文明餐桌"行动倡议书》《规范经营倡议书》和《文明经营承诺书》，提出了"引领文明消费、节约"的倡议，以及"加快幸福郫县建设步伐""强化内部管理，提高节约意识"等意见，努力创建"科学膳食、绿色饮食、文化美食"的新风尚。此外，还组织餐饮企业切实练好内功，加大电子商务应用、实施信息化技术等手段实现转型升级，打造品质餐饮。郫都荣胜和火锅、天房阁私房菜、铜壶苑休闲庄充分利用互联网手段，不断打造新的盈利链条和销售卖点，受到了成都市有关主管部门的奖励。2017年，郫都区商务局聘请海外友人、创客代表、行业精英、美食达人、网络红人共十名群体代表担任"美食郫都推广大使"，助推川菜文化服务消费。2018年，郫都区商务局还全力推进了望丛祠、三道堰景区等5个重点景区、景点、商圈、商业街区餐饮标识精细化和标准化的推广示范工作。

II. Catering Businesses Keeping Transforming and Upgrading Themselves

As the economy develops rapidly, the demands for better food and life of the people continues to grow. Therefore, the catering businesses keep transforming and upgrading themselves in various ways with the encouragement and guidance of the Government of Pidu District, gaining good effects. In 2009, called up by the government, a total of 372 million yuan was invested by the catering businesses in the district (then Pixian County) for better dining environment, cultural atmosphere, cuisine creation, service skills, etc., which enabled the businesses to accept the food industry transferred from downtown area and to attract more and more consumers with its characteristic ecology, culture, rural feelings, specialty dishes, etc. Among these businesses, Xiyuyuan, Taiheyuan Restaurant, Sanyuan Farm, Hongxing Restaurant, Ziyun 98, etc. invested a total of 61.45 million yuan to upgrade their dining environment, equipment and external decorations. In 2014, Mengtongquan Country Club and Wangyangge Hotel upgraded their dining halls, hotel rooms, catering environment and roads; Rongshenghe Hotpot, Shuziyuan, Ande's Furong Steamed Bun and other enterprises set 5 branches in downtown Chengdu, Mianzhu and Xindu, promoting the standardization, chain management and brand building of the catering businesses. In 2015, Pixian County Catering Trade Association took active part in organizing market researches and multi-level and multi-channel seminars in accordance with the development situation of the industry, in order to adapt the new trends and accelerate the development of the whole industry. They also advocated healthy, scientific and reasonable consuming behaviors and encouraged the research of conservation-oriented specialty dishes and streamlined the layout of dining tables, all of which demonstrated an ecology-centered and environmentally friendly orientation. Moreover, the association put out *The Proposal for the Catering Industry to Conduct the "Civilized Table" Movement*, *The Proposal for Normative Management* and *The Commitment Letter of Good Faith*, and came up with the ideas of "encouraging civilized consumption and conservation", "speeding up the construction of Happy Pixian County" and "strengthening internal management and the consciousness to save food", etc., aiming for a new trend of "scientific, green and cultural diet". The association also guided the catering businesses to develop themselves from within, increase the application of E-commerce and transform and upgrade themselves via IT technologies, thus the quality of the industry could be improved. Rongshenghe Hotpot, Tianfangge Specialty Cuisine and Tonghuyuan Happy Farmhouse in Pidu all make full use of the Internet to explore new profit points and expand branches, thus realize sustainable development and have been rewarded by the Government of Chengdu. In 2017, the Commerce Bureau of Pidu District selected 10 representatives, including some foreign friends, entrepreneurs, industry leaders, gourmet masters and Internet celebrities to be the "Ambassadors of Pidu's Food", in order to promote the consumption of local Sichuan cuisine and relevant cultural services. In 2018, the Commerce Bureau launched the detailed and standardized promotion of 5 key scenic areas including Wangcong Shrine and Sandaoyan Scenic Spot.

2020年以来，郫都区相关政府部门在餐饮产业方面重点发展消费新形态、打造消费新场景，吸引了国际美食、特色首店、老字号门店、川西农耕桑蚕火锅、星级场景农家乐等不同业态的餐饮企业参与其中，开启了川菜美食体验游、望丛祭祀祈福游、乡村振兴体验游等消费新体验；打造并推出了郫都区美食体验新场景，尤其是以青年人最为喜爱的新消费场景为载体，精心打造了“团聚缘来·安逸郫都”青年美食派对，评选出青年社交消费新场景，同时推出了“安逸郫都·美食地图”及“安逸郫都·精品旅游路线”等内容，通过多元化的经济布局和多重消费形态构建，并结合川菜文化，搭建产业经济合作平台，进一步提升了“安逸郫都”的品牌价值，释放出更多的餐饮消费活力。

Since 2020, relevant government departments have been working hard on the new patterns and new scenes of consumption in the catering service industry, attracting many catering businesses of different types, including some international delicacies, characteristic No.1 chain stores, time-honored stores, Western Sichuan embroidery-themed hotpot, star-rated happy farmhouses, etc., which brings the consumers brand-new experiences of tasting Sichuan food, praying at the Wangcong Shrine, and learning about the rural revitalization. There are also new scenes for experiencing the delicacies. In particular, the Destined Teenager Delicacy Party, the Delicacy Map, the Recommended Tour Routes with the theme of "Comfortable Pidu", etc. have built a platform for economic cooperation among the industries, combined the culture of Sichuan cuisine with the activities, further increased the value of the brand of "Comfortable Pidu", as well as activated the catering consumption via multi - element economic structures and consuming forms.

三、特色餐饮品牌逐渐形成

在餐饮经营、管理过程中，品牌的价值含量不但可以提升餐饮企业的知名度，提高产品的美誉度，更是企业开拓市场、参与竞争的有力武器。长期以来，郫都区相关政府部门和行业协会十分重视特色餐饮品牌的打造，并为此采取了餐饮名菜、名店、名宴、名厨等多种评定形式，造就了一批知名餐饮品牌。

III. Characteristic Catering Brands Gradually Taking Shape

The brands play an important role in the operation and management of catering businesses, which cannot only raise the prestige of the businesses but also expand their market and enhance their competitiveness. For a long time, relevant government departments of Pidu and industrial associations have been attaching great importance to the cultivation of catering brands. They have created a series of famous catering brands by electing famous dishes, shops, feasts, cooks, etc.

在名菜、名店、名厨评定方面，例如2009年，郫县餐饮同业公会牵头组织了“郫县第二届餐饮名菜和首届餐饮名店”评定活动，共评出33道“郫县名菜”和14家“郫县餐饮名店”。此次评定活动，是郫都餐饮发展史上的一个重要举措，既是对郫都餐饮业发展的总结，也是郫都餐饮业发展的新起点，有利于引导各餐饮企业树立品牌意识、强化市场竞争意识，推动郫都的餐饮实力更上一级台阶。此后，郫都开展了更为多样的评选和烹饪比赛活动。2017年和2019年，郫都区商务局先后开展了两届“名菜、名厨、名店”评选工作，邀请烹饪大师、专家评委及大众评委组成评委团，此次评选出的餐饮名店有红星饭店、印象泰和园、铜壶苑等；名小吃有芙蓉包、周鹅油烫鹅、罗记豆瓣抄手、徐记酥肉豆花等；名菜有犀浦鲶鱼、杨鸡肉、百姓烩菜、家常豆瓣鱼等；烹饪名师及大师有官燎、童逊、陈洪光等。2021年，在成都市开展的“成都名菜”遴选中，郫都区有犀浦鲶鱼、芙蓉包等十余个品种入选。

On the one hand, for the election of famous dishes, shops and cooks, Pixian County Catering Trade Association organized The Second Famous Dishes Election and The First Famous Catering Shops Election in 2009, which yielded 33 "Famous Dishes of Pixian County" and 14 "Famous Catering Shops of Pixian County". The election above is an important milestone for the catering business of Pidu, which summarizes the development of the catering business as well as marks a new start for the business community. It raises the awareness of the businesses to create brands and participate in the market competition, promoting the strength of the whole business to the next level. Later, they held more various elections and cooking contests in Pidu. In 2017 and 2019, the Commerce Bureau of Pidu District successively held the first and the second elections for "famous dishes, cooks and shops". The judge group consisting of master chefs, expert judges, and public judges elected famous shops including Hongxing Restaurant, Taiheyuan Restaurant, Tonghuyuan Happy Farmhouse, etc., famous snacks including Hibiscus Buns, Zhou's Oil Scorched Goose, Luo's Wontons in Chili Bean Sauce, Xu's Tofu Pudding with Crispy Pork Topping, etc., famous dishes including Xipu Catfish, Yang's Chicken, Multi-Ingredient Folk Stew, Home-Cooking Chili Bean Paste Flavored Fish, etc., as well as famous cooks including Guan Liao, Tong Xun, Chen Hongguang, etc. In 2021, more than 10 brands from Pidu District were elected as "Chengdu's Famous Dishes", including Xipu Catfish and Hibiscus Buns.

在名宴的打造方面，“郫县豆瓣宴”是其代表作。2015年，郫县餐饮同业公会在围绕“打造郫县餐饮特色品牌，交流郫县经典川菜技艺，推进郫县餐饮与川菜、文化、旅游产业融合发展”的工作中，组织郫都餐饮企业开展郫都特色餐饮的创新活动，并将“郫县豆瓣筵席”的研发、制作和比赛纳入重点流程，通过挖掘、汇集与郫县豆瓣有关的经典川菜来精心组合地方特色宴席，以此促进郫都川菜文化、川菜产业与旅游产业的发展。2016年，蜀都川菜公司与四川旅游学院签订战略合作协议，研发、创制了“郫县豆瓣风情宴（传统版、时尚版）”，受到广泛好评，并于2016年荣获第十三届成都美食旅游节“川菜十佳宴席奖”，在之后的2017年，又荣获中国饭店协会授予的“中国地方特色名宴”称号。

In terms of the famous feasts, the Bean Paste Feast is the representative. In 2015, Pixian County Catering Trade Association organized the research contest of “Pixian Chili Bean Paste Banquet”, an innovative event of specialty catering, in order to find out the classic Sichuan cuisine related to Pixian chili bean paste and to assemble them into a characteristic feast for better development of the culture and industry of Sichuan food and relevant tourism industry. In 2016, Shudu Sichuan Food Company and Sichuan Tourism University signed a strategic cooperation agreement to create the “Pixian Chili Bean Paste Banquet (Traditional Version and Fashionable Version), which has been widely praised. In 2016, it was awarded with the “Top 10 Sichuan Feast” during the 13th Chengdu Food and Tourism Festival; in 2017, it was awarded the title of “Characteristic Local Feast of China” by China Hospitality Association.

此外，郫都区还积极参与由成都市政府相关部门组织的美食活动，2018年，在成都市政府主办、成都市商务局等承办的2019“过节耍成都·寻找最年味（第二季）”活动中，郫都区荣获2018“十大最具年味区域”和“十大最具年味活动”奖项，此外，还在奥地利设立了“郫县豆瓣·川菜原辅料海外推广中心”，有效地助推了川菜产业的国际化发展，进一步提高了以郫县豆瓣为核心的郫都美食在国际上的品牌影响力。

Moreover, Pidu District has taken active part in the events organized by the Government of Chengdu. In 2018, Pidu District won the prizes of “Top 10 Districts of the Best Spring Festival Feeling” and “Top 10 Activities with the Best Spring Festival Feeling” during the 2019 “Spend Festivals in Chengdu and Find the Best Spring Festival Feeling (Season 2)” organized by the Government of Chengdu and held by the Commerce Bureau of Chengdu. Besides, Pidu District has set the “Overseas Promotion Center of Ingredients of Sichuan Cuisine • Pixian Chili Bean Paste” in Austria, which further expanded the international influence of Pidu’s food represented by the Pixian chili bean paste.

四、美食节会层出不穷

郫都区通过大力开展美食节会推广活动，极大地提升了郫都区的饮食知名度和影响力。2009年，郫都承办了“2009中国（成都）国际美食旅游节郫县会场”活动，其主题是“吃在成都、寻川菜之根，味在郫县、展川菜之魂”。2013年，郫都参与主办了“2013·郫县豆瓣博览会”，主题是“展示豆瓣文化，体验川菜味乡”，先后举办了安德川菜汇美食体验街开街仪式、川菜文化体验馆开馆仪式、豆瓣宝贝暨川菜美食推广大使选拔大赛总决赛、豆瓣文化川菜坝坝筵、“郫县制造”优质品牌推介会暨调味品企业贸易洽谈会现场签约仪式和川菜厨艺大赛等活动。2015年，郫都区举办了“郫县味道·豆瓣菜品厨艺比赛”，以“体验豆瓣文化，展示川菜味乡”为主题，以传承、挖掘、弘扬“郫县味道·豆瓣菜品”为驱动。2018年，郫都区商务局借助省、市资源，搭建会展平台，组织区内企业参加“2018欧洲·成都美食文化节”“第十五届成都国际美食旅游节”等多项会展活动，在比利时、奥地利、捷克等国开展2018“郫都造”及川菜产业欧洲推广活动，展示以郫县豆瓣为主的“郫都造”川菜调味品、川菜文化、非遗文化。2020年，郫都区以“川味世界·安逸郫都”为主题，举办了“2020成都国际美食节郫都区体验会场暨第八届郫县豆瓣博览会”，开展了“安逸郫都·十大美食体验新场景评选”“夜耍郫都·夜间绿道美食市集生活节”“乐享郫都·青年社交场景营造”“大V 1+1”寻味之旅等主题营销活动，推出了“安逸郫都·美食地图”“安逸郫都·精品旅游路线”及郫都“首批文明消费餐饮示范店”等内容。这些美食节会，已成为推广宣传郫都美食文化的重要平台与窗口。

郫县豆瓣宴（雷勇／摄）

IV. Gourmet Festivals Springing up One after Another

A good deal of gourmet festivals and promotion activities has been launched by Pidu District Government, so as to increase the popularity and impact of Pidu cuisine. In 2009, Pidu District Government organized activities of "parallel sessions of 2009 China (Chengdu) International Food and Tourism Festival in Pixian", with the theme of "food in Chengdu embodying the core of Sichuan cuisine, while its sole lies in Pixian's condiments". In 2013, Pidu hosted the "2013 Pixian chili bean paste Fair", with the topic of "showing the bean paste culture and learning about original taste of Sichuan cuisine". Then, Pidu organized the opening ceremony of Ande Sichuan Cuisine Experience Street, opening ceremony of Sichuan Cuisine Culture Experience Hall, the final of selection contest of Bean Paste Baby, the Sichuan Cuisine Promotion Ambassador, Bean Paste Culture: Sichuan Babayan Feast (with nine dishes served separately as the guests arrive in succession), Signing Ceremony of "Manufactured in Pixian" Quality Brand Introduction Event and Trade Fair of Condiment Enterprises as well as Culinary Competition of Sichuan Cuisine. In 2015, Pidu hosted the "Taste of Bean Paste: Culinary Competition of Dishes Cooked with Bean Paste", with the theme of "learning about the bean paste culture and showing original taste of Sichuan cuisine", and the aim to carry forward, seek, and promote the "Taste of Pixian: Dishes Cooked with Bean Paste". In 2018, Bureau of Commerce of Pidu District built the fair platform and organized the enterprises of Pidu to participate in many events including "2018 Europe and Chengdu Gourmet Festival" and "the 15th Chengdu International Gourmet Festival", with the help of provincial and municipal resources. In addition, the events like "Manufactured in Pidu" and "Sichuan Cuisine Industry Promotion Events in Europe" of the year 2018 were launched in Belgium, Austria, Czech Republic, etc., showing the Sichuan cuisine condiment, Sichuan cuisine culture and intangible cultural heritages "from Pidu" with the bean paste as the core. In 2020, with the subject of "Comfortable Pidu: World of Sichuan Cuisine", Pidu organized "Pidu Experience Session of 2020 Chengdu International Gourmet and the 8th Pixian chili bean paste Fair", and hosted the marketing campaigns like "Comfortable Pidu: Selection of Top Ten Culinary Experience New Scenes", "Touring in Pidu at Night: Nightly Gourmet Fair along Greenway", "Enjoying the life in Pidu: Creating Social Scenes for the Youth", and "Trip to Seek the Delicacy by Internet Celebrities", as well as advocated the events like "Comfortable Pidu: the Map of Delicacy", "Comfortable Pidu: Recommended Tour Routes" and the "First Batch Demonstration Restaurants in Pidu for Civilized Consumption" and so on. Such gourmet festivals have already become the key platform and channel to promote and publicize Pidu's cuisine culture.

五、饮食文化场馆及川菜产业城建设独树一帜

郫都区在饮食文化场馆建设上同样取得了显著成效，到2021年9月，已拥有川菜博物馆、川菜文化体验馆、中国川菜博览馆三大场馆，全方位展示了川菜丰富的历史文化、国内外发展状况。

V. Food Culture Venues and Sichuan Cuisine Industry City Developing a School of Its Own

Pidu District has made great achievements in the food culture venue building. It has owned Sichuan Cuisine Museum, Sichuan Cuisine Cultural Experience Museum and China Sichuan Food Expo until September, 2021, demonstrating enriched historic culture of Sichuan cuisine in all rounds and its development situation both home and abroad.

此外，郫都区还重点建设了全国首个以地方菜系命名的产业城，即中国川菜产业城。该产业城由成立于2005年的川菜产业园区提升而来，是成都市绿色食品产业生态圈重点发展的6个产业功能区之一，其发展定位是川菜技艺融合创新创造中心主承载区、郫县豆瓣全产业链基地主承载区、火锅底料研发制作中心主承载区、川味美食国际供应链主承载区、特色休闲食品研制基地协同发展区等，在此基础上主要发展以郫县豆瓣为核心的复合调味品和休闲食品两大主导产业，聚力建设"全国川菜产业高地，世界川菜文化中心"。至2021年，川菜产业城已有"中华老字号"企业2家、中国驰名商标7件、省市著名商标27件、四川省名牌产品31个，拥有国家级重点农业产业化龙头企业4家、国家高新技术企业5家、绿色食品A级证书23个，以及国家、省、市企业（工程）技术中心、院士（专家）创新工作站42家，包括全国调味品行业首家国家企业技术中心，已成为成都建设"世界旅游名城""世界美食之都"的一个重要载体。如今，川菜产业城在生产上"串点成链、聚链成圈"，已构建了完整的产业生态圈创新链，实现了从研发设计、生产制造、流通消费等环节的全链衔接；在生活上建立了川菜小镇，开发了工业旅游，着力补齐城市服务功能的短板；在生态上加快发展绿色产业，推进生态场景的叠加，满足各类人群的多种需求，形成了生活、生态、生产"三生"共融的城市功能，逐步实现由"园"到"城"的华丽蜕变。

What's more, the industrial city named after regional cuisines is the key construction project of Pidu, namely China's Sichuan Cuisine Industrial City, the first of its kind in China. The industrial city is upgraded from Sichuan Cuisine Industrial Park built in 2005. It is one of six industrial functional areas mainly developed in the green food industry ecosystem of Chengdu. It is deemed as the supporting region for Sichuan cuisine technology fusion innovation center, whole industry chain base of Pixian chili bean paste, hotpot flavoring research and development and manufacturing center and international supply chain of Sichuan cuisine as well as coordinated development area for special snack food. Based on that, the compound condiment with Pixian chili bean paste as its core and snack food are two major industries. Arduous efforts have been made to build "a highland of Sichuan cuisine industry all over China and a center of Sichuan cuisine culture in the world". By 2021, Sichuan Cuisine Industrial City has already had 2 China time-honored enterprises, 7 China renowned brands, 27 provincial and municipal famous brands, 31 famous products of Sichuan, 4 national key agricultural industrialized leading enterprises, 5 national high-tech enterprises, 23 Grade A Certificates for Green Food and 42 enterprise (engineering) technology centers at national, provincial and municipal level and innovative workstations of academicians and experts. It also includes the first national enterprise technology center in the condiment industry, becoming an important supporting region for Chengdu to build "world famous tourist city" and "world gourmet capital". Now, Sichuan Cuisine Industry City is connecting points into chains and clustering chains into circles in production, building a complete industrial ecosystem and full connection from research and development, design, production, manufacturing, circulation and consumption. In terms of life service, Sichuan cuisine town has been established, industrial tourism has been developed, and efforts have been made to complement the weak point of city service function. In terms of ecology, development of green industries has been accelerated, the superposition of ecological scenes has been promoted, which meet the composite needs of all kinds of people, form the urban function of life, ecology, production, ecology and life service, and gradually realize the magnificent transformation from an Industrial Park to an Industry City.

第四节 郫都饮食文化发展探源

Section Four Origin of Pidu Food Culture

九天开出一成都，千流万脉源望丛。望帝杜宇将都城迁至郫邑，教民农桑；从帝鳖灵治水兴蜀。由望从二帝开创的“天府之国”，为郫都饮食文化的发展铺垫了良好的物质基础。经过几千年的漫长发展，时至今日，郫都以一粒小小的豆瓣为发轫，衍生出了一个川菜产业园，直至一座川菜产业城。郫都饮食文化之所以能够历经数千年绵延不断，并形成今天庞大的产业集群，究其原因，主要得益于“天时、地利、人和”，即优越的地理环境与物产资源，厚重的历史和文化积淀，强大的政策支持与社会推动。

“Chengdu has been built since the beginning of time, and all of its history can be dated back to the era of the ancient kings like King Wang and King Cong.” King Wang, Du Yu, moved the capital of Ancient Shu Kingdom to Piyi and taught his folks farming and silkworm rearing skills; King Cong, Bie Ling, made great contribution to the prosperity of the kingdom by taming the flood. The two kings laid a solid foundation for the “Land of Abundance”, which in the meantime provided rich materials for the development of Pidu’s food culture. With thousands of years passing by, the tiny bean halves of Pidu has born the fruit of today’s industrial park of Sichuan cuisine, making Pidu the district of Sichuan cuisine industry. The secret that Pidu food culture could last for thousands of consecutive years and end up with today’s huge industrial clusters lies in the advantageous geographical conditions and abundant materials, heavy accumulation of history and culture, as well as strong support from the government and the society.

一、优越的地理环境与物产资源

郫都地处天府之国核心区，土地肥沃、物产丰富，历来都是四川地区的重要粮食产区。晋代常璩《华阳国志·蜀志》载：“蜀川人称郫、繁曰膏腴”。宋代乐史《太平寰宇记》引《益州志》言：“谓繁县之地土地肥良，号称小郫。”发展至今，郫都已成为全国优良的粮油基地，也是当今的创业热土、宜居佳处。

I. Advantageous Geographical Conditions and Abundant Materials

Located at the heart of the “Land of Abundance”, Pidu has always been an important grain production area of Sichuan thanks to its fertile land and abundant materials. According to *The Records of Huayang • Chronicles of Shu Kingdom* written by Chang Qu in the Jin Dynasty, “the people in Shu believe that Pi (today’s Pidu) and Fan (located at today’s Xindu) counties are rich places”. *The Record of Overall Geographical Conditions* written by Le Shi in Song Dynasty quoted *The History of Yi Prefecture* that “Fan county boasts fertile land, thus it is called ‘Small Pi’.” Today, Pidu has become an excellent place of grain and oil production, a hub of start-ups, as well as a good place to live in.

郫都位于川西平原腹心地带，自然条件得天独厚，气候温暖湿润，无严寒酷暑；境内地势平坦，又为都江堰的首灌区，拥有纵横交错的排灌体系，江河众多，四季常青。郫都最具代表性的壤土是灰色潮土性水稻土，耕地质量居于国内上乘。郫都得天独厚的地理环境，孕育了悠久的农耕文明，造就了丰富的物产资源。数千年来，勤劳智慧的郫都人民遵循时令、辛勤耕耘，出产了众多品质优良的特色食材。在郫都，五谷杂粮、禽畜河鲜、瓜果蔬菜蔚为大观，其中，不乏远近驰名的云桥圆根萝卜、唐元韭黄、新民场生菜、汇菇源金针菇、中延榕珍杏鲍菇、广福韭菜、豆芽、唐昌板鸭、太和牛肉、郫县豆瓣、德源大蒜等。2006年，“国家杂交水稻工程技术研究中心成都分中心”在郫都区犀浦镇正式运行，袁隆平院士专程来成都分中心揭牌，并在该中心亲手种下了“泰隆3号”和“二优8815”杂交水稻，开始了袁隆平超级杂交水

郫都区唐昌镇平乐村（杨健/摄）

稻“种三产四”丰产工程的郫都实践。至2018年，郫都区全年粮食作物播种面积约5 650公顷，以水稻、小麦为主，兼种红薯、土豆、玉米等粮食作物，总产量5.56万吨，其中大春粮食作物播种面积5 227公顷，产量4.14万吨，小春粮食播种面积422公顷，产量0.19万吨；油料作物3 613公顷，以油菜为主，总产量0.95万吨；蔬菜生产面积18 224公顷（含复种），总产量77.97万吨，总产值20亿元；食用菌生产5 490万袋（折合面积约368公顷），品种达20个，既有香菇、平菇、金针菇、木耳等普通品种，也有杏鲍菇、羊肚菌、茶树菇、灵芝、猴头菇、滑菇、草菇等珍稀品种，总产量2.776万吨，总产值17 272万元。丰富的物产资源为郫都饮食文化及产业发展奠定了坚实的物质基础。

Located at the heart of Chengdu Plain, Pidu enjoys a flat terrain and a warm and humid climate without severe cold and heat; as one of the first-irrigated areas of Dujiangyan, Pidu boasts a crisscross drainage and irrigation system and many rivers, making it evergreen throughout the year. The most representative soil in Pidu District is the gray fluvo-aquic paddy soil, and the quality of its arable land is superior in the country, It is the advantageous geographical and environmental conditions that gave birth to such a long farming civilization and created a wealth of product resources. For thousands of years, the diligent and brave ancestors in Pidu have been working hard to produce so many high-quality and characteristic materials of food as seasons passed by. Here you can find everything from various kinds of grains, poultry, aquatic products, fruits, vegetables to Yunqiao spring radish, Tangyuan chives, Xin Minchang lettuce, Huiguyuan needle mushroom, Zhongyanrongzhen king oyster mushroom, Pixian chili bean paste, Guangfu chives and bean sprout, Deyuan garlic, Tangchang dried salted duck, Taihe beef and other localspeciality that are known far and wide. In 2006, “China National Hybrid Rice Research and Development Center / Chengdu Hybrid Rice Research Center” was formally established in Xipu Town, Pidu District and inaugurated by Academician Yuan Longping, who planted the “Tailong 3” and “Erfu 8815” hybrid rice, marking the beginning of Academician Yuan’s production-increasing project of the super hybrid rice (which aims at producing grains that should have been produced by 4 acres of arable land with 3 acres of arable land). As of 2018, the annual sown area of grain crops in Pidu District has reached 5,650 hectares, mainly rice and wheat, supplemented by sweet potatoes, potatoes, corns and food crops, with a total output of 55,600 tons. Among them, the sown area of major spring-planted crops (mainly rice) is 5,227 hectares with a yield of 41,400 tons; the sown area of the minor spring-planted crops

(mainly wheat) is 422 hectares, with a yield of 1,900 tons; the sown area of oil crops, mainly rapeseed, is 3,613 hectares, with a total output of 9,500 tons; the sown area of vegetables reaches 18,224 hectares (including the repeatedly sown area), with a yield of 779,700 tons and production value of 2 billion yuan; the production of edible mushrooms (including ordinary varieties such as shiitake mushroom, enoki mushroom, needle mushroom, black fungus, etc. and rare ones like king oyster mushroom, toadstool, chestnut mushroom, lingzhi, monkey head mushroom, pholiota nameko, straw mushroom) is 5.49 million bags (converting into 368 hectares), with a total output of 27,760 tons and production value of 172.72 million yuan. The rich materials laid a solid foundation for Pidu's food culture and the catering business.

二、厚重的历史和文化积淀

郫都作为古蜀文明的发祥地之一，自古以来就是钟灵毓秀之区，悠久的历史、深厚的人文积淀和丰富的文化内涵，不但滋养出当地的人文情怀和民俗风情，也催生了多姿多彩的美食文化。

II. Accumulation of History and Culture

Pidu, the important birthplace of the ancient Shu civilization, has always been a place endowed with the fine spirits of the universe. With such a long history and so many people of talents in succession, the heavy accumulation of history and culture of Pidu also drew forth a colorful food culture.

郫都的早期古城遗址及出土的一系列石器、陶器文物都表明，在距今约4 000年前，郫都就已形成城邦，拥有了悠久的古老文明。从先秦至明清及近现代，郫都诞生了一大批重要的历史文化名人，包括道家文化承上启下的划时代传承人严君平；汉代大儒、文学家、语言学家扬雄；汉代大司空何武；宋代文学家张俞等。严君平充实和深化了老子的宇宙观和“无为而治”的政治思想，使老子的道家学说更加博大精深。严君平的弟子扬雄不仅撰写了《太玄》《法言》《方言》《训纂篇》等著述，还写出了敲开汉成帝皇宫大门的《蜀都赋》，文中不仅详述了四川地区丰富的物产资源，还生动描绘了当时富庶人家举办宴会时的盛况。此外，历史上曾到过郫都的文化大家也很多，如晋代“竹林七贤”之一、发明郫筒酒的郫县县令山涛，唐代杜甫，宋代陆游、范成大，近代文学艺术名人张大千、张天翼、熊佛西、陈白尘等，他们都在郫都留下了许多文化印迹，包括诗、词、楹联、赋、文、碑记等。郫都史前遗址和文物古迹众多，如望丛祠、杜鹃城遗址、古城遗址、扬雄墓、唐昌文庙等。郫都出土的汉代文物极为丰富，有全国闻名的说唱俑、《盐井》画像砖、摇钱树、东汉铜马、东汉石棺等。珍贵的青铜器、陶器等文物，不仅反映了郫都在汉代的生产劳动、神话传说、生活场景、娱乐生活等各个方面，而且说明当时“郫邑”的文化艺术已处于较高水平。郫都的悠久历史，不仅留下了大量的文化遗存、名人轶事，还形成了丰富多彩的民俗传统文化。“祭望丛・拜杜鹃”“三道堰龙舟赛”“春台会”等民俗文化活动至今仍年年可见。由这方热土孕育出的民俗文化，充分展示了其种类繁多、特色鲜明、形式多样的文化特征。总的来说，郫都在其漫长发展过程中所积淀的人文资源和丰厚的文化内涵，都为郫都饮食文化的传承和繁盛，提供了连绵不绝的灵感源泉与精神动力。

The ancient city ruins of Pidu and the stone artifacts and potteries excavated there all indicate that there has been a city state in Pidu 4000 years ago, which boasted an ancient civilization. From pre-Qin era, Ming, Qing dynasties to modern days, many important historical and cultural celebrities were born in Pidu, including Yan Junping the epoch-making inheritor between the past and the next of Taoism, Yang Xiong the Han-Dynasty scholar, litterateur and linguist who combines profundity with virtue, He Wu the Dasikong in Han Dynasty, Zhang Yu a litterateur in Song Dynasty, etc. Yan Junping enriched and deepened Laozi's outlook on the universe and political thoughts of “governing by doing nothing that goes against nature”, making Laozi's Taoist doctrine more broad and profound. Yang Xiong, as Yan's student, not only wrote pieces of work including *Taixuan*, *Fayan*, *Fangyan*, *Xunzuanpian*, but also the *Story of Chengdu* that described the abundant resources and spectacular feasts in Sichuan, bringing him appreciation of the Emperor Cheng of the Han Dynasty. Moreover, many men of letters have been to Pidu, including Shan Tao, one of the “seven sages of the bamboo grove” in Jin Dynasty and the county mayor of then Pixian County who invented Pitong wine; Du Fu, poet of Tang Dynasty; Lu You, poet of Song Dynasty; Fan Chengda, litterateur and poet in Song Dynasty; Zhang Daqian, Zhang Tianyi, Xiong Foxi, Chen Baichen and many other celebrities. They left the cultural heritage, including poems, ditties,

唐昌文庙（唐昌镇供图）

couplets, odes, essays and tablet inscriptions. There are numerous prehistorical sites as well as cultural relics and historic monuments, like Wang Cong Shrine, the Dujuancheng Ruins, Ruins of Ancient City, Yang Xiong's Tomb and Confucius Temple in Tangchang. The cultural relics in the Han Dynasty unearthed in Pidu are quite rich. The well-known comedian figurine, portrait brick of *Salt Well*, Money Tree, bronze horse and sarcophagus of the Eastern Han Dynasty, precious bronze ware and pottery not only reflect production labor, myths and legends, scenes of life and entertainment life in all respects, but also demonstrate that the culture and arts of Pidu were at a higher level at that time. The long history of Pidu left a huge amount of cultural relics and anecdotes of famous persons. More importantly, it helped to form enriched folk and traditional cultural activities. The cultural customs like "Worshiping King Wang, King Cong and the Cuckoo", "Dragon Boat Race of Sandaoyan" and "Temple Fair" still survive today. As typical examples of Pidu folk culture, they are rooted in this land for thriving with various types and distinctive features. Diversity is their cultural characteristics. Generally speaking, the profound humanitarian deposit and enriched cultural connotation formed in the long-term development of Pidu have injected vitality and intellectual impetus into Pidu's food culture for its prosperity.

三、强大的政策支持与社会推动

中华人民共和国成立后，尤其是改革开放以来，郫都的饮食文化与产业发展均取得了长足进步，除了依托优越的地理环境和物产资源、厚重的历史和文化积淀外，还得益于以下三个方面。

III. Strong Policy Support and Social Promotion

After the foundation of the People's Republic of China, especially since reform and opening up, Pidu's food culture and industry have made great progress due to excellent geographic advantages and material resources as well as solid historical and cultural foundation. Three aspects as follows also contribute to its development.

（一）政府政策的大力支持与行业协会的积极推动

长期以来，四川省各级政府都先后制定和出台了一系列促进四川饮食文化和餐饮产业发展的政策措施，从而为郫都当地饮食文化和餐饮产业的发展，提供了必要的宏观政策基础。1999年11月，四川省人民政府发布《关于大力发展川菜产业有关问题的通知》，提出要切实加强对川菜发展工作的领导，大力扶持和发展川菜龙头企业、城乡餐饮业和川菜加工业，加大川菜产业对内、对外开放力度。2013年，四川省商务厅制订和颁布了《四川省川菜产业发展规划（2013—2015）》。2016年，成都市政府颁布了《关于进一步加快成都市川菜产业发展的实施意见》，提出以“千亿级、四中心”为总体目标，加快成都市川菜产业发展，把成都建成全球川菜标准制定和发布中心、全球川菜原辅料生产和集散中心、全球川菜文化交流和创新中心、全球川菜人才培养和输出中心。2018年，四川省人民政府又制订和颁布了《四川省促进川菜走出去三年行动方案（2018—2020年）》《关于大力推动农产品加工园区发展的意见》。2020年，四川省人民政府从推动现代服务业的高度，制订了《关于加快构建“4+6”现代服务业体系推动服务业高质量发展的意见》，提出推动川菜传承与创新，振兴老字号川菜，实施川菜菜品及品牌文化创新工程，促进川派餐饮与商业贸易、文体旅游等融合发展；打造一批特色商业街、美食街、川菜小镇、酒镇酒庄；开展以川菜、川酒、川茶为重点的川派餐饮品牌行系列活动；支持成都建设国际美食之都。这一系列政策、措施，不仅表明四川省政府、成都市政府长期以来的高度重视，也切实为包括郫都区在内的整个四川饮食文化及餐饮产业的发展，提供了有力的政策保障。

1. Enhanced Support from Government Policies and Proactive Promotion from Industrial Association

For a long term, governments at all levels in Sichuan Province formulated and launched a series of policies and measures to promote its development of food culture and catering industry, a macro policy based on which the local food culture and catering industry in Pidu have kept growing. In November, 1999, the People's Government of Sichuan Province published *Notice on Issues about Rapid Development of Sichuan Cuisine Industry*, bringing up to effectively strengthen the leadership of Sichuan cuisine development, vigorously support and develop corporate champions, rural and urban catering businesses and processing industry of Sichuan cuisine, as well as make the industry more open both to domestic and foreign markets. In 2013, Sichuan Provincial Department of Commerce formulated and issued *Sichuan Province Development Program of Sichuan Cuisine Industry (2013-2015)*. In 2016, Chengdu Municipal Government issued *Implement Opinion on Further Speeding up Sichuan Cuisine Industry Development in Chengdu*. It put forward a general goal "Industry Worth Hundreds of Billion Yuan with Four Centers". It also stipulated to expedite Sichuan cuisine industry development in Chengdu and make Chengdu a global center of Sichuan cuisine for standard formulation and issuance, raw material production and distribution, cultural exchange and innovation as well as talent cultivation and output. In 2018, the People's Government of Sichuan Province laid down and published *Sichuan Province Three-Year Action Plan to Drive Sichuan Cuisine Going Out (2018-2020)* and *Opinions on Enhancing Development of Agricultural Product Processing Park*. In 2020, Sichuan Government worked out *Opinions on Establishing "4+6" Modern Service System to Promote High Quality Development of Service Industry*, which implicated to promote Sichuan cuisine heritage and innovation, revitalize time-honored brand, implement cuisine and brand culture innovation projects, promote its integrated development with commercial trade, recreational activities and tourism; create a batch of specialty shopping streets, food streets and Sichuan cuisine or wine towns; launch a series of Sichuan catering brand activities focusing on Sichuan cuisine, wine and tea; support Chengdu to be an international gourmet capital. These policies and measures not only demonstrated the long-term attention of provincial and municipal governments, but also provided a strong policy guarantee for food culture and catering industry development of the whole Sichuan including Pidu.

在四川省政府及成都市政府的领导下，郫都区委、区政府，以及相关部门、行业协会结合郫都优势，高度重视郫都饮食文化和川菜产业的发展，进一步提供强有力的政策指导和支持力度，开展一系列促进和推广活动。首先，郫都区委、区政府为了以郫县豆瓣为基础建设川菜产业园、推动川菜产业发展，制订了《成都市郫都区促进服务业加快发展的支持政策》《成都市郫都区服务业2017~2022发展规划》等政策和规划，设立了专门的管理机构——川菜产业园管理委员会。该机构聚焦川菜产业，先后制订了《川菜产业振兴与高质量发展方案》《成都中国川菜产业功能区建设方案》等配套方案，指导川菜产业园区及功能区发展和生态圈建设。郫都区商务局等部门也积极为川菜产业、餐饮产业的发展提供支持，2018年，郫都区商务局开展了望丛祠、三道堰景区海骏达城等5个重点景点、商圈、商业街区的餐饮标识精细化及标准化推广示范工作。其次，郫都区餐饮同业公会、郫都区餐饮旅游商会等协会，积极组织、承办了一系列促进和推广活动。郫都区餐饮同业公会不仅精心组织和开展了郫都名菜、名小吃、名店、名厨评定，还动员会员企业紧抓元旦、春节传统消费旺季，推出团圆饭、年夜饭等系列活动，充分发挥了其桥梁、纽带作用。

Under the leadership of Sichuan and Chengdu Government, Pidu District CPC Committee and District Government, related departments and industry associations highly prioritize the development of food culture and Sichuan cuisine industry. They make use of Pidu's advantages to further offer enhanced policy guidance and support as well as launch promotion activities. First, Pidu District CPC Committee and District Government established Sichuan Cuisine Industrial Park based on Pixian chili bean paste to accelerate the development of Sichuan cuisine industry. Policies and programmes such as *Supporting Policies for Promoting the Rapid Development of Service Industry in Pidu District of Chengdu, 2017-2022 Development Plans for the Service Industry in Pidu District of Chendu* were made. A specialized management institution was thus formed, namely Management Committee of Sichuan Cuisine Industrial Park. The institution focused on Sichuan cuisine industry and formulated supporting solutions like *Sichuan Cuisine Industry Revitalization and High Quality Development*, *Development Scheme of Chinese Sichuan Cuisine Industry Function Zone in Chengdu*, which guided development of Sichuan Cuisine Industrial Park and Function Zone as well as related ecosystem. Commerce Bureau of Pidu

District among others proactively provided support for Sichuan cuisine and catering industry. For example, in 2018, the bureau launched the popularization and demonstration of fine management and standardization of catering logos in 5 key scenic spots, business districts and commercial blocks including Wang Cong Shrine, Haijunda City, and Sandaoyan. Second, Pidu District Catering Trade Association and Pidu Catering and Tourism Association among others proactively organized and held a series of promotion activities. Pidu Catering Trade Association not only painstakingly carried out the evaluation of Pidu famous dishes, snacks, shops and chefs, but also mobilized member enterprises to make good use of shopping seasons like Christmas, New Year's Day and Chinese New Year, during which, the enterprises launched activities such as family reunion dinner and New Year's Eve dinner. Therefore, the association played an important role to promote Pidu food culture and industry development as a bridge and link.

（二）院校人才与技术支撑

郫都区高校资源丰富，区内有电子科技大学、西南交通大学、西华大学、成都技师学院、郫都区职业技术学校、成都川菜产业学校等众多院校，拥有丰富的科研资源和高素质人才资源，在校师生20余万名，每年都会培养一大批川菜、食品类各类人才。郫都区通过深化校地融合，广泛集聚创新资源，引导众多高校、科研院所、专业机构等与企业开展产、学、研合作项目，建立产、学、研合作机构和博士后工作站，共建产业技术研究院、小试中试车间等。目前，郫都区川菜产业城内拥有国家、省、市（工程）技术中心和院士（专家）创新工作室40余家。其中，全国调味品行业首家国家企业技术中心落户郫都，此外还拥有郫县豆瓣产业技术研究院、川菜调味品产业技术研究院、川菜产业研究院等。不仅如此，郫都区还通过《“郫都菁英”产业人才计划若干政策》，对引进的包括食品饮料、文化创意等众多领域高层次人才进行扶持奖励；通过实施“引博入企”“星期天工程师”工程，探索校地企园深度融合，组建专家人才智库等措施，引进大量专业人才。雄厚的人才与技术资源优势，为郫都饮食文化与餐饮产业的可持续发展，提供了重要的人力与技术支撑。

2. The Support of Talents and Technologies from Colleges and Universities

There are a number of institutions of higher-learning, such as University of Electronic Science and Technology of China, Southwest Jiaotong University, Xihua University, Chengdu Technical College, Pidu Vocational and Technical College, Chengdu Sichuan Cuisine School, etc. as well as abundant scientific research resources and lots of high-quality personnel in Pidu. With a total of more than 200 thousand teachers and students in these institutions, a large number of personnel graduate and take up Sichuan food industry each year. Through deepening the cooperation between the local higher-learning institutions with the businesses and integrating the innovation resources widely, the Pidu District Government guides the higher-learning institutions, scientific research institutions and professional institutions to conduct industry-university-research cooperation projects with the businesses. Meanwhile, they will build the industry-university-research cooperation institutions and the postdoctoral workstation, co-create the industrial technology research institute, exploratory test and pilot scale test workshops as well. At present, there are over 40 engineering technology centers at national, provincial and municipal levels and innovation workrooms of the academician or expert within the Sichuan Cuisine Industrial Park in Pidu District. Among them, the first national enterprise technology center for the condiment industry has been established in Pidu. There also are Pixian chili bean paste Industry Technology Institute, Sichuan Cuisine Condiment Industry Technology Institute and Sichuan Cuisine Industry Institute, etc. In addition, the Pidu District Government supports and awards the high-level talents from all fields including food and beverage as well as the cultural creativity via its *Several Policies about the Industrial Talents Plan "The Elite in Pidu"*. Also, the government has brought in many professional personnel through the measures of implementing the projects of "Introducing the Doctors to the Enterprises" and "Sunday Engineer", exploring the deep integration of the local higher-learning institutions and the enterprises, and creating the think tank of professional personnel. Such an advantage of rich personnel and technology

resources has provided important support in terms of manpower and technology for the sustainable development of food culture and catering industry in Pidu.

（三）餐饮食品企业的不懈努力

郫都饮食文化与餐饮产业的繁荣发展，离不开餐饮食品企业的不懈努力。首先，郫都是中国农家乐的发源地，各类农家乐、乡村酒店遍及全区，这些餐饮企业是郫都餐饮市场繁荣发展的中流砥柱，如“中国农家乐第一家”徐家大院，鹃城一绝杨鸡肉，掩映在竹林深处的竹里湾，以及泰和园、红星饭店等。这些餐饮企业积极主动作为，从大胆创办到转型升级，从个体经营到企业规模化、品牌化经营，始终以农业资源为载体，将观光、体验、休闲与餐饮紧密结合，既满足了人们对美好生活的需求，又同时促进了自身发展，丰富了郫都饮食文化的内涵。其次，郫都区还形成了以郫县豆瓣为核心的川菜调味品特色产业集群，并吸引了乳制品及饮料、休闲食品、营养保健食品等产业集聚，它们在不断开展技术与产品更新的同时，还积极参与各类展销会、推介会，并利用各种媒体多途径宣传，这些举措，都有效提升了自身产品及郫都饮食文化与产业的知名度。第三，一些餐饮食品企业还积极投身到川菜文化的保护与传承活动中，如郫县豆瓣股份有限公司依托产业优势、品牌优势，深入挖掘川菜文化、郫县豆瓣文化资源，建设川菜美食体验街、川菜文化体验馆、蜀都特产商城及中国川菜博览馆等，全力打造川菜文化主题旅游景区，使郫都成为四川省内川菜文化场馆最为丰富的区域。可以说，通过郫都广大餐饮食品企业的不断努力，不仅提升了自身的综合实力、促进了餐饮产业的繁荣与发展，而且还推动了郫都饮食文化的传承、传播、创新和发展。

中国农家乐第一家——徐家大院

3. The Tireless Efforts of the Catering and Food Businesses

The prosperity and development of the food culture and catering industry in

郫都工业园区（刘兴银/摄）

Pidu is the result of the tireless efforts of catering and food businesses. Firstly, Pidu is the birthplace of Chinese happy farmhouse, where there are various kinds of happy farmhouses and rural hotels. These catering businesses act as the mainstay for the prosperity and development of the catering market in Pidu. For example, the Countyard of the Xu's, the first happy farmhouse in China, Yang's Chicken which is famous in Pixian, Zhuliwan Hotel hiding deep in the bamboo forest, and Taiheyuan Restaurant and Hongxing Restaurant founded in 1990s have experienced from the process of bold foundation and private operation at the beginning to upgrading and large-scale and brand-oriented operation. Then they have combined the sightseeing, experiencing, recreation and catering together, meeting people's demands for a beautiful life, thus driving the development of these businesses and enriching the cultural connotation of Pidu food culture. Secondly, a featured industrial cluster of Sichuan cuisine condiments with "Pixian chili bean paste" as the core has been formed, bringing together the industries like dairy products, beverage, snack food, nutritious and health-care food. They are continually conducting the technology and product renewal on the one hand, and engaging actively in trade fairs, introduction and marketing events and launching publicity via various media to increase the popularity of their own products and the food culture and industry in Pidu on the other hand. Thirdly, some catering and food businesses are actively protecting and inheriting the culture of Sichuan cuisine. Taking the Pixian chili bean paste LLC as an example, the company explores the culture of Sichuan cuisine, the cultural resources of Pixian chili bean paste deeply, builds the Sichuan Food Experience Street, Sichuan Cuisine Cultural Experience Museum, Shudu Specialty Store and China Sichuan Food Expo and creates the tourist attraction themed on Sichuan Cuisine Culture, making Pidu the place with the most cultural venues of Sichuan cuisine in Sichuan Province. So to speak, the continuous efforts of these businesses make for the improvement of comprehensive strength of catering and food industry in Pidu, the prosperity and development of the catering industry, as well as the inheritance, spreading and innovation of Pidu food culture.

第二章 CHAPTER TWO

名特食材 得天独厚

Unique Famous and Special Ingredients

郫都区地处川西平原腹心地带，土地肥沃、江河纵横、气候温润、四季常青，得天独厚的地理环境，孕育了悠久而灿烂的农耕文明。数千年来，勤劳智慧的郫都人民遵循时令，辛勤耕耘，培育出众多品质优良的特色食材。五谷杂粮、禽畜水产、瓜果蔬菜应有尽有，蔚为大观，其中不乏远近驰名的名优特产，如郫县豆瓣、云桥圆根萝卜、唐元韭黄、德源大蒜、唐昌板鸭、太和牛肉等。这些丰美食材，既为川菜产业的发展提供了重要的物质基础，也满足了广大民众对饮食生活的美好需求。

Pidu District is located in the center of Western Sichuan Plain. The soil is fertile, rivers are densely covered, and climate is moist with evergreen all the year round here. Advantageous geo-environmental conditions have nurtured splendid and time-honored agriculture civilization. For thousands of years, industrious and talented Pidu people who follow the seasons and work hard in fields have produced large number of high-quality and distinctive ingredients. There are all kinds of cereals, livestock, aquatic products, fruits and vegetables, which present a splendid sight. Among them, there are famous specialties, such as Pixian bean paste, Yunqiao round carrot, Tangyuan Chinese chive, Deyuan garlic, Tangchang dried salted duck, Taihe beef, etc. Abundant delicious food have provided important material base for the development of Sichuan cuisine industry, and meet people's growing demands for a better diet life.

第一节 地理环境

Section One Geographical Environment

一、位置与气候

郫都区隶属四川省成都市，位于成都市西北，介于东经103°42'～104°2'，北纬30°43'～30°52'之间，东北与彭州市、新都区接壤，东南与金牛区毗邻，南面与青羊区相连，西南与温江区对接，西北与都江堰市相交。辖区面积438平方千米，下辖7个街道和3个镇、155个村（社区），常住人口1 390 913人（第七次全国人口普查结果）。2016年12月撤销郫县，设立成都市郫都区，全面纳入成都市“11+2”中心城区管理，正式成为成都建设“全面体现新发展理念城市”中心城区。2020年，全区GDP实现655.6亿元，经济综合实力连续21年进入全省“十强”。郫都区交通外联内畅、方便快捷，有成灌快铁、成都地铁2号线、成都地铁6号线、有轨电车共4条轨道交通，外加红光大道、成灌高速公路（羊西线）、西华大道、西源大道、西区大道5条平行快速通道纵向直达市中心，成都绕城高速公路（成都四环路）和成都第二绕城高速公路（成都六环路）横贯境内。此外，郫都区还建有334千米的绿道、135个城市公园及小游园，构成了居民“可进入、可参与、景观化、景区化”的城市慢行系统，与四通八达的交通网络彼此交错，相互融通。

I. Location and Climate

Pidu District, which belongs to Chengdu City, is located in the northwest of Chengdu, between 103°42' and 104°2' E, and 30° 43' and 30° 52' N. It adjacent to Pengzhou city and Xindu District in the northeast, Jinniu Districtin the Southeast, Qingyang District in the South, Wenjiang District in the Southwest, Dujiangyan City in the Northwest. Covering an area of 438 square kilometers, Pixian District governs 7 streets, 3 towns and 155 villages (communities), with a permanent population of 1,390,913 (the results of the Seventh National Census). In December 2016, Pixian County merged into Chengdu City and designated as today's Pidu District , and has been fully incorporated into "11+2"central city management of Chengdu. Since then, Pidu District has been officially become the central urban area of Chengdu that reflects the new development concept. The GDP of Pidu District in 2020 is 65.56 billion yuan. The comprehensive economic strength has entered the "top ten" of Sichuan province for 21 consecutive years. The people's government of Pidu District is at No. 998, Middle Wangcong Road, Pitong Street. Traffic in Pidu District is convenient. Chengguan express, Chengdu Metro Line 2, Chengdu Metro Line 6 and 4 trams directly to the city center. Hongguang Avenue, Chengguan highway (Yangxi avenue), Xihua avenue, Xiyuan avenue, Xiqu avenue are 5 parallel fast lanes go straight to center of Chengdu. Chengdu G42 and SA2 cross Pidu district. Meanwhile, there are 334 km green-way, 135 city parks and street gardens, which have formed urban non-motorized traffic system with the characteristics of accessibility, participation, landscape and scenic area. That system and all-round traffic network has complemented each other in Pidu District.

纵横交错的郫都交通（康耘/摄）

郫都区属亚热带季风性湿润气候，冬无严寒、夏无酷暑、雨量充沛。年平均气温16℃，一月平均气温5℃，八月平均气温26℃左右，年度极端最高气温35.3℃、极端最低气温-4.0℃；年降水量979.4毫米，日照1 014.0小时，具有春早、夏长、秋雨、冬暖、无霜期长、冬季多雾、日照偏少和四季分明的特点，具有成都平原地区“上风上水”的气候优势。

Pidu District is sub-tropical humid monsoon climate. The winteris not severely cold, the summer does not have the intense summer heat, and rainfall is abundant here. The annual average temperature is 16℃; the average temperature in January is 5℃; the average temperature in August is around 26℃; the extremely highest temperature in a year is 35.3℃; the extremely lowest temperature is -4.0℃. Annual precipitation here is 979.4mm and annual sunshine duration is 1,014.0 hours. With the characteristics of early spring, long summer, rainy autumn, warm winter, long frost-free period, foggy winter, less sunshine and distinct seasons, Pidu District has the climate advantage of up-wind and up-water in Chengdu plain.

二、水文与地貌

郫都区地处都江堰自流灌区之首，承担了成都市区90%以上饮用水的供水任务。其地表水均为都江堰宝瓶口内江分出的水系，在仰天窝闸门分出的蒲阳河、柏条河、走马河、江安河等四大河流进入郫都区，或分或合，又形成蒲阳河、走马河—清水河、沱江河、柏条河、徐堰河、毗河、府河、江安河等八大干渠，俗称“八河并流”。区境内河道总长158千米，又以这些干渠为动脉，派生出66条支渠、116条斗渠、219条农渠和密如蛛网的毛渠，形成纵横交错的排灌体系，造就了郫都区“水旱从人，时无饥馑”的自然禀赋。

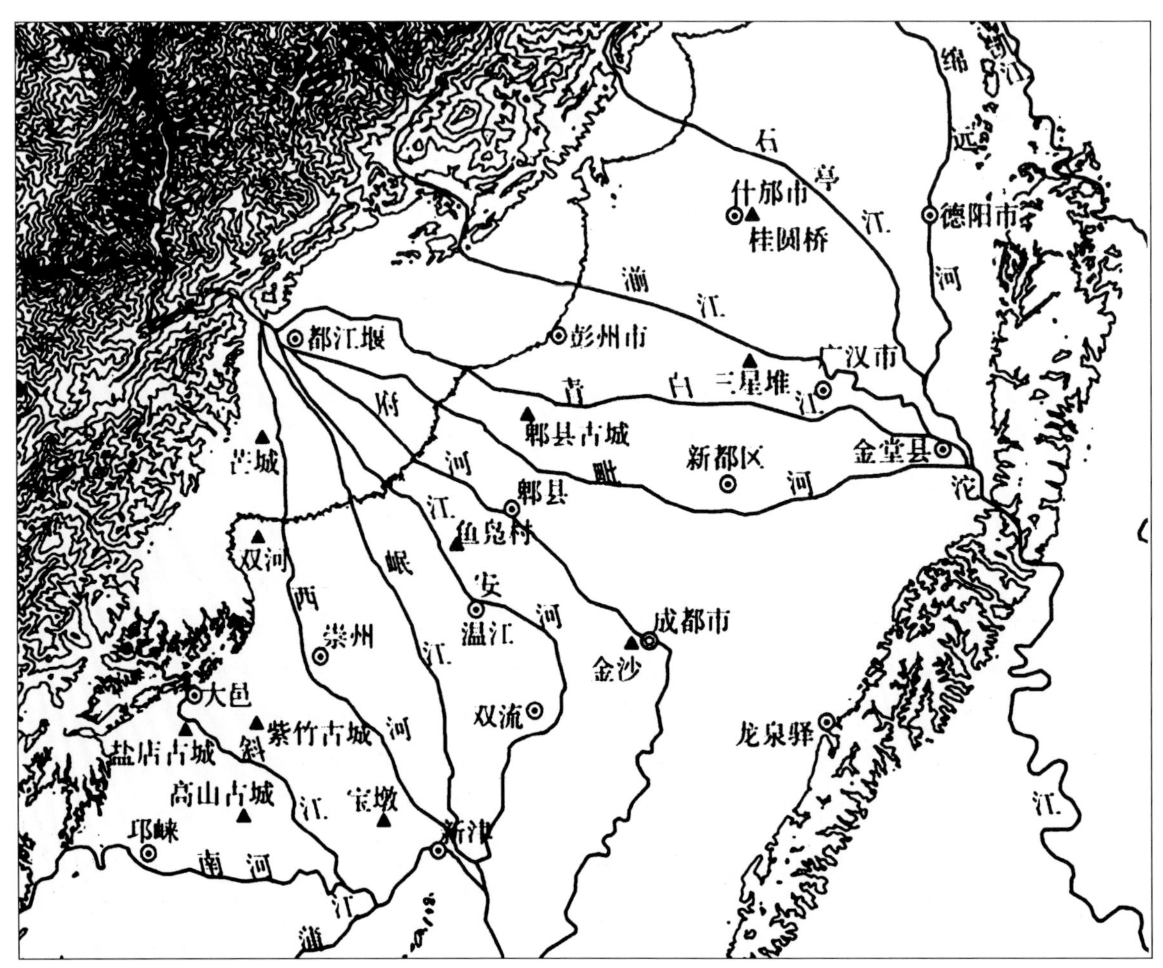

都江堰自流灌区水系分布图

水利资源丰富的郫都区（ 邹灿均/摄 ）

II. Hydrology and Geomorphology

Pidu District is at the top of Dujiangyan gravity irrigation district and take 90% responsibility of providing drinking water for Chengdu urban area. Its surface water all belong to river system separated from the Neijiang River at Precious-bottle-neck, Dujiangyan. Puyanghe River, Baitiaohe River, Zoumahe River and Jiang'anhe River separatedfrom Yang Tianwo gate flow into Pidu District. Those 4 rivers separate or combine to form 8 trunk canals, such as Puyanghe River, Zoumahe-Qinshuihe River, Tuojianghe River, Baitiaohe River, Xuyanhe River, Pihe River, Fuhe River and Jiang'anhe River, which are commonly known as 8 parallel flows. The total length of river course is 158 km. Taking these trunk canals as arteries, there are 66 branch canals, 116 lateral canals, 219 field ditches and sub lateral canals like spider web, which have formed crossed irrigation and drainage system and created the natural endowments of "the irrigation is controllable, the people live here never know how it feels to be hungry" in Pidu District.

郫都区地貌类型属四川盆地西平原区，具有川西坝区的典型特点，是岷江冲洪积扇状平原，由西北向东南倾斜，略似一只五指并拢、由西北伸向东南的手掌，相对高差为121.8米，具有“大平小不平”的特点。因古河道的冲击和近代河流的冲刷切割，形成众多扇形状展开，微地貌呈凸凹状的条堤形地，相对高度不超过2米；西北部浅丘台地横山子，是区内唯一的山丘。

As the fan-like Minjiang alluvial plain and inclined from northwest to Southeast, the geomorphic type of Pidu District belongs to western plain of Sichuan basin, which has the typical character of western Sichuan plain. It is slightly like a palm with five fingers close together and extending from northwest to Southeast. The relative height difference is 121.8m, which is characterized by "most area is flat and small part uneven". Because of impact of ancient river and scour of modern rivers, many fan-shaped levees with relative height no more than 2 meters are formed in Pidu. Heng Shanzi, the shallow high platform in the northwest, is the only hill in this region.

郫都区气候温润、水利发达、地势平坦、土地肥沃，为众多优质食材的生产提供了天然、优质的自然环境。郫都是成都平原农耕文明的源头之一，自望帝教民务农伊始，数千年来，郫都民众充分借助自然资源优势发展农业生产，创造性地开发出“水旱轮作”“立体种植”“稻田养鱼”“林盘综合利用”等农业劳作方式，培育出了丰美的优质食材，形成了悠久、深厚的林盘农耕文化。如今，郫都区的林盘农耕文化已成为第五批全国重要农业文化遗产，并已形成优质粮油、无公害蔬菜、珍稀食用菌、花卉苗木四大特色产业布局。郫都农业现代化步伐的不断加快，对郫都区饮食文化的发展也产生了重要影响。

Mild climate, developed water conservancy, flat terrain and fertile land, provide a natural high-quality environment for food production and make Pidu District one of the sources of agricultural civilization in Chengdu Plain. Since the King Wang Du Yu taught people the knowledge of agriculture, people in Pidu District have taken advantages of natural resources to develop agriculture for thousand years. They have creatively developed agricultural work methods such as "paddy-upland rotation", "stereo-planting", "raising fish in paddy field" and "comprehensive utilization of forest plate", etc. Rich and high-quality food have been produced, Pidu Linpan farming culture has been long-standing. Nowadays, Linpan farming culture has become the 5th batch China national important National agricultural heritage. Agriculture of Pidu has formed 4characteristic industrial layout , which are quality oil and grain, pollution-free vegetable ,rare edible mushroom、flowers and seedlings. Agriculture here is being modernize quickly and has an important effect on food culture in Pidu District.

第二节 名特食材

Section Two Famous and Special Ingredients

郫都区自然地理条件优越，农业生产发达，食材资源丰富，名优产品众多。目前，全区“三品一标”产品达248个，其中有有机农产品163个，无公害农产品56个，绿色食品24个，地理标志证明商标4个，农产品地理标志1个。郫都区现已成为全国生态建设示范区、全国农产品加工示范基地、四川省首批生态农业建设试点县、四川无公害农产品基地和无规定动物疫病区示范区。近年来，郫都区农业生产规模化、现代化、生态化、标准化、产业化步伐不断加快，有效地推动了农业与商贸、文化、创意、旅游、康养、体育产业的融合发展，在成都加快建设全球川菜原辅料生产和集散中心、促进中国川菜产业城一、二、三产业联动发展、推动乡村振兴等方面均发挥了重要作用。由于郫都区特色食材众多，限于本书篇幅，下文主要选取获得国家地理标志产品称号、列入非物质文化遗产保护名录，以及传统名特产或种植面积与产量突出等极具代表性的名特食材进行介绍。需要指出的是，郫县豆瓣作为郫都区最重要的名特调味品，其传统制作技艺已入选国家级非物质文化遗产名录，鉴于其在川菜中的重要地位与作用，将在第三章中专文阐述，这里不再介绍。

Natural geographical conditions of Pidu District are endowed, agricultural production is well developed, food resources are rich, and many famous and high-quality products are plentiful. At present, Pidu District has 248 high quality products in total, which contains 163 organic agricultural products, 56 pollution-free agricultural products and 24 green food, 4 geographical indication certification trademarks, 1 agricultural product with geographical indication. Now, Pidu District has become a national demonstration area of ecological construction, a national demonstration base of agricultural products processing, one of the first pilot counties of ecological agriculture construction in Sichuan Province, and the demonstration area of pollution-free agricultural products& non-specified animal epidemic. Recent years, the pace of large-scale production, modernization, ecology, standardization and industrialization of agricultural production in Pidu District is accelerating continuously, which promotes the integrated development of agriculture with commerce, trade, culture, creativity, tourism, health, and sports industries. Pidu District has played an important role in accelerating the construction of Chengdu as a global production and distribution center of raw materials for Sichuan cuisine, promoting the joint development of the primary, secondary and service industries of Sichuan cuisine industry city of China and stimulating rural revitalization. Due to the large number of characteristic food ingredients in Pidu Region and limited space, this book mainly selects the representative food ingredients which protected by National Geographical Indication Indications of Agricultural Products and listed in the intangible cultural heritage protection list, as well as traditional specialties in prominent planting area and output. What should be pointed out is that Pixian Bean paste is the most important and special condiment in Pidu District. Its traditional making technique is listed in the national intangible cultural heritage list. Its unique characteristics built an important status in Sichuan cuisine. Pixian Bean paste will be expatiated in Chapter Three.

一、云桥圆根萝卜

云桥圆根萝卜为郫都区著名特产，是国家农产品地理标志保护产品，种植历史已达千年，且形成了与之相关的乡风民俗。在1983年版的郫都区《新民场乡志》中，有“白露点圆根子萝卜”的农事习俗活动记载，1989年版的《郫县志》也载有“春不老萝卜”的有关文字。

I. Yuqiao Spring Radish

Yuqiao spring radish is a famous local speciality of Pidu District, and the products protected by National Geographical Indications of Agricultural Products. The cultivation of Yuqiao spring radish is more than one thousand years and has its related customs. There is a record of farming custom of “planting spring radish in White Dew” in the *Chronicles of Xinminchang* published in 1983. In the *Chronicles of Pixian* published in 1989, there is a record of “Chun Bulao radish”(not blossom in Spring, and not hollow).

云桥圆根萝卜

云桥圆根萝卜外形扁圆、大小均匀、表皮纯白、光滑细嫩、脆嫩多汁、回甜不辣、营养丰富，一直是百姓餐桌上的美味家常食材，既可鲜食、煮食、凉拌，又可干制、腌渍，更是制作老鸭汤、牛肉炖萝卜等菜肴的上等食材。

Yunqiao spring radish has a oval shape, uniform size, pure white peel, smooth and tender. It is rich in nutrition, and tastes crisp & juicy, sweet & not spicy. It has always been a delicious home food on table. It can not only be eaten fresh, boiled, cold and dressed with sauce, but also can be dried and pickled. It is also a top-grade ingredient for making dishes such as Duck Soup, Stewed Beef with Radish and so on.

自20世纪70年代开始，云桥圆根萝卜逐步走出郫都推向全国。近年来，郫都区大力加强云桥圆根萝卜的生产、开发，强化萝卜种植繁育技术规范，不断优化萝卜品质，业已步入现代化、产业化、品牌化的发展之路，以云桥圆根萝卜为主要食材的深加工制品，已成为旅游休闲、佐餐馈赠的佳品，远销全国30多个省、市。目前，云桥圆根萝卜种植面积达1 340公顷，年产量80万千克，产值近亿元。

Since 1970s, Yuqiao spring radish has been stepped out of Pidu and introduced to the whole country. Take the road of modernization, industrialization and brand development, Pidu District has enhanced the production of Yuqiao spring radish, strengthen the technical specifications of planting and breeding, and constantly optimize the quality of this radish in recent years. At present, Yunqiao spring radish has a planting area of 1,340 hectares, an annual output of 800 thousand kg and an output value of nearly 100 million yuan. The deep-processing products mainly made of Yuqiao spring radish have become ideal gift for tourism and leisure, and are sold to more than 30 provinces and cities in China.

唐元韭黄

二、唐元韭黄

唐元韭黄为郫都区著名特产，是国家地理标志产品。唐元韭黄种植历史悠久，至今已有300多年历史。唐元韭黄色泽黄白如玉，具有鲜、香、脆、嫩、回味甜的特征；营养价值丰富，富含多种维生素、胡萝卜素，以及硒、磷等矿物质，具有温中健脾、活血化瘀、增强体力等食效。唐元韭黄食用方便，大量应用于菜肴和面点制作，常见品种有韭黄肉丝、韭黄煎蛋、韭黄包子等，其深加工制品——“韭黄酒”已投入市场。

II. Tangyuan Chinese Chive

Tangyuan Chinese chive is a famous local speciality of Pidu District, and the product with National Geographical Indication. Tangyuan town has 300 years' history of cultivating Chinese chive. Tangyuan Chinese chive has a yellow and white color just like jade, with the characteristics of fresh, fragrant, crisp, tender and sweet aftertaste. It is rich in nutrition and various vitamins, carotene, selenium, phosphorus and other minerals. It has the function of warming the spleen, promoting blood circulation, removing blood stasis and enhancing physical strength. Tangyuan Chinese chives are easy to eat and are widely used in dishes and pastries, such as shredded pork with Chinese chives, fried eggs with Chinese chives, steamed stuffed bun with Chinese chives. Its deep processed product—“Chinese Chive Wine” has been put into market.

唐元韭黄因其品质优良而在郫都区大量种植，并形成规模化的种植基地。2002年，唐元韭黄荣获“西南农业博览会名优农产品”金奖；2003年和2004年，唐元韭黄先后取得《无公害农产品基地证书》和《无公害农产品证书》；2006年，唐元韭黄基地通过了省级农业标准化示范项目验收，并于当年成功申报为“国家级农业标准化示范项目”；2008年，唐元韭黄通过中国绿色食品发展中心的“绿色食品”认证；2011年，唐元万亩韭黄基地被四川省农业厅认定为首批“四川省现代农业万亩示范区”。近年来，唐元韭黄产业不断发展，已成为郫都区的特色农业支柱产业，种植面积达2 312公顷，为西南地区最大的韭黄生产基地，年产韭黄2.1万余吨，占据成都韭黄市场70%以上的市场份额。此外，唐元韭黄还远销日本、韩国、加拿大等地，年出口量达200多吨。

Tangyuan Chinese chive is widely planted because of its excellent quality, and has a planting base in Pidu District already. In 2002, Tangyuan Chinese chive won the gold award of "Famous and Excellent Agricultural Products of Southwest Agricultural Expo"; In 2003 and 2004, Tangyuan Chinese chive planting base and products obtained pollution-free agricultural product base certificate and product certificate successively ; In 2006, Tangyuan Chinese chive base passed the acceptance check of Provincial Agricultural Standardization Demonstration Project, and successfully applied for National Agricultural Standardization Demonstration Project that year; In 2008, Tangyuan Chinese chive was certified to the "Green Food" by China Green Food Development Center; In 2011, Tangyuan ten thousand mu Chinese chive base was recognized as the first batch of "10000 mu Demonstration Area of Modern Agriculture in Sichuan Province" by Sichuan Provincial Department of Agriculture. In recent years, Tangyuan Chinese chive industry develops continuously and has become a characteristic agricultural pillar in Pidu District. The planting of Tangyuan Chinese chive reaches 2,312 hectares, which is the largest Chinese chive production base in Southwest China. The annual output is more than 21,000 tons, accounting for more than 70% of the market share of Chengdu Chinese chive market. At the same time, Tangyuan Chinese chive has also been exported to Japan, South Korea, Canada and other places, with an annual export volume of more than 200 tons.

三、德源大蒜

德源大蒜为郫都区著名特产，是国家地理标志保护产品。是郫都区德源街道百姓世代种植的农产品，早在清代乾隆年间就已名声斐然，至今已有近300年历史。德源大蒜具有蒜瓣体大、色泽光亮、口味香辣、胶汁浓郁等特点，蒜素含量高于普通大蒜，有解毒杀菌、抗病毒、降压、降血脂等作用，是制作川菜的上等调味品。

德源大蒜

III. Deyuan Garlic

Deyuan garlic is a famous local speciality of Pidu District, and the product with National Geographical Indication. Deyuan people has plant garlic for generations. Deyuan garlic has a history of nearly 300 years and has got its fame since Qianlong period of Qing dynasty. Deyuan garlic has the characteristics of large garlic petals, bright color, spicy taste and thick gum juice. The content of allicin is higher than ordinary garlic. It has the functions of detoxification, sterilization, anti-virus, blood pressure reduction and blood lipid reduction. It is the top-grade condiment for Sichuan cuisine.

如今，德源街道是成都市最大的种子蒜种植基地，也是成都最大的优质蒜薹原产地之一，享有“西部蒜乡”的美誉。德源街道着眼绿色大蒜基地化建设，积极推行标准化种植工艺，大力实施现代化农业生产模式，大蒜规范种植面积达670公顷，年产大蒜2万余吨，产值超过4亿元，形成了优质蒜薹、优质种子蒜基地的规模效应。

Called as the western garlic hometown, Deyuan is the biggest garlic seeds breeding base, and the largest place origin of high-quality garlic sprout in Chengdu nowadays. Deyuan has enhanced the construction of green garlic base, promoted the standard planting technique, implemented modern agricultural producing mode. As a result, the standard garlic planting area is 670 hectares. The annual output of garlic is more than 20,000 tons and the output value is more than 400 million yuan. Deyuan achieves the scale effect of high-quality garlic sprout and high-quality garlic seed base both inside and outside Sichuan Province.

四、新民场生菜

新民场生菜为郫都区著名特产，是国家地理标志产品。生菜原产于欧洲地中海沿岸，自20世纪80年代初引入郫都新民场镇种植。在本地得天独厚的地理环境孕育下，生菜品质上佳、推广速度快、产量增长迅速，一年可种四季，成为当地农民增收的主要经济作物之一。如今，新民场生菜种植面积达10 569公顷，年产量达10万吨，产值超过2亿元。

IV. Xinminchang Lettuce

Xinminchang lettuce is a famous local speciality of Pidu District, and the product with National Geographical Indication. Lettuce originated in European Mediterranean coast. It has been introduced into Xinminchang of Pidu District in 1980s. In the endowed geo-environent, lettuce grow rapidly with high quality. The climate permits 4 harvests every year and makes lettuce one of the main cash crop of local farmers. Nowadays, there are 10,569 hectares of lettuce fields. Annual output can reach 100,000 tons and the output value exceeds 200 million yuan.

新民场生菜

新民场生菜外形美观、色泽翠绿、外叶舒展、心叶半包、爽脆化渣、微苦回甜、清爽解腻，水分丰富，富含维生素B、莴苣素、碳水化合物及多种矿物质，具有清热、解暑、减肥等食疗作用。新民场生菜食用方便，既可直接生食，也可做成蔬菜沙拉，还可炒制或煮汤。

Xinminchang lettuce looks beautiful with green color. Its outer leaves stretch and central leaves half wrapped. The taste is crisp and fresh, a little bitter, sweet in aftertaste, with large water content. It is rich in vitamin B, lettuce element, carbohydrates and minerals, with function of clearing heat, relieving summer heat and losing weight. Xinminchang lettuce can not only be eaten raw directly, but also be made into vegetable salad, fried dishes or boiled soup dishes.

五、广福韭菜

广福韭菜是郫都区著名特产，因主要产于安德街道的广福村而得名。韭菜又称丰本、草钟乳、起阳草、懒人菜等，原产于中国，已有3000多年的栽培历史。韭菜营养价值丰富，富含维生素C、维生素B_1、维生素B_2、烟酸、胡萝卜素、碳水化合物、矿物质及丰富的纤维素，对促进肠道蠕动、预防心血管疾病均有一定的食疗作用。

V. Guangfu Leek

Guangfu leek is a famous local speciality of Pidu District. The name comes from its place of production, Guangfu Village of Ande Subdistrict. Leek, also known as fengben, Cao Zhongru, Qi Yangcao, Lan Rencao, etc., is native in China and has a cultivation history of more than 3,000 years. Leek has rich nutrition. It is full of vitamin C, vitamin B_1, vitamin B_2, niacin, carotene, carbohydrate, mineral substance, and cellulose, which can promote intestinal peristalsis and prevent cardiovascular disease.

广福韭菜

如今，韭菜已成为广福村的特色产业，全村韭菜种植面积达140公顷。广福村不仅推出了以韭菜为主打食材的创新菜点，如韭香奶油餐包、翡翠小米饭时蔬、蒜香韭酱法国香包、香煎培根牛排佐川味韭菜酱等，还不断加快韭菜产业化发展，推进韭菜深加工，开发韭菜系列产品。位于村中的四川润禾家园科技公司，是一家专门从事韭菜产品研发及深度加工的专业化企业，其产品有韭菜青粉、韭菜挂面、韭菜绿酒等。

Currently, leek has become the feature industry of Guangfu village. The leek fields there are 140 hectares. Guangfu village has created new dishes made of leek, such as leek cream steamed bun, millet mixed with leek, bread with leek sauce, fried bacon and bread with Sichuan style leek sauce. Meanwhile, Guangfu village is accelerating the industrialization of leek, promote the deep processing of leek, and develop new products with leek. Sichuan Runhejiayuan Technologies Co., Ltd located in the village is an enterprise which focuses on the research and development of leek processing. It has produced leek flour, leek noodle and leek green wine, etc.

六、苕菜

苕菜，又名野豌豆、野苕子、薇菜、巢菜，为豆科植物野豌豆的嫩叶或茎叶。宋代时，苏东坡好友巢元修将四川野豌豆种带到湖北黄州东坡的田间地头上随意播撒，不但满足了苏东坡的思乡之情，还满足了他们的口福之欲。苏东坡向黄州人介绍此菜时，便称其为“元修菜”。苕菜为一年或二年生草本植物，茎细柔蔓生，几无毛或被稀柔毛，有棱，高10~30厘米，托叶一边有线形齿或4~5个线形齿，多生长于田边及灌木林间。苕菜营养价值丰富，含有碳水化合物、钙、磷、维生素、膳食纤维和较高的蛋白质，具有补肾调经、祛痰止咳、清热利湿、和血散瘀等作用。

苕菜

VI. Shaocai

Shaocai also known as wild pea, wild sweet potato, Osmunda japonica and chaocai, is the tender leaf or stem leaf of wild pea. In Song Dynasty, Chao Yuanxiu, the close friend of Su Dongpo planted shaocai at random in fields in Huangzhou, Hubei province. Shaocai did not only relieve Su Dongpo's homsickness, but also satisfy their appetite for food. When Su Dongpo introduced it to Huangzhou people, he called it Yuanxiu Cai. Shaocai is an annual or biennial herb with thin and soft stems, few glabrous or thinly pilose, ribbed. It is 10 ~ 30 cm high, linear teeth or 4 ~ 5 linear teeth on one side of stipules, growing in the edge of fields and shrubs. Shaocai is nutritious. It is full of carbohydrate, calcium, phosphor, vitamin, dietary fiber and protein. It has the functions of tonifying kidney and regulating menstruation, eliminating phlegm and relieving cough, clearing heat and dampness, and dispersing blood stasis.

明代李时珍《本草纲目》载：“薇，生麦田中，原泽亦有”，“即今野豌豆。蜀人谓之巢菜。蔓生，茎叶气味皆似豌豆，其藿作蔬、入羹皆宜。”苕菜为野生食材，在郫都区农田边广泛分布，是郫都区民众日常食用的美味食材。每年清明节前后，人们到田间采摘苕菜，可新鲜食用，亦可制成干菜。新鲜苕菜气味清香，可炒、可烧，可羹、可汤；干制苕菜同然清香有存，可入馔，可熬粥。郫都区用苕菜制成的菜肴较多，代表性菜品有米汤苕菜、苕菜狮子头、干煸苕菜等。

Compendium of Materia Medica written by Li Shizhen in the Ming Dynasty recorded, “Shaocai, born in the wheat field, also exists in bush”, “that is today's wild pea, which Shu people call it chaocai. Its stem and leaf has the flavor of pea, and is suitable for vegetables and soup.” As a wild food ingredient, shaocai is widely distributed along the fieldsin Pidu district and is a delicious ingredient for people. Around Qingming Festival every year, people go to pick shaocai in fields, which can be eaten fresh or made into dried vegetables. Fresh shaocai smells good and can be fried and cooked, or made into soups and stews. The dried shaocaistill has a delicate fragrance,and can be mixed into porridge. There are lots of dishes made of shaocai, including Rice Soup with Shaocai, Braised Lion's Head Meatballs with Shaocai, Dry-Fried Shaocai, etc.

优质水稻

七、优质水稻

郫都区土地肥沃、气候温润，水稻种植历史十分悠久，稻田养鱼共生系统、水旱轮作系统早在2000多年前就已出现。宋代李昉、李穆、徐铉等编纂的《太平御览》卷九二六载："《魏武四时食制》曰：郫县子鱼，黄鳞赤尾，出稻田，可以为酱。"如今，郫都区已成为国家优质水稻种植试验基地。2017年，袁隆平杂交水稻科学园落户郫都区菁蓉小镇，2020年5月正式对外开放。项目一期占地1.68公顷，建筑面积约8 000平方米，户外配套高标准农田134公顷，已获批成为国家现代农业科技示范展示基地。2021年，将有4 018个试验品种的秧苗被移栽进试验田，试验品种总数创历年新高。

VII. High Quality Rice

Pidu District has fertile land, warm climate and a long history of rice planting. The fish-rice symbiotic system and the paddy-dry land rotation system appeared more than 2,000 years ago. Chapter 926 of *Tai Ping Yu Lan* compiled by Li Fang, Li Mu, and Xu Xuan in the Song Dynasty recorded, "*Arrangement of Food throughout the Year* said that growing in paddy field, the baby fish of carp with yellow scale and red tail can be used for carp roe." Nowadays, Pidu District has become the national high quality rice planting base. In 2017, Yuan Longping Hybrid Rice Science Park has been established in Jingfu Town, Pidu District, which has been opening for the public since May, 2020. The first phase of the project covers an area of 1.68 hectares, with a construction area of 8,000 square meters, and 134 hectares of outdoor supporting high standard farmland. It has been approved as the National Modern Agricultural Science and Technology Demonstration Base. In 2021, 4,018 experimental rice seedlings will be transplanted into the experimental field, and the total number of experimental varieties will record a new high over the years.

郫都区以袁隆平杂交水稻科学园建设为契机，着力打造"种业硅谷"，创新开展"稻鱼""稻菜"和"稻蒜"的有机绿色种养结合新模式。同时结合林盘文化、川西文化和水稻文化，着力打造新时代的林盘院落景观，积极探索"种稻致富"、稻米品牌塑造、青少年农耕文化研学旅行基地建设，走出一条"种稻致富"的一、二、三产业融合互动的乡村振兴郫都模式。

Pidu District has taken the establishment of Yuan Longping Hybrid Rice Science Park as opportunity to build "Silicon Valley of Seeds", and to explore new mode of organic green planting and breeding through combination of rice & fish, rice & vegetable and rice & garlic. Meanwhile, Pidu District has combined Linpan culture, western Sichuan culture and rice planting culture to create a new Linpan-courtyard landscape, to explore the construction of " to acquire wealth by planting rice ", rice brand building and agricultural culture study tour base for teenagers. With the integration and interaction of primary, secondary and tertiary industries, Pidu District find its own path of rural revitalization through rice planting.

八、金针菇

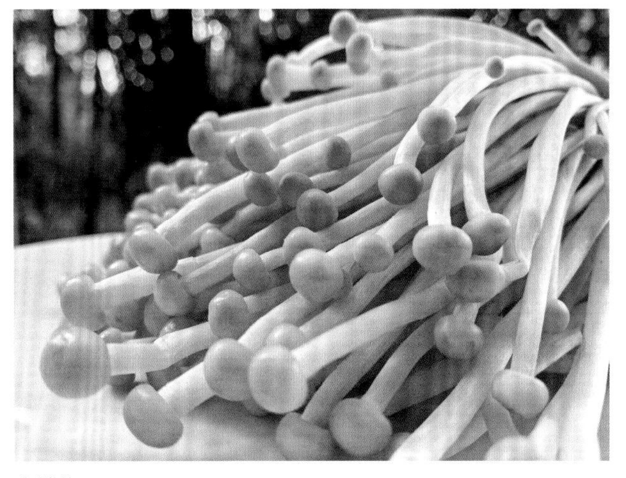
金针菇

VIII. Golden Needle Mushroom

金针菇，又称冬菇、朴蕈、绒毛柄金钱菌等，在我国广泛分布，栽培历史悠久。早在公元800年，金针菇就作为一种食药两用菌而开始人工栽培。金针菇具有菌盖滑嫩、柄脆、营养丰富、味道鲜美等特点，富含蛋白质、多种维生素，以及钙、磷、铁等多种矿物质和人体所需的18种氨基酸，特别是精氨酸和赖氨酸含量高于普通菇类，能促进儿童智力增长，日本人称其为“增智菇”。

Golden needle mushroom, also known as Donggu, Puji, tremella villosa, etc., is widely distributed in China and has a long cultivation history. As early as 800 A. D., golden needle mushroom began to be artificially cultivated both for food and medicine. The pileus of golden needle mushroom is smooth and tender; the stipe is crisp, and the taste is delicious. It is rich in protein, a variety of vitamins, calcium, phosphorus, iron and other minerals and 18 kinds of amino acids. In particular, the content of arginine and lysine of golden needle mushroom is higher than that of ordinary mushrooms, which can promote children's intellectual growth. Hence, Japan calls it "wisdom increasing mushroom".

金针菇味道鲜美，质地滑嫩且脆，在烹饪和食品中应用广泛，既可鲜食，用于拌、炒、炝、熘、烧、炖、煮、蒸等菜肴，也可做汤，还可以做成初级加工产品，如金针菇罐头、金针露、低糖金针菇脯、金针菇蜜饯等。

Golden needle mushroom is widely used in cooking as it's delicious, tender, smooth and crisp. It can be eaten fresh, make it cold and dressed with sauce, stir frying, steaming, and making soup, etc. It can also be used as a primary processing product, like canned golden needle mushroom, golden needle mushroom dew, preserved golden needle mushroom with less sugar, preserved golden needle mushroom.

位于郫都区唐昌镇的成都汇菇源公司，是郫都区引进的重点高端农业项目，也是西南地区最大的鲜菇生产基地、全国首家黄色金针菇工业化生产基地。该公司不仅大量生产金针菇，也是集食用菌生产、研发、销售为一体的企业，日产鲜菇50吨，年产值超亿元，产品已销往全国各地。

Chengdu Huiguyuan Company, located in Tangchang Town, Pidu District, is a key high-end agricultural project introduced by Pidu District. It is also the largest fresh mushroom production base in Southwest China and the first industrialized production base of golden needle mushroom in China. The company not only produces golden needle mushroom in large quantities, but also integrates production, research and sale. The daily output of fresh mushrooms is 50 tons, and the annual output value exceeds 100 million yuan. Their products have been sold all over the country.

九、杏鲍菇

杏鲍菇是集食用、药用、食疗于一体的珍稀食用菌，其子实体单生或群生，菌盖直径2～12厘米，菌柄长4～12厘米，粗0.5～3厘米，呈棍棒状至球茎状，光滑，无毛，近白色，中实，肉白色，肉质细纤维状。杏鲍菇的营养成分十分丰富，植物蛋白含量高达25%，含有人体所需的18种氨基酸及多糖、寡糖，具有防癌、抗癌、提高人体免疫力、促进消化等作用，被称为“平菇王”“干贝菇”。

杏鲍菇

IX. King Oyster Mushroom

King oyster mushroom, known as "king of Pleurotus ostreatus" and "scallop mushroom", is a precious edible fungus which can be taken as food and medicine. It is solitary or in group, the pileus diameter is 2-12cm, and the stipe is 4-12cm long and 0.5-3cm wide, rod-shaped or bulbous, smooth, hairless, nearly white. King oyster mushroom is white inside and looks fibrous. It is rich in nutrients, with plant protein content up to 25%, 18 kinds of amino acids, polysaccharides and oligosaccharides. It has the functions of anti-cancer, enhancing human immunity and promoting digestion.

郫都区的杏鲍菇主要由成都中延榕珍菌业有限公司生产，是郫都区农业现代化、食材工业化生产的典型代表。该公司是四川省农业产业化重点龙头企业、川台农业合作示范基地，位于郫都区战旗村，日产杏鲍菇60余吨，年产值1.2亿元，产品份额占成都市场80%以上、西南市场70%以上。中延榕珍杏鲍菇，菌肉肥厚、质地脆嫩，菌柄组织致密、结实、乳白，口感独特，既可作为蔬菜直接烹饪食用，也可进行食品深加工，做成杏鲍菇软罐头系列产品，市场开发前景广阔。

King oyster mushroom of Pidu District is mainly produced by Chengdu Zhongyan Rongzhen Mushroom Co., Ltd. Located in Zhanqi village, Pidu district, the company is a key leading enterprise of agricultural industrialization in Sichuan Province and a demonstration base of Sichuan-Taiwan agricultural cooperation. The daily output of king oyster mushroom is over 60 tons, and the annual output value is 120 million yuan. It has taken more than 80% market share in Chengdu and 70% market share in Southwestern China. Zhongyan Rongzhen’s king oyster mushroom is plump, tender and crisp. It is tight in stipe with white color. Unique taste of king oyster mushroom brings a prospect market. It can not only be directly cooked and eaten as vegetables, but also be processed into canned products. It has a broad market development prospect.

十、唐昌豆腐乳

唐昌豆腐乳为郫都区著名的传统特产，因产于郫都区唐昌镇而得名。豆腐乳又名腐乳，在我国制作、食用历史悠久。公元5世纪，北魏时期的书中就有“干豆腐加盐成熟后为腐乳”的记载，距今已有1000多年历史。清代赵学敏编著的《本草纲目拾遗》记述道：“豆腐又名菽乳，以豆腐腌过酒糟或酱制者，味咸甘心。”清代四川罗江人李调元整理刊印的《醒园录》，详细记载了豆腐乳及衍生品的多种制法，包括豆腐乳法一种、酱豆腐乳法两种、糟豆腐乳法三种。豆腐乳营养价值丰富，锌和B族维生素含量较多，对预防老年痴呆有重要作用，其蛋白质的含量是豆腐的

唐昌豆腐乳

两倍，且易消化吸收，具有养胃健脾、增加食欲、预防积食等功效。

X. Tangchang Fermented Bean Curd

Tangchang fermented bean curd is a famous local speciality of Pidu District, which is named because it is produced in Tangchang Town. Fermented bean curd, also known as soy cheese, has a long history of making and eating in China. In the 5th century A. D., there was a record in the books of the Northern Wei Dynasty that "adding salt into dried bean curd ,later it becomes fermented bean curd ", which has a history of more than 1,000 years. In the book *Supplement to Compendium of Materia Medica* written by Zhao Xuemin in the Qing Dynasty recorded, "Tofu is also known as bean milk. It tastes salty outside and sweet inside after pickled with distillers' grains or sauce." *Xingyuan Record* published by Li Tiaoyuan from Luojiang, Sichuan in the Qing Dynasty recorded the way of making fermented bean curd and various derivatives in detail, including 1 method of making fermented bean curd, 2 methods of making preserved fermented bean curd, 3 methods of making fermented bean curd with distilled grain. Fermented bean curd is nutritious and full of zinc, vitamin B, which can prevent Alzheimer's disease. Protein in fermented bean curd is twice in Tofu. It is easy to digest and absorb, with functions of nourishing the stomach and spleen, increasing appetite and preventing food accumulation, etc.

唐昌豆腐乳主要用黄豆、蚕豆仁、小麦粉、辣椒、菜籽油、粮食白酒、香辛料等多种食材经过传统工艺发酵而成，成品规格统一、色泽红亮、口感滑嫩，芳香馥郁、风味独特，既可作为佐餐小菜直接食用，也可作为调味料制作美味菜肴，如腐乳蒸腊肉、腐乳蒸鸡蛋、腐乳炖鲤鱼等。如今，唐昌豆腐乳积极实施现代化、品牌化发展，在唐昌镇战旗村建立了四川浪大爷食品有限公司，所生产的红油豆腐乳畅销省内外。

Tangchang fermented bean curd is made of soybean, broad bean, wheat flour, chilly, colza oil, Chinese liquor, spices, etc. through traditional technique. It has the unified specifications, bright red color, tender taste, pleasant smell and special flavor. It can not only be eaten directly as a side dish, but also be used as a seasoning to make dishes, such as Steamed Bacon with Fermented Bean Curd, Steamed Egg with Fermented Bean Curd, Stewed Carp with Fermented Bean Curd, etc. At present, Tangchang fermented bean curd is being modernized and branded. Fermented bean curd with red chili oil produced by Sichuan Lang Daye Food, Co. Ltd in Zhanqi village, Tangchang town, sold well both inside and outside the province.

第三章 CHAPTER THERE

郫县豆瓣 川菜之魂
Pixian Chili Bean Paste, Soul of Sichuan Cuisine

郫县豆瓣是辣椒随移民进入四川地区后，又一个移民文化与智慧的结晶。郫县豆瓣诞生于清代中期，制作工艺考究，具有色泽红褐、油亮光润、回味悠长等特点。将它运用于川菜烹饪之中，与仅用辣椒调味而表现为单纯的辣味不同，郫县豆瓣可使菜肴的辣味呈现出另外一番辣而不燥、酱香浓郁、入口醇厚的特性。郫县豆瓣的诞生，不仅丰富了川菜独有的麻辣味、家常味等复合味型的风味特色，而且成为回锅肉、麻婆豆腐、毛肚火锅、过江豆花等众多经典川菜的重要调味品。可以说，郫县豆瓣是近现代川菜领域最具标志性、基础性与核心地位的调料之一，故有“川菜之魂”之美誉。

Pixian chili bean paste is the quintessence of immigrant culture and wisdom after chili has been introduced to Sichuan with immigrants. It was born in the Mid-Qing Dynasty and has exquisite production technology. It has the characteristics of red and brown in color, oily and shiny, spicy with rich soy sauce flavor, moderate viscosity, mellow and long aftertaste, etc. It not only enriches the unique spicy & homely flavor characteristics of Sichuan cuisine, but also becomes an important seasoning for many classic Sichuan dishes, such as Twice-Cooked Pork, Mapo Tofu, Beef Tripe Hot Pot, Silken Tofu, etc. So to speak that Pixian chili bean paste is one of the most important symbolic, basic and core seasonings of modern Sichuan cuisine, known as the “Soul of Sichuan Cuisine”.

第一节 郫县豆瓣的历史及制作工艺
Section One The History and Workmanship of Pixian Chili Bean Paste

郫县豆瓣诞生于清代，历经民国时期，至新中国成立后得到快速发展，从传统制作工艺到现代生产工艺，在确保风味、品质不变的基础上，注重传承与创新的相互结合，逐步走上规模化、产业化和国际化的发展之路，知名度和美誉度也得以不断提升。

Pixian chili bean paste was born in the Qing Dynasty, promoted in the period of the Republic of China, and developed rapidly after the foundation of the People's Republic of China. From traditional production to modern technology, on the basis of invariable flavor and quality, with inheritance and innovation, Pixian chili bean paste constantly moving towards to large-scale, industrialization and internationalization. Popularity and reputation of Pixian chili bean paste are continuously improved.

一、郫县豆瓣的诞生与发展历程

晒场上的郫县豆瓣

据原益丰和酱园后人陈述宇、酱园管账李同德，原绍丰和酱园后人陈述启，原元丰源酱园管账张悦民共同撰写的《郫县豆瓣今昔》，以及郫县地方志编纂委员会编纂的《郫县志》记载，清代初年，四川因战乱而致人口锐减、土地荒芜、百业凋敝。自清代康熙年间开始实行移民政策，由此拉开了“湖广填四川”这一人口大迁徙的序幕。其间，福建汀州府永定县孝感乡翠亨人陈逸仙移民入川，见郫都土地肥沃、气候宜人，便在县城南门外一千米处落户，以制作酱油、麸醋、盐渍辣椒，到清代咸丰年间，由陈家后人陈守信将现文调巷陈家祠堂与南大街打通，设立益丰和酱园。陈守信，号益谦，取其益为号首，其年正值咸丰年间，以“丰”为时记，“和”则顺应天、地、人，且有“家和万事兴”之意，酱园由此得名。陈守信扩大生产，却发现盐渍辣椒易出水、不易保存，遂在祖辈的基础上潜心研究数年，加入适当比例的发酵胡豆瓣子，既能增加黏稠度、香醇度，又便于长期贮存，特别是日晒夜露后酿成的辣豆瓣酱更受欢迎。于是，真正意义上的郫县豆瓣就此诞生，其独特的制作技艺也基本定型。

I. The Birth and Development Process of Pixian Chili Bean Paste

According to the records of *History of Pixian Chili Bean Paste*, which was co-authored by Chen Shuyu, successor of old Yifenghe paste shop, Li Tongde, accountant of old Yifenghe paste shop, Chen Shuqi, successor of old Shaofenghe paste shop, and Zhang Yuemin, accountant of old Yuanfengyuan paste shop, as well as the records of *Chronicles of Pixian*, at the beginning of the Qing Dynasty, the population of Sichuan was sharply decreased because of the war, lands was laid waste and the region suffered from economic depression. Since the beginning of Kangxi Period, the immigrants from other provinces including Hubei and Hunan moved to Sichuan driven by the immigration policy at that time. During that period, Chen Yixian from Xiaogan Township, Yongding County, Tingzhou City, Fujian Province settled down in Sichuan. Seeing the fertile land and pleasant climate in Pidu, he settled 1 kilometer away from the South Gate of the county to make soy sauce, bran vinegar, and salted pepper. In Xianfeng Period, Chen Shouxin a descendant of the Chen family, thirled the Chen family ancestral temple in Wendiao Lane with the South Street and set up Yifenghe paste shop. Chen Shouxin, named Yiqian. He took Yi as the first word for paste shop. Feng meant the year of Xianfeng. "Feng" and "He" implied heaven, earth and people, and the meaning of "a peaceful family will prosper". Therefore, the paste shop is named. Chen Shouxin expanded production and found that the salting chili was easy to water out and difficult to be preserved. After years' research on the basis of ancestors, Chen shouxin found that adding an appropriate proportion of fermented bean petals can not only increase the viscosity and mellow degree, but also facilitate long-term storage. In particular, the spicy bean paste made after weather exposure is very popular. Therefore, the real Pixian chili bean paste was born, and its unique workmanship was basically finalized.

在陈守信的经营下，益丰和酱园的生产规模不断扩大，后又把西街的顺天酱园合并，统一生产、销售。最初雇工匠十人左右，年产豆瓣酱数千斤，主要销往县城及近邻场镇饮食店铺和城关居民食用。陈守信于清代光绪年间去世，益丰和酱园由其第六子陈竹安经营，产量逐步上升到三四万斤。清光绪三十一年(1905年)，彭县人（今彭州市）弓鹿宾在“陕西帮”的支持下，到郫都城东街开设了元丰源酱园，改变了益丰和独家经营的局面。1931年，陈竹安之侄陈文换在城南外街李家花园开设绍丰和酱园，主要酿制郫筒

陈守信研究郫县豆瓣

酒，同时兼营郫县豆瓣，由此形成益丰和、元丰源、绍丰和三足鼎立的格局。

Under the operation of Chen Shouxin, Yifenghe expanded the production scale of bean paste, and continue merged the Shuntian paste shop in West Street to produce and sell uniformly. At first, there were about 10 workers, with an annual output of thousands of tons of bean paste, which was mainly sold to the catering shops in the county, nearby towns and the residents living those ares. During Guangxu Period, Chen Shouxin died. Yifenghe paste shop was operated by his sixth son Chen Zhu'an, and the output of bean paste increased to 15,000 or 20,000 kg. In the 31st year of Guangxu Period (1905 A. D.), Gong Lubin from Pengxian County, with the support of the "Shaanxi Gang", opened "Yuanfengyuan" paste shop in Eastern Street of Pidu, which broke the monopoly of Yifenghe. In 1931, Chen Wenhuan, the nephew of Chen Zhu'an, opened Shaofenghe paste shop in Li' Garden, outer street of southern city, mainly brewing Pitong Wine and making Pixian chili bean paste as well. Thus, Pixian chili bean paste has formed a tripartite pattern of Yifenghe, Yuanfengyuan and Shaofeng.

郫县豆瓣走出四川始于民国初年。此时，外地客商到郫都来做生意，常购买郫县豆瓣馈赠亲友，于是，郫县豆瓣的名声越传越远，销量逐步扩大，大多经成都销往重庆、湖北、贵州、陕西等地。民国4年（1915年），四川军政府派员去西藏犒赏川军，在益丰和与元丰源各订购豆瓣三四万斤，两家酱园互比质量，如期交货，都用荷叶和油纸作为内封，外面用竹篓包装，所有豆瓣酱全用人力挑运，三个多月后才运到驻藏川军营地，产品色味不变，深受官兵欢迎。四川军政府特此传令嘉奖两家酱园并发给奖牌。到20世纪30年代，郫县豆瓣还因质量上乘被四川省劝业会评为酿造优等品。但进入20世纪40年代后，由于管理不善、社会动荡等原因，郫县豆瓣的生产受到极大影响。

Pixian chili bean paste has been sold out of Sichuan at the beginning of the Republic of China. At this time, travelling businessmen who did business in Pidu purchased Pixian chili bean paste as gifts for relatives and friends. The reputation of Pixian chili bean paste spread further and the sales expanded. Most of Pixian chili bean paste has been sold to Chongqing, Hubei, Guizhou, and Shaanxi etc. via Chengdu. In the fourth year of the Republic of China (1915 A. D.), Sichuan Military Government purchased about 15,000 or 20,000 kg bean paste in Yifenghe and Yuanfengyuan respectively. It was sent to Tibet to reward the Sichuan army. Those two shops compared with each other on quality and delivery goods on schedule. Both of them were sealed bean paste with lotus leaves and oil paper, and packed with bamboo baskets outside. Those bean pastes were transported manually, and reached to the Sichuan military camp in Tibet more than three months later, which kept the original flavor and was very popular among soldiers. The Sichuan Military Government hereby awarded the two bean paste shops and issue medals. Hence the military government awarded those two shops medals. In 1930s, Pixian chili bean paste was named as excellent brewed goods by Sichuan Chamber of Commerce because of its high quality. However in 1940s, the production of Pixian chili bean paste has been affected deeply by poor management and social instability.

中华人民共和国成立后，郫县豆瓣的生产得以逐步恢复，并进入到快速发展期。1955年公私合营后，益丰和、绍丰和、元丰源三家酱园合并为四川省地方合营郫县酱园，后更名为国营郫县豆瓣厂。在政府支持下，扩建厂房、增添设施，郫县豆瓣的传统生产工艺逐步向现代生产工艺发展，产量不断上升，郫县豆瓣也随之有计划地销往全国各地，并随着川菜的发展走向海外。改革开放后，郫县豆瓣开始销往日本、泰国、前南斯拉夫、美国等国家和我国香港地区，供当地川菜馆使用。

After the founding of the People's Republic of China, the production of Pixian chili bean paste gradually recovered and entered a period of rapid development. In 1955, Yifenghe paste shop, Shaofenghe paste shop and Yuanfengyuan paste shop merged into one joint state-private enterprise, and later renamed state-owned Pixian Chili Bean Paste Factory. With the support of the government, the traditional production process of Pixian chili bean paste has gradually developed to modern production process, and the output has been increasing. Pixian chili bean paste has been sold all over China, and has gone overseas with the development of Sichuan cuisine. After the reform and opening up, Pixian chili bean paste has been sold to Japan, Thailand, former Yugoslavia, the United States and other countries, as well as Hong Kong region of China, for the use of local Sichuan restaurants.

1999年，四川省郫县豆瓣股份有限公司成立，并经中华人民共和国对外贸易经济合作部批复取得进出口经营权，同年12月，由国家国内贸易局认证“鹃城牌”郫县豆瓣为中华老字号。2008年，“郫县豆瓣”传统制作技艺被列入国家非物质文化遗产保护名录。近年来，郫县豆瓣、丹丹、饭扫光、鑫鸿望等郫县豆瓣生产企业，不断对生产设备和制作工艺进行升级换代，同时还对产品结构进行了大幅调整和完善，彻底打破了之前仅仅生产烹饪型郫县豆瓣的单一模式，着力开发了一系列创新产品，包括调味系列、即食系列、佐餐系列、火锅底料等全新产品，销售范围遍布全国各地，并出口至数十个国家和地区。伴随着郫县豆瓣产业链的不断延伸，也同时带动了原料种植、食品包装、运输、竹编等相关产业的发展。改革开放以后，随着川菜的大发展、大繁荣，郫县豆瓣在川菜中的使用更加广泛，再加上生产经营不断走向规模化、产业化和国际化，营销推广力度持续加强等原因，其国内外知名度、美誉度和影响力均得到空前提升。

In 1999, Sichuan Pixian Chili Bean Paste Incorporated Company was established and approved by the Ministry of Foreign Trade and Economic Cooperation to obtain the import and export operation right. In December of the same year, the Ministry of Domestic Trade of the People's Republic of China certified “Juancheng” Pixian Chili Bean Paste as a time-honored brand in China. In 2008, “Pixian chili bean paste” traditional production technology was listed in the National Intangible Cultural Heritage protection list. In recent years, the production equipment and production process of Pixian chili bean paste production enterprises such as Pixian Douban, Dan Dan, Fansaoguang, Xinhongwang, etc. are constantly upgraded, and the product structure is constantly adjusted and improved. From the single cooking bean paste, it has developed multi-purpose and multi series products, including seasoning, instant foods, condiment, hot-pot seasoning, etc. which are sold all over the country and exports to dozens of countries and regions. The continuous extension of Pixian chili bean paste industrial has also driven the development of raw material planting, food packaging, transportation, bamboo weaving and other related industries. In addition, after the reform and opening up, with the great development of Sichuan cuisine, Pixian chili bean paste was more widely used in Sichuan cuisine. Due to the continuous large-scale production, industrialization and internationalization, and the continuous strengthening of marketing promotion, Pixian chili bean paste improved its popularity, reputation and influence both at home and abroad.

二、郫县豆瓣的制作工艺

郫县豆瓣是以辣椒、蚕豆、面粉、食盐等为原料，再经加工、发酵制成，其制作工艺分为传统制作工艺和现代化生产工艺，两者的主要差异在于生产设备和工具的不同。传统制作工艺使用的生产设备和工具主要是竹木器具和陶制酱缸；现代生产工艺使用的生产设备和工具大多是机器及现代化、智能化设备，机械化、工业化、智能化程度高，但两者的制作工艺流程则基本相同，即先制作蚕豆醅、辣椒醅，再将它们混合发酵制成郫县豆瓣。其具体工艺流程如表3—1。

II. Workmanship of Pixian Chili Bean Paste

Pixian chili bean paste is made by processing and fermentation with chili, broad bean, flour and salt, etc. Its workmanship can be divided into traditional workmanship and modern production technics. The main difference between them lies in the difference of production equipment and tools. Traditional workmanship mainly uses bamboo and wooden equipment, fictile paste jar. Modern production technics mainly use machines and modern intelligent equipment, with high level of mechanization, industrialization and intellectualization. However, the basic procedures are almost the same. Firstly, workers make fermented broad bean and fermented chili, and then mix them to make Pixian chili bean paste by fermenting. The specific procedure is in Figure 3-1.

（一）郫县豆瓣的传统制作工艺

郫县豆瓣是郫都得天独厚的自然环境与郫都人匠心浇注共同作用的结果。郫县豆瓣的生产非常注重温度、阳光和水质条件的相互作用，传统制作工艺考究，突出强调翻、晒、露三道工序，它们相辅相成，缺一不可，即所谓“温、光、水、翻、晒、露”六字真诀。

1. Traditional Workmanship of Pixian Chili Bean Paste

Pixian chili bean paste is the result of endowed environment and craftsmanship of Pidu people. The production of Pixian chili bean paste attaches special importance of temperature, sunshine and water quality. The traditional workmanship is very fastidious and emphasizes three processes of weather exposure: stirring, drying and dewing. Those necessary procedures supplement each other. They complement each other and are indispensable, that is, the so-called six word true secrets of making Pixian chili bean paste — temperature, light, water, stirring, drying and dewing.

郫都区属亚热带季风气候，雨量充沛，相对湿度在83%左右，气候温润，无酷暑、无严寒，无霜期长，年平均温度16℃，最高平均温度35.3℃，最低平均温度4℃，这些自然条件，为郫县豆瓣的酿造提供

表3—1 郫县豆瓣的制作工艺流程

Figure3-1 Technological Process of Making Pixian Chili Bean Paste

郫县豆瓣的传统制作工艺

了非常适宜的温度环境。郫都区“春早、夏长”，日照率为24%～32%，年平均太阳辐射总量少，紫外线强度偏低，为环境中微生物群落的结构稳定提供了适宜的光照条件，营造出适合酿造郫县豆瓣的独特工艺环境。郫都区处于都江堰岷江水系自流灌溉区上游地带，是成都市饮用水源重点保护区域，水源无污染，地表土层由第四系沉积物发育而成，土层深厚，富含磷、钾、钙、镁等矿物质及微量元素，为郫县豆瓣的酿造提供了优质水源。

Pidu District is located in the sub-tropical monsoon climate, with abundant rainfall and relative humidity of about 83%. The climate is warm and moist, without shrive heat and cold, and long frost free period. The annual average temperature is 16℃, the maximum average temperature is 35.3℃, and the minimum average temperature is 4℃, which provides a suitable temperature environment for making Pixian chili bean paste. Spring is early and summer is long, relative sunshine duration here is 24%-32%. The annual average total amount of solar radiation is low and the ultraviolet intensity is low, which provides suitable light conditions for the stability of microbial community structure in the environment and creates a unique process environment for the brewing of Pixian chili bean paste. In another aspect, Pidu district is located in the upstream of the gravity irrigation area of Dujiangyan, Minjiang River system. It is the key protection area of drinkinml water sources in Chengdu. The water source is pollution-free. The surface soil layer is developed from quaternary sediments, and rich in minerals such as phosphorus, potassium, calcium, magnesium, and trace elements, which has provided high-quality water source for Pixian chili bean paste brewing.

具体而言，郫县豆瓣的传统制作工艺主要包括以下两个阶段：

第一阶段是蚕豆醅、辣椒醅的制作。蚕豆也称胡豆，蚕豆醅的制作方法，首先是蚕豆瓣的脱壳、烫瓣，即精选上等二流板蚕豆，先用石磨碾压脱壳，再用热水漂烫后放入冷水中冷却、浸泡，然后滤掉多余的水分；其次是拌面粉、制曲、发酵，即按照一定的比例在蚕豆瓣中加入精面粉拌匀，放入竹篓中接种米曲酶，之后移入曲房，使之自然发酵，形成成熟的甜瓣子。辣椒醅的制作方法是精选二荆条海椒，经过去蒂、除杂、清洗等工序，沥干水分后剁碎，再添加食盐混合均匀，放入槽桶中密封储存，通过盐渍发酵。

Specifically, the traditional workmanship for making Pixian chili bean paste mainly includes the following two stages.

The first stage is making fermented broad bean and fermented chili. Broad bean is also called as lima bean. The first step is the shelling and scalding, that is, select the first-class & second-class beans, roll them with a stone mill, scald them with hot water, cool them in cold water successively, soak them, pick them up and filter out the water; The second is flour mixing, koji making and fermentation, that is, add refined flour into beans in proportion, well mixed and put it into a bamboo basket, inoculate rice koji enzyme and to the koji room to make it ferment naturally and form mature sweet bean. Fermented chili is to select erjingtiao pepper, go through the process of edulcoration and cleaning, crushing, salt adding, and put them into a tank for sealed storage to make them salted and fermented.

第二阶段是蚕豆醅、辣椒醅的混合发酵阶段，其中的关键是“翻、晒、露”三道工序。一是翻：将辣椒醅和甜瓣子按照特定比例配合后放入陶制缸中，每日以特制木杵上下翻动，木杵从缸周入、缸心出，每缸翻搅15次以上，将物料拌和均匀，同时使其释放出酿制过程中产生的杂味，导入新鲜空气，保障发酵质量。二是晒：在白天充分利用阳光提供的能量，促进酱醅中功能微生物的生长繁殖，并保持一定的发酵温度，为各种酶类发挥作用创造适宜条件。三是露：在无雨的夜晚，随着昼夜温差所产生的露珠凝结，将空气中的有益微生物沉降到酱醅表面，丰富酿制过程中的微生物种类，通过酱醅温度降低带来的变温发酵，为微生物代谢产物的多样性创造条件，共同促进郫县豆瓣独特风味物质的形成。

The second stage is the mixed fermentation stage of fermented broad bean and fermented chili, in which there are three key processes — stirring, drying and dew. Stirring: fermented chili and sweet beans are mixed in a specific proportion and put into a pithos. Workers use a special wooden pestle stir the paste up and down every day. The wooden pestle enters from the cylinder periphery to center. Each pithos is stirred more than 15 times to mix the materials well, release the miscellaneous flavor generated in the brewing process, and let fresh air in, which can ensure the fermentation quality. Drying: make full use of the energy provided by sunlight during the day to promote the growth and reproduction of functional microorganisms in bean paste, and maintain a certain fermentation temperature to create suitable conditions for various enzymes. Dewing: during the night without rain, with the condensation of dew caused by the temperature difference between day and night, beneficial microorganisms in the air settle on the surface of fermented bean paste can enrich the microbial species in the fermentation process. The variable fermentation caused by the reduced temperature of

郫县豆瓣的现代生产工艺（刘贵明/摄）

bean paste creates conditions for the diversity of microbial metabolites, and promotes the formation of unique flavor of Pixian chili bean paste.

（二）郫县豆瓣的现代生产工艺

采用传统工艺生产郫县豆瓣，由于工艺复杂、生产周期长、生产成本较高等原因，产量会因此受到很大限制，从而制约了其规模化、产业化发展进程。改革开放以后，尤其是进入21世纪以来，随着人们饮食需求的提高和川菜在国内外的繁荣发展，以及工业4.0、物联网、人工智能等高科技的兴起，郫县豆瓣的知名度和需求量也得到不断提高，如果仅仅依靠传统工艺进行生产，其产量将不能满足消费者日益增长的需求，因而必须不断引入新机器、新设备与新技术，通过现代生产工艺进行大规模工业化生产，这样才可与市场需求相吻合。

2. Modern Production Technics of Pixian Chili Bean Paste

Due to complicated process, long production cycle and high production cost, the output of Pixian chili bean paste was limited, which restricts its development on scaling up and industrialization. After the reform and opening up, especially since the beginning of the 21st century, with the improvement of people's dietary demands, the prosperity of Sichuan cuisine at home and abroad, as well as the rise of high technologies such as industry 4.0, internet and AI, the popularity and demand of Pixian chili bean paste have continuously increased. The traditional workmanship cannot meet the growing needs of consumers in terms of its output. New machines, new equipment and new technologies have been continuously introduced to carry out mass-production and industrialization, which help to match the market demand.

1. 蚕豆醅的加工

制作蚕豆醅，其工艺流程依然是蚕豆瓣的脱壳、烫瓣、拌面粉、制曲、发酵。其中，常用的脱壳方法有两种：一种是湿法脱壳，一般采用2%的NaOH溶液浸泡后脱壳，其特点是瓣粒比较完整，但需注意脱皮及浸泡时间不宜过长，否则会使豆瓣变得僵硬，制曲发酵以后也不易软解，影响成品质量。二是干法脱壳，即使用脱壳机对蚕豆进行批量脱壳，可大大提高劳动效率，减轻劳动强度，改善卫生条件。制曲时多采用人工添加米曲霉等菌种，再配置自动发酵设备，通过自动通风、控温、控湿等手段来保障成曲质量，采用这些新技术，可避免传统制曲工艺往往会掺杂其他杂菌而带来的安全隐患。米曲霉在制曲过程和发酵初期均有着十分重要的作用，必须让其在此过程中占据主要优势，避免其他杂菌在制曲、发酵阶段造成污染。

1. Processing of Fermented Broad Bean

The technological process of making fermented broad bean is still shelling, scalding, flour mixing, koji making and fermentation. There are two ways of exuviation. One is wet shelling. Normally, 2% NaOH solution is used for soaking and exuviation. Beans processed by this way can keep complete shape, but it is often necessary to pay attention to the peeling and soaking time, otherwise the beans will become stiff, and it is not easy to soften after koji making and fermentation, which will affect the quality of products. Another way is dry shelling. Sheller is used for mass exuviation of broad beans, which improves the labor efficiency greatly, alleviates the labor intensity and improves the sanitary conditions. During koji making, manual addition of Aspergillus oryzae is mostly used, and new automatic fermentation equipment is configured to automatically ventilate, temperature control, humidity balance for koji making, which can ensure the

蚕豆醅制作工艺

quality of koji making and eliminate the potential safety hazards caused by other miscellaneous bacteria in traditional koji making. Aspergillus oryzae plays an important role in the koji making process at the beginning stage of fermentation. It must occupy the main superiority in this stage to prevent pollution caused by other miscellaneous bacteria in the fermentation.

辣椒醅的加工工艺（严建国/摄）

2. 辣椒醅的加工

制作辣椒醅，其工艺流程主要是辣椒去蒂、除杂、清洗、破碎、盐渍、发酵等。其中，破碎时选用专用粉碎机粉碎椒坯。盐渍一般是将辣椒碎放入大型盐水循环发酵池中，并配置抽水泵，将发酵池下层的水抽取后浇淋在辣椒碎表面，使盐水上下循环，促进盐渍均匀。此后即可覆盖池面，封存发酵。

2. Processing of Fermented Chili

The process of making fermented chili is picking off stalk, edulcoration, cleaning, crushing, salting and fermentation, etc. Workers use shredding machines to crush the chili. During the process of salting, crushed chili is generally put into a large salt water circulating fermentation tank, and a water pump is configured to extract the water at the lower layer of the fermentation tank and pour it on the surface of the crushed pepper, so as to circulate the salt water up and down and promote salting evenly. After that, the tank can be covered and sealed for fermentation.

3. 混合发酵

蚕豆醅和辣椒醅的混合发酵，大多采用大型发酵池或自动发酵设备。发酵池上常配置现代自动大棚和可移动天窗，以实时自动监控发酵晒场的温度、湿度、含氧量，调节光照度、强制通风，同时用机器自动翻搅，从而实现“白天翻、夜晚露、晴天晒、雨天盖”的良好发酵效果，极大地提高了生产效率及产量。

3. Mixed Fermentation

The mixed fermentation of fermented broad bean and fermented chili mostly adopts large fermentation tank and automatic fermentation equipment. The fermentation tank is often equipped with modern automatic shed and movable skylight to automatically monitor the temperature, humidity and oxygen content of the fermentation in real time, as well as the illumination intensity and forced ventilation. Meanwhile, the machine is used to stir automatically, so as to realize the good fermentation of “stirring during day time, dewing in night, drying in sunny day, sheltering in rainy day”. Thus the output and productivity of Pixian chili bean paste have been increased greatly.

总之，郫县豆瓣的现代生产工艺传承“晴天晒、雨天盖，白天翻、夜晚露”的古法技艺精髓，又融入了现代化加工技术与设备，同时在各个生产环节实施质量监测与管理，不仅能有效地提高生产效率、降低劳动强度，而且还能更好地保证产品质量的稳定性和安全性，为郫县豆瓣的规模化、产业化、国际化发展奠定坚实基础。

In brief, the modern production technics of Pixian chili bean paste has inherited the core of ancient workmanship —

"drying in sunny day, sheltering in rainy day, stirring during day time, dewing in night". Integrating modern technology and equipment and carrying out quality monitoring and management in all production links can not only effectively improve production efficiency and alleviate labor intensity, but also ensure the stability and safety of product quality, which has laid a solid foundation for the mass production, industrialization and internationalization of Pixian chili bean paste.

第二节 郫县豆瓣的风味奥秘及在川菜中的地位与作用

Section Two The Flavor Secrets of Pixian Chili Bean Paste and Its Position and Function in Sichuan Cuisine

郫县豆瓣以辣椒、蚕豆、面粉、食盐等为原料，通过前后两个发酵阶段制作而成，风味极具特色。伴随着郫县豆瓣在川菜烹饪中的广泛运用，从而造就了部分川菜菜品别具一格的独特风味，郫县豆瓣也因此成为川菜烹饪中的一款标志性调味料。

Through pre-fermentation and post-fermentation stages, Pixian chili bean paste is made of chili, broad bean, wheat flour, salt, etc. It has the characteristics of red and brown in color, oily and shiny, spicy with rich soy sauce flavor, moderate viscosity, mellow and long aftertaste, etc. Therefore, it has been widely used in Sichuan cuisine, making Sichuan dishes in a unique flavor and become an indispensable symbolic seasoning of Sichuan cuisine.

一、郫县豆瓣的风味奥秘

从滋味和气味来看，郫县豆瓣的风味特征主要是咸鲜醇厚、辣而不燥、酱香浓郁，并且随着后发酵时间的适当增加，其酱香风味更加浓郁，其成因，主要来自原料在生产环节中所形成的多种微生物的生理、生化反应。

I. The Flavor Secrets of Pixian Chili Bean Paste

From the taste and smell, the flavor characteristics of Pixian chili bean paste are mainly salty, fresh and mellow, spicy and in rich paste flavor. With the appropriate longer post-fermentation, its flavor becomes stronger. The reason is mainly from the physiological and biochemical reactions of various microorganisms in raw materials and production processes.

在郫县豆瓣生产过程中不同的发酵阶段，由于多种微生物的生理、生化反应不同，从而形成了其独特的风味体系，氨基酸、有机酸、脂肪酸和糖类物质是造就其滋味的主要成因，虽然后期发酵产生的香味成分含量极微，却对其气味的形成产生了很大作用。其中，郫县豆瓣的咸味主要来源于添加的食盐，其鲜味主要来自蛋白质的分解产物，即原料中的蛋白质在发酵过程中所产生的蛋白酶、肽酶、谷氨酸胺酶，经水解后生成多种游离氨基酸，其中谷氨酸含量最高，由此产生了郫县豆瓣的鲜味。此外，郫县豆瓣的醇厚与它自身所拥有的极微弱的甜味与酸味有密切关系。其极微弱的甜味来自三个方面：一是原料中的淀粉水解所生成的葡萄糖和麦芽糖；二是原料中的蛋白质分解所生成的部分呈甜味的氨基酸；三是原料中的油脂水解生成的味道微甜的甘油。其微弱的酸味来源于混合发酵过程中产生的乳酸、醋酸等有机酸。郫县豆瓣的辣味来源于辣椒，辣椒在经盐渍及发酵后，其所含的3−甲基−1−丁醇会相应减少。据研究表明，这可能是造成辣椒坯辛辣刺激味损失的原因，从而使得郫县豆瓣形成辣而不燥的风味特征。

Due to the physiological and biochemical reactions of a variety of microorganisms in different fermentation stages,

Pixian chili bean paste forms its unique flavor.Amino acid, organic acids, fatty acids and sugars are the main sources of its flavor. The content of flavor components produced by post- fermentation is few, but playing a great role in the formation of its flavor. Salty taste of Pixian chili bean paste mainly comes from the salt. Its delicate flavor mainly comes from the protein decomposition, that is, the protein in the raw material is hydrolyzed to a variety of free amino acids after the action of protease, peptidase and glutamine synthetase produced in this process, of which the content of glutamate is the highest, resulting in the delicate flavor of Pixian chili bean paste .Moreover, the mellowness of Pixian chili bean paste is closely related to its tiny sweet and sour taste. Its extremely tiny sweetness comes from three aspects; namely, glucose and maltose produced by the hydrolysis of starch in the raw material, some sweet amino acids produced by the decomposition of protein in the raw material, tiny sweet glycerol produced by the hydrolysis of oil in the raw material. Its tiny sour taste comes from organic acids created in mixed fermentation, like lactic acid and acetic acid. Its spicy taste comes from chili. 3-methyl-1-butanol in Chili will be reduced after salting and fermenting, which could be the reason for the loss of pungent taste of chili. Hence Pixian chili bean paste is spicy but not pungent.

郫县豆瓣香味丰富，尤其是酱香突出而浓郁，这主要是源于它拥有众多的香味成分，包括醇类、醛类、酯类、酚类、有机酸类、含硫化合物、呋喃酮类等。据研究表明，郫县豆瓣中的主要呈香物质是4-乙基愈创木酚，而醇类、酯类、醛类等挥发性组分都是混合发酵期才形成的。郫县豆瓣在其发酵过程中，后期呈香物质会不断增多，增加的主要成分是杂环类化合物，说明风味化合物的检出数量会随着后熟时间延长而略有增加。其中，酱香味与3-甲硫基丙醛、糠醛和川芎嗪紧密相关。郫县豆瓣的成香机制较为复杂，大体可归纳为四类：一是由原料成分所产生；二是由米曲霉代谢产物生成；三是由耐盐酵母、细菌的

表3-2 郫县豆瓣特征挥发性物质演变及其香型特性变化

发酵时段	原料	特征挥发性物质		香型特性变化
		发酵前	发酵后	
发酵前期	辣椒	3-甲基-1-丁醇、正己醇、壬醛、(E)-2-壬烯醛、愈创木酚和4-乙基苯酚等	乙醇、芳樟醇、苯乙醇、乙酸、3-羟基-2-丁酮、苯甲醛、苯乙醛、乙酸乙酯等	发酵后，辣椒原料中的特征性挥发性物质成分浓度均有不同程度降低。发酵后，在辣椒胚中新生成了重要风味成分
	蚕豆	乙醇、正己醇、1-辛烯-3-醇等	乙醇、正己醇、1-辛烯-3-醇、3-甲基丁醛、2-甲基丁醛、3-甲硫基丙醛和苯乙醛、乙酸乙酯等	发酵后，醇类物质赋予产品醇香、果香和蘑菇香；醛类物质赋予产品麦芽香、酱香和蜂蜜香等；酯类物质赋予产品浓郁的花果香和甜香味
发酵后期	辣椒醅+蚕豆醅	其挥发性物质是辣椒醅和蚕豆醅发酵后的总和（见上）	2-甲基-1-丁醇、3-甲基丁醛、2-甲基丁醛、苯甲醛、乙酸乙酯、4-羟基-5（2）-乙基-2（5）-甲基-3（2H）-呋喃酮(HEMF)等	混合发酵后，郫县豆瓣中多种风味成分浓度显著增加，它们主要赋予郫县豆瓣麦芽香、果香、烟熏香和酱香等香气特征。此外，HDMF和HEMF可赋予郫县豆瓣独特的焦糖香味。

Table 3-2 Evolution of Volatile Aroma Compounds and Changes of Aroma Characteristics of Pixian Chili Bean Paste

Fermentation Stage	Raw Material	Volatile Aroma Compounds		Changes of Aroma Characteristics
		Before Fermentation	After Fermentation	
Early Stage	Chili	3 - methyl - 1-butyl alcohol, n-Hexanol, nonanal, (E)-2-Nonenal, guaiacol, 4-Ethylphenol, etc.	Ethanol, linalool, Phenethyl alcohol, acetic acid, 3-hydroxy-2-butanone, benzaldehyde, Phenylacetaldehyde, Ethyl acetate, etc.	After fermentation, the changes of aroma characteristics in chili will drop on different levels. Two alcohols will reduce significantly. New important flavor will be created in chili embryo after fermentation.
	Broad Bean	Ethanol, n-Hexanol, 1-octene-3-alcohol, etc.	Ethanol, n-Hexanol, 1- octene-3-alcohol, 3-Methylbutyraldehyde, 2-Methylbutyraldehyde, 3-methylthiopropanal, Phenylacetaldehyde, Ethyl acetate, etc.	After fermentation, alcohols will bring mellow aroma, fruit aroma and mushroom aroma to the products; aldehydes material will bring malt aroma, paste aroma and bee honey aroma to the products; ester will bring thick flower and fruit aroma , and sweet aroma to the products.
Later Stage	Fermented Chili and Broad Bean	Volatile substances are the sum of those created after fermentation of chili and broad bean. (vide supra)	2-Methyl-1-butanol, 3-Methylbutyraldehyde, 2-Methylbutyraldehyde, benzaldehyde, Ethyl acetate, ethyl isovalerate, Ethyl hexanoate, 4-hydroxy-5(2)-ethyl-2 (5)-methyl -3 (2H)-furanone(HEMF), etc.	After mixed fermentation, concentration of diverse flavor components in Pixian chili bean paste will be increased significantly. Those components bring malt aroma, fruit aroma, smoking aroma and paste aroma to Pixian chili bean paste. In addition, HEMF will bring special caramel aroma to the paste.

代谢产物生成；四是由非酶化学反应生成。在整个生产过程中，郫县豆瓣的呈香物质会不断改变和增加，卢云浩、何强等人较为详细地研究出郫县豆瓣特征挥发性呈香物质演变及其香型特性变化，具体情况见表3-2。从表中可以看出，郫县豆瓣发酵后所具有的特征挥发性呈香物质十分多样，使得产品具有醇香、酸香、花香、果香、焦糖香、烟熏香及酱香等多种香型特征，香气中的香味非常丰富。结合感官评价，其中的酱香尤为突出。

Pixian chili bean paste has rich aromas and flavors, especially the paste aroma, which is mainly due to its many aroma components like alcohols, aldehydes, esters, phenols, organic acids, sulfocompound, furanone, etc. According to research, the main aroma component of Pixian chili bean paste is 4-ethyl guaiacumwood phenol. The volatile ingredients, like alcohols, aldehydes, esters, are formed in mixed fermentation stage. The aroma substances of Pixian chili bean paste increased continuously in the process of fermentation, mainly heterocyclic compounds, indicating that the amount of flavor compounds increased with the extension of time. Sauce flavor is closely related to 3-methylthio, furfural and tetramethylpyrazine. The mechanism of aroma formation is complex, which can be divided into 4 categories generally; namely, from the raw materials, from metabolite of Aspergillus oryzae, from metabolite of salt-tolerant yeasts and bacteria, created by non-enzymatic reactions. In the whole process of production, aroma substances of Pixian chili bean paste are constantly changing and increasing. Evolution of volatile aroma compounds and the changes of aroma characteristics of Pixian chili bean paste were studied in details by Lu Yunhao and He Qiang, which are in Table 3-2. We can see that the fermented Pixian chili bean paste has a variety of volatile aroma substances, which makes the products have a variety of aroma characteristics, such as mellow, sour, flower aroma, fruit aroma, caramel aroma, smoke and paste aroma. Based on sensory evaluation, paste aroma is the most outstanding.

二、郫县豆瓣在川菜中的地位与作用

郫县豆瓣因其独特的风味，而在川菜烹饪中得到广泛运用，川菜中的许多名菜都与它息息相关，在川菜的独特味型中也能常常看到它的身影。近现代以来，郫县豆瓣已成为川菜标志性、基础性与核心地位的烹饪调料之一，其重要价值主要体现在以下两个方面。

II. The Position and Function of Pixian Chili Bean Paste in Sichuan Cuisine

Pixian chili bean paste is widely used in Sichuan cuisine because of its unique flavor. Many famous Sichuan dishes cannot be cooked without it. It helps to form the unique flavor in Sichuan cuisine. In modern times, Pixian chili bean paste has become one of the most important symbolic, basic and core seasonings of Sichuan cuisine. Its status and function are as follow.

（一）近现代川菜独特味型的标志性复合调料

在菜肴烹制过程中，调味是决定菜肴风味的关键，而调料是其重要的物质基础。根据调料的呈味方式，常分为单一味调料和复合味调料。其中，复合味调料的使用则更有利于方便、快捷地确定菜肴风味，因而广受欢迎，并成为当代调味品产业发展的一个重要方向。其实，早在宋代就已经出现了一种复合味调料，名为“一了百当”。据陈元靓《事林广记》“一了百当”条载：“甜酱一斤半，腊槽一斤，麻油七两，盐十两，川椒、马芹、茴香、胡椒、杏仁、姜、桂等份，为末。先以油就锅熬香，将料末同槽，酱炒熟，入器。”此后，可以随时“修馔”，“料足味全，甚便行饔”，即做菜时酌量加入，尤其适宜外出时做菜使用。元代时，复合味调料逐渐增多，仅在倪瓒所著的《居家必用事类全集》中，就记有“鲙醋”的制法及“芥辣醋”“五辣醋”等复合味调料，常用于多种菜肴及汤品的风味调制。

1. Symbolic Compound Seasoning of the Unique Flavor of Modern Sichuan Cuisine

Seasoning as an important material basis, is the key to determine the flavor of dishes. According to the flavor feeling, seasonings are usually divided into single flavor seasoning and compound flavor seasoning. The use of compound flavor seasoning is more helpful to fix the flavor of dishes easily and quickly, which is widely welcomed by people, and becomes the important developing trend of modern seasonings industry. In fact, one compound seasoning named as “Yi Liao Bai Dang” (all matters end when the main matter ends) was invented in the Song Dynasty. *Shi Ling Guang Ji* written by Chen Yuanliang has recorded the making of “Yi Liao Bai Dang”. “Prepare 1.5 jin of sweet paste, 1 jin of vinasse, 7 liang of sesame oil, 10 liang of salt, and powders of Sichuan chili, Cuminum cyminum, fennel, pepper, almond, ginger, laurel. Firstly boil the oil in pot; secondly add above materials to fry; and then keep the completed paste in Jar”. After that, people can “fix the dishes” at any time, and “it's full of materials and flavors, and is especially convenient”. That is, when cooking, use it as appropriate, especially suitable for cooking when travelling. In the Yuan Dynasty, compound seasonings gradually increased. *The Complete Collection of Household Necessities* written by Ni Zan has recorded the ways of making “Kuai vinegar”, “mustard-spicy vinegar”, “five-spicy vinegar”, etc. Those seasonings were used for making diverse dishes and soups.

在四川，至清代中叶以后，随着辣椒的引入，尤其是郫县豆瓣作为复合味调料诞生之后，因其逐渐运用于川菜制作之中，并凭借其特有风味而大大丰富了川菜的烹饪味型。作为一种复合味调料，郫县豆瓣比单一味调料具有更多的优势，其色泽红褐、质地油润、瓣子酥脆，极微弱的甜味、酸味，乃至多种多样的香味，都使得郫县豆瓣在咸鲜的基础上，更加凸显出味道醇厚、辣而不燥、酱香浓郁的特性。郫县豆瓣在川菜调味中的广泛使用，不仅极大地丰富了川菜的风味特征，也在很大程度上确立和提升了其核心调味品的地位，如麻辣味型、家常味型、鱼香味型、酸辣味型等川菜烹饪中的独特味型，大多都将郫县豆瓣作为一种重要的复合调味料参与其中。根据商务部发布的行业标准《川菜烹饪工艺》（SB/T 10946-2012）载，麻辣味型是由花椒、辣椒油、干辣椒、郫县豆瓣、食盐、白糖、酱油、芝麻油等调料调制而成，色泽红亮、麻辣咸鲜、香味浓郁；家常味型主要是以郫县豆瓣、食盐、酱油、豆豉、甜面酱等调料调制而成，色泽红亮、咸鲜微辣，具有多种调料的混合香味；鱼香味型则是由食盐、白糖、酱油、醋、泡辣椒、郫县豆瓣、姜、蒜、葱、辣椒油、芝麻油等多种调料调制而成，具有色泽红亮，且咸、酸、甜、辣各味平衡等特征。在这些独特的川菜味型中，郫县豆瓣的主要作用，除了增色，更是增辣、增香、增鲜，以及增加醇厚的复合味感。在四川民间和餐饮行业中，曾经还将一种味型称为“豆瓣味”，就是源于豆瓣酱是其主要调料，且用量较大，这类菜品因具有咸鲜微辣、略带甜酸、豆瓣酱香味突出而得名。不仅如此，一些川菜菜品的命名也直截了当冠之以“豆瓣”字样，如豆瓣鲜鱼、豆瓣肘子、豆瓣子姜等，其共同点都是在制作菜品时会使用到郫县豆瓣。由此可言，郫县豆瓣是铸就川菜独特味型当之无愧的标志性复合调料。

After the Mid-Qing Dynasty, with the introduction of chili into Sichuan, Pixian chili bean paste has been invented as compound seasoning and been used in Sichuan dishes gradually. Pixian chili bean paste has enriched the special flavor of Sichuan dishes with its characteristics of salty& mellow, spicy but not pungent, and its paste aroma. As a compound seasoning, Pixian chili bean paste has its advantages over single seasoning. Its color is red, brown and oily, its petals are crisp, thick and mellow, and its extremely tiny sweetness, acidity and a variety of aroma make Pixian chili bean paste more mellow, spicy but not pungent on the basis of salty and fresh. Its paste flavor is richer. When used in Sichuan cuisine, it helps to enrich the flavor characteristics of the unique flavor, and enhance the core flavor of dishes. For example, the unique flavor of Sichuan cuisine, such as spicy flavor, home-made flavor, fish-smelling taste flavor, sour and spicy flavor, are taking Pixian chili bean paste as an important composite seasoning. According to the occupation standard *Sichuan Cuisine Cooking Techniques (SB/T 10946-2012)* issued by the Ministry of Commerce, hot and spicy flavor is seasoned by Sichuan pepper, chili oil, dry chili, Pixian chili bean paste, salt, sugar, soy sauce, sesame oil. It has bright reddish color, hot and spicy tastes, and thick aroma. Home-made taste type is seasoned by Pixian chili bean paste, salt, soy sauce, fermented soya bean, sweet soybean paste. It's red in color, salty and less spicy, thick and mellow. Fish-smelling type is seasoned by salt, sugar, soy sauce, vinegar, pickled pepper, Pixian chili bean paste, ginger, garlic, green onion, chili oil, sesame oil. It has bright reddish color, balanced flavor of salty, sour, sweet, and spicy. In those special taste types, Pixian chili bean paste increases the color, spicy taste, aroma, umami and mellow compound taste. Among these unique flavors of Sichuan cuisine, main function of Pixian chili bean paste is not only to brighten color, but also to increase compound flavor of spicy, fragrant, fresh and mellow, so as to ensure the core flavor of these unique flavor. In addition, there is another type called as “Douban Flavor”. The main seasoning is Pixian chili bean paste, which has the paste aroma of Douban, salty and less spicy, a little sweet and sour. Some Sichuan dishes cooked with Pixian chili bean paste use Chinese characters of Douban in their names, such as Chili Bean Paste Flavored Fish, Chili Bean Paste Flavored Pork Knuckle, Chili Bean Paste Flavored Tender Ginger. It could be said that Pixian chili bean paste is the important symbolic compound seasoning for special flavor of Sichuan cuisines.

（二）近现代部分川菜菜品的基础性复合调料

至清代中后期，通过不断地兼收并蓄，川菜逐渐形成较为完整的风味体系，共有筵席菜、大众便餐菜、家常风味菜、三蒸九扣和风味小吃五大类。其中，大众便餐菜是指清末至民国时期城乡“四六分饭铺”制作的菜肴，其特点是烹制快速、经济实惠，在风味上以麻辣味厚的菜式居多，而豆瓣酱的参与，对此类菜肴的风味形成与确立起到了重要的助推作用。因此，豆瓣酱常常用于许多大众便餐菜的烹调之中，如水煮牛肉、麻婆豆腐、鱼香肉丝等，都属于这类菜肴，郫县豆瓣均为其基础性复合调料之一。时至今日，在四川的大众化餐馆中，诸如大蒜烧鲶鱼、土豆烧甲鱼、苦笋烧肉、水煮鱼、水煮牛蛙等烧菜及水煮系列菜，都会使用郫县豆瓣来调味。进入冬季，在四川各地烧菜馆的炉灶上，一锅锅色泽红亮的烧菜，品类丰富、麻辣飘香，持续诱惑着过往的行人，在这当中，郫县豆瓣自然功不可没。家常风味菜，是指广大城乡居民常在家中制作和食用的菜肴，要求取材方便、操作易行，在风味上以适口为要，豆瓣酱也是烹调此类菜肴的重要复合调料，如回锅肉、盐煎肉、过江豆花、蚂蚁上树、麻辣肥肠鱼、干煸鳝丝等，在烹调过程中，这类菜肴都无一例外地使用了郫县豆瓣。

2. Basic Compound Seasoning for Some Modern Sichuan Cuisine

In the middle and late Qing Dynasty, Sichuan cuisine gradually formed a relatively complete flavor system, including banquet dishes, popular light meals, home -made flavor dishes, San Zheng Jiu Kou dishes and flavor snacks. Among them, popular light meals refer to the dishes made by "Siliufen Restaurants" in urban and rural areas from the late Qing Dynasty to the Republic of China. Those dishes are rapidly cooked, economical, hot and spicy in flavor. The salty and mellow flavor, spicy but not pungent, and rich paste flavor of bean paste play an important role in the formation of the spicy flavor of dishes, so it is widely used in cooking popular light meals. Boiled Beef in Chili Sauce, Mapo Tofu and Fish Flavored Pork Slices all belong to popular dishes, in which Pixian chili bean paste is used as one of the basic compound seasonings. Up to now, Pixian chili bean paste used as seasoning in lots of braised dishes and water boiled dishes in Sichuan restaurants, such as Braised Catfish with Garlic, Braised Shell Fish with Potatoes, Braised Pork with Bitter Bamboo Shoots, Water Boiled Fish, Water Boiled Bullfrog. In winter, on the stoves of braised dishes restaurants all over Sichuan, there are diverse braised dishes with bright reddish color, hot and spicy taste, with thick fragrance. Those braised dishes are tempting the pedestrians' appetites and taste buds, and make them to enjoy the food inside. Pixian chili bean paste does play an important role in those dishes. Home-made flavor dishes refer to dishes made and eaten by families, which materials are easily obtained and cooking methods are common. They are palatable in flavor. Chili bean paste is also an important composite seasoning for those dishes. Home-made flavor dishes such as Twice-Cooked Pork, Stir-Fried Salty Pork, Silken Tofu, Vermicelli with Spicy Minced Pork, Spicy Fish and Pork Intestines, and Dry Fried Eel Slivers are all cooked with Pixian chili bean paste.

特别值得一提的是麻辣火锅，即红汤火锅。它的汤卤在传统制作中也离不开郫县豆瓣，其制法是锅置中火上，牛油烧至五至六成热，入郫县豆瓣煵酥，加姜末、辣椒粉、花椒炒香，注入牛肉汤烧开，再倒至砂锅中，然后将砂锅置旺火上，入绍酒、豆豉、醪糟汁烧开，打尽浮沫即成卤水。其特点是色泽棕红、汤鲜香浓、麻辣味厚，而对于汤卤的红艳，郫县豆瓣和辣椒粉的使用贡献了很大力量。如今，麻辣火锅受到越来越多国内外消费者的喜爱，随着需求量的不断攀升，火锅汤卤调料的制作业已进入标准化、规模化、工业化的制作阶段，涌现出了许多著名的火锅底料品牌，其工业化生产的火锅底料也大多离不开郫县豆瓣。根据市场调查，在火锅底料配方中标明使用了“郫县豆瓣”的有臻鲜火锅、好人家火锅、小郡肝火锅、麻辣空间火锅、大红袍火锅、新希望火锅、桥头火锅等；标明使用了“豆瓣酱”的则有名扬火锅、海底捞火锅、小龙坎火锅、大龙燚火锅、川娃子火锅等，其采用的“豆瓣酱”，其实大多也是“郫县豆瓣”。此外，在干锅、冷锅鱼、串串香、麻辣烫、冒菜等火锅衍生品的烹调过程中，也几乎少不了郫县豆瓣的加入。

It is worth mentioning that spicy hot pot, which is called as red soup pot, is also inseparable from Pixian chili bean paste of its soup brine. The preparation method is to put the pot on a medium heat, burn the butter until it is 50% to 60% of the temperature, add crispy bean paste, stir minced ginger, chili powder and pepper, add beef soup to boil, and then pour them into the casserole on high heat, add yellow wine, fermented soya bean and fermented glutinous rice juice, and clean the floating foam to form brine. It is characterized by brown red color, fresh and fragrant soup, and thick spicy taste.

加有郫县豆瓣的火锅底料

Pixian chili bean paste and chili powder have contributed much to its flavor. Nowadays, spicy hot pot is very popular all over the world. With the increase of demands, the seasonings for basic soup have been mass-produced, standardized, industrialized. There are lots of famous hotpot brands, which formula is depending on Pixian chili bean paste. According to market research, hotpot brands which have marked "Pixian chili bean paste" in their formulas for basic soup seasonings are Zhen Xian Hotpot, Hao Ren Jia Hotpot, Xiao Jun Gan Hotpot, Ma La Kong Jian Hotpot, Da Hong Pao Hotpot, New Hope Hotpot, Qiao Tou Hotpot, etc. Hotpot brands which have marked "chili bean paste" in their formulas for basic soup seasonings are Ming Yang Hotpot, Hai Di Lao Hotpot, Shoo Loong Kan Hotpot, Da Long Yi Hotpot, Chuan Wa Zi Hotpot, etc. The "chili bean paste" they used here are mostly Pixian chili bean paste. In addition, Pixian chili bean paste is almost indispensable in cooking different hot pots, such as dry pot, cold pot fish, Chuan Chuan Xiang, Ma La Tang spicy steam-pot and other famous Sichuan dishes.

郫县豆瓣在大众便餐菜、家常风味菜和火锅这三大川菜类别中的大量使用，再次验证了郫县豆瓣是川菜烹饪中的一种基础性复合调料与核心调料。

The widely use of Pixian chili bean paste in Sichuan's popular light meals, home flavor dishes and hot pot proves once again that Pixian chili bean paste is a basic compound seasoning and core seasoning in Sichuan cuisine.

第三节 郫县豆瓣产品的应用

Section Three Application of Pixian Chili Bean Paste

郫县豆瓣自诞生以来，不仅在许多经典川菜的烹调中能见到它的身影，郫都人还潜心研发了许多新的郫县豆瓣产品，并运用到川菜创新中。其中，以郫县豆瓣为核心调料创制出的"郫县豆瓣宴"，主题十分鲜明，既彰显了非常浓郁的地方饮食风貌，也从另一个层面丰富了川菜别具一格的风味特色和文化内涵。

Pixian chili bean paste has been widely applied in Sichuan cuisine since its invention. It can be seen not only in classic dishes, but also used to formulate new seasonings, which are a driving force for the innovation of Sichuan cookery. In addition, "Pixian Chili Bean Paste Banquet" has been developed, enriching the flavors of Sichuan cooking and the depth and width of Sichuan culinary culture.

郫县豆瓣与经典川菜

Pixian Chili Bean Paste and Classic Sichuan Dishes

所谓经典川菜，是指在较长时间内广为流传，民众认可度很高，能够充分体现烹饪技术特色和文化内涵，具有典范性、代表性意义的川菜菜种。郫县豆瓣作为辣椒与胡豆瓣、食盐融合发酵的智慧结晶，成了许多经典川菜不可或缺的调味品。

Classic Sichuan dishes refer to the typical Sichua foods that have been widely accepted for a relatively long period of time and are representative of Sichuan cooking techniques and culture. Pixian chili bean paste, a fermented combination of broad beans, salt and chili peppers which were not introduced into Sichuan until the Qing Dynasty, has been an inseparable condiment for a number of classic Sichuan dishes.

食材配方

猪肘1个（约1 000克） 瓢儿白200克
蒜苗100克 郫县豆瓣50克 姜片15克
姜米10克 葱段20克 食盐5克 糖色15克
料酒20毫升 白糖20克 酱油毫升
醋10毫升 味精2克 清水1 000毫升
水淀粉30克 食用油100毫升

成菜特点

色泽红亮、咸鲜微辣、略带酸甜、肉质软糯。

制作工艺

1. 猪肘洗净，焯水后备用；蒜苗切成蒜苗花。
2. 猪肘入汤碗中，放入清水、姜片、葱段、食盐、料酒、糖色，入笼蒸至软糯后捞出装入凹盘中。
3. 瓢儿白入沸水中焯熟，捞出摆放在肘子四周。
4. 炒锅置火上，入食用油烧至100℃时，放入郫县豆瓣炒香至油呈红色，之后入姜米、一半蒜苗花炒香，注入蒸猪肘的原汤（200毫升）及白糖、酱油、醋烧沸，下水淀粉收汁至浓稠，最后加入另一半蒜苗花及味精搅匀，起锅后浇淋在猪肘上即成。

Pork knuckle, which features lean meat, abundant skin and tendon, rich in colloid, is a popular food among people in Sichuan. The most popular dishes include Dongpo Pork Knuckle, Simmered Pork Knuckle with Red Dates, etc. This dish, using Pixian chili bean paste as the main seasoning, helps improve the appetite.

Ingredients

One pork knuckle (about 1,000g); 200g bok choy; 100g bagy leeks; 50g Pixian chili bean paste; 15g ginger, sliced; 10g ginger, finely chopped; 20g scallion, segmented; 5g salt; 20ml Shaoxinml Shaoxing

豆瓣肘子

Chili Bean Paste Flavored Pork Knuckle

肘子皮厚、筋多、胶质重，瘦肉多且呈卷子形，常带皮烹制，是四川百姓常吃常新的美食，最著名的有东坡肘子、红枣煨肘等。豆瓣肘子是以郫县豆瓣为主要调料制成味汁来调味，既开胃，又解腻。

cooking wine; 15g caramel; 20g white sugar; 35ml soy sauce; 10ml vinegar; 2g MSG；1,000ml water; 30g cornstarch batter; 100ml cooking oil

Features

lustrous reddish brown in color; slightly spicy and sweet and sour in taste; soft and glutinous in texture

Preparation

1. Rinse the pork knuckle and blanch. Chop the baby leeks.
2. Place the knuckle in a bowl, add water, ginger slices, scallion, salt, cooking wine and caramel, and steam till soft. Remove the knuckle from the bowl and transfer to a deep platter.
3. Blanch the bok choy till just cooked, remove and lay around the knuckle.
4. Heat a wok over the flame, add the oil and heat to 100℃. Blend in the Pixian chili bean paste, and stir fry till the oil becomes reddish and fragrant. Add the chopped ginger and half of the leeks, and continue to stir till aromatic. Pour in 200ml stock, blend in the white sugar, soy sauce and vinegar, and bring to a boil. Add the batter to thicken the sauce, blend in the rest of the leeks and MSG, and pour over the knuckle.

盐煎肉

Stir-Fried Salty Pork

盐煎肉又称『生爆盐煎肉』，是四川家常风味菜的代表作之一，被视为回锅肉的姐妹菜。盐煎肉与回锅肉的区别是不带皮、生炒，调料中要加入豆豉，成菜较干香。

食材配方

猪去皮二刀肉200克　蒜苗100克　郫县豆瓣30克　豆豉6克　食盐0.5克　酱油2毫升　白糖2克　味精1克　食用油50毫升

成菜特点

色泽棕红、咸鲜微辣、干香滋润。

制作工艺

1. 猪肉切片；蒜苗切成马耳朵形。
2. 锅置火上，入食用油烧至160℃时，放入肉片、食盐炒至干香，再入郫县豆瓣炒香至油呈红色，之后下豆豉炒香，最后入酱油、白糖、蒜苗、味精炒匀后装盘。

Stir-Fried Salty Pork is a typical home-style dish in Sichuan, a twin of Twice-Cooked Pork. Unlike Twice-Cooked Pork. This dish is a stir fry using pork belly without skin, and fermented soy beans are added.

Ingredients

200g skinless Erdao (second cut) pork; 100g baby leeks; 30g Pixian chili bean paste; 6g fermented soy beans; 0.5g salt; 2ml soy sauce; 2g white sugar; 1g MSG; 50ml cooking oil

Features

reddish brown in color; salty, delicate, savory and spicy in taste

Preparation

1. Slice the pork. Cut the baby leeks into horse-ear segments.
2. Heat a wok over the flame, add the oil and heat to 160℃. Blend in the pork and salt, and stir to evaporate the water contents. Add the Pixian chili bean paste, and continue to stir fry until the oil becomes reddish. Add fermented soy beans, and stir till aromatic. Add the soy sauce, white sugar, baby leeks and MSG, blend well and transfer to a serving dish.

蚂蚁上树

Vermicelli with Spicy Minced Pork

蚂蚁上树是川菜中的传统名菜，因成菜中的肉末会贴附在粉丝上，形似蚂蚁爬树而得名。

食材配方

水发红薯粉丝200克　猪碎肉100克
郫县豆瓣40克　食盐1克　酱油10毫升　味精1克
葱花15克　鲜汤100毫升　食用油70毫升

成菜特点

色泽红亮、质地爽滑、咸鲜微辣。

制作工艺

1. 锅置火上，入食用油烧至100℃时，下猪碎肉、食盐、酱油（2毫升）炒香。
2. 锅置火上，入食用油烧至120℃时，下郫县豆瓣炒香，掺入鲜汤，再下猪碎肉、红薯粉丝、酱油烧至入味，待汁水将干时，放入味精、葱花炒匀，起锅装盘即成。

A traditional classic of Sichuan cuisine, this dish is so named because pork mince clings to vermicelli like ants climbing up the tree.

Ingredients

200ml water soaked sweet potato vermicelli; 100g pork mince; 40g Pixian chili bean pate; 1g salt; 10ml soy sauce; 1g MSG; 15g scallion, finely chopped; 100ml stock; 70ml cooking oil

Features

reddish and lustrous in color; smooth and tender in texture; savory and slightly spicy in taste

Preparation

1. Heat oil in a wok to 100℃, add pork mince, salt and 2ml soy sauce, and stir fry until fragrant.
2. Heat oil in a wok to 120℃, and stir fry Pixian chili bean pate until fragrant. Add stock, pork mince, vermicelli and soy sauce, braise until the sauce almost dries up, add MSG and chopped scallions blend well and transfer to a serving dish.

水煮牛肉是川菜中的传统名品，相传起源于四川省自贡市。“水煮”是川菜独特的烹饪方法，不能望文生义，“水煮”并非用清水煮制，而是用麻辣味的汤汁煮制，然后再用热油浇淋双椒末而成。

食材配方

牛肉200克　蒜苗50克　芹菜100克　青笋尖150克　干辣椒20克　花椒5克　郫县豆瓣80克
食盐3克　酱油8毫升　料酒15毫升　鲜汤350毫升　水淀粉70克　食用油180毫升

成菜特点

色泽红亮、麻辣鲜香、肉质细嫩、蔬菜清香。

制作工艺

1. 牛肉横切成薄片，加入食盐、料酒、水淀粉拌匀；蒜苗、芹菜切成长约10厘米的段；青笋尖切成长约10厘米的片；干辣椒、花椒入锅炒香，晾凉后剁细成双椒末。
2. 将蒜苗、芹菜、青笋尖入锅，加食盐炒至断生后装入碗中垫底。
3. 锅置火上，入食用油烧至100℃时，放入郫县豆瓣炒香，掺入鲜汤，入酱油、料酒、牛肉煮至成熟后倒入碗中，撒上双椒末，淋上少许180～200℃的热油即成。

Boiled Beef in Chili Sauce is a traditional dish originating in Zigong of Sichuan. Water boiling is a cooking method peculiar to Sichuan. Instead of being boiled in plain water, ingredients are first boiled in spicy sauce and poured over with sizzling oil.

Ingredients

200g beef; 50g baby leeks; 100g celery; 150g asparagu lettuce tips; 20g dried chilies; 5g Sichuan pepper; 80g Pixian chili bean paste; 3g salt; 8ml soy sauce; 15ml Shaoxinml Shaoxing cooking wine; 350ml stock; 70g cornstarch batter; 180ml cooking oil

Features

red in color; spicy in taste; soft beef and crunchy vegetables

Preaparation

1. Cut the beef into thin slices, and mix well with salt, cooking wine and batter. Cut the baby leeks and celery into 10cm segments. Cut the asparagus lettuce tips into 10cm slices. Dry roast the dried chilies and Sichuan pepper, leave to cool and finely chop the mixture.
2. Stir fry the baby leeks, celery and asparagus lettuce in a wok till just cooked, and transfer to a serving bowl.
3. Heat a wok over the flame, add the oil and heat to 100℃. Blend in the Pixian chili bean paste, and stir fry till fragrant. Pour in the stock and blend in the soy sauce, cooking wine and beef. Boil till the beef is cooked through, and pour the contents of the wok into the serving bowl. Sprinkle with chopped chilies and Sichuan pepper, and pour over 180-200℃ oil.

水煮牛肉
Boiled Beef in Chili Sauce

家常姜汁热窝鸡是一道大众化的传统川菜，制作工艺较为简单，在四川地区的普及度较高，极具开胃之功，特别适合家庭烹制。

食材配方

熟鸡肉400克　瓢儿白200克　郫县豆瓣30克　姜米40克　蒜米10克　葱花15克　食盐1克　白糖3克　味精1克　胡椒粉1克　酱油5毫升　醋30毫升　芝麻油5毫升　鲜汤300毫升　水淀粉15克　食用油60毫升

成菜特点

色泽红亮、咸鲜酸辣、姜味浓郁。

制作工艺

1. 熟鸡肉斩成3厘米见方的块；瓢儿白入沸水中焯水后装入圆盘的四周。
2. 锅置火上，入食用油烧至120℃时，放入郫县豆瓣、姜米、蒜米、葱花炒香，掺入鲜汤，放入鸡块、食盐、白糖、胡椒粉、酱油，用中火烧5分钟至鸡块入味，最后加入味精、醋、芝麻油、水淀粉，收浓汁水成清二流芡后起锅装盘即成。

家常姜汁热窝鸡

Home-Style Ginger Chicken

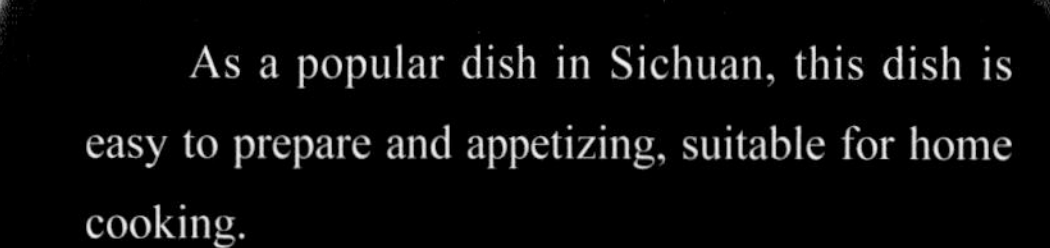

As a popular dish in Sichuan, this dish is easy to prepare and appetizing, suitable for home cooking.

Ingredients

400g cooked chicken; 200g bok choy; 30g Pixian chili bean paste; 40g ginger, finely chopped; 10g garlic, finely chopped; 15g spring onions, finely chopped; 1g salt; 3g white sugar; 1g MSG; 1g pepper; 5ml soy sauce; 30ml vinegar; 5ml sesame oil; 300ml stock; 15g cornstarch batter; 60ml cooking oil

Features

reddish brown in color; delicate, sour and spicy in taste with strong ginger fragrance

Preparation

1. Cut the chicken into 3cm cubes. Blanch the bok choy and lay around the rims of a serving plate.
2. Heat a wok over the flame, add the oil and heat to 120℃. Blend in the Pixian chili bean paste, ginger, garlic and spring onions, and stir fry till fragrant. Pour in the stock and add the chicken, salt, white sugar, pepper and soy sauce. Braise over medium heat for five minutes, and blend in the MSG, vinegar, sesame oil and batter. Transfer to the serving dish.

雪魔芋烧鸭

Braised Duck with Snow Conjak Jelly

雪魔芋是将魔芋豆腐经冷冻后制成，用清水涨发后呈海绵状，质地松软，富有弹性。用雪魔芋与鸭肉同烧，风味独特，是川菜的传统经典品种。

食材配方

鸭肉400克　雪魔芋50克　蒜苗节15克　郫县豆瓣30克　泡仔姜片20克　蒜片10克　花椒2克　食盐1克　酱油5毫升　料酒20毫升　味精2克　鲜汤1 000毫升　水淀粉15克　食用油100毫升

成菜特点

色泽红亮、质松多汁、咸鲜微辣。

制作工艺

1. 将鸭肉斩成长约6厘米、宽约2厘米的条，入沸水中焯水后捞出，沥干余水备用。
2. 雪魔芋用清水浸泡至透，切成长约5厘米、粗约1.8厘米的条，入沸水中焯水后捞出，浸泡于温水中。
3. 锅置火上，入食用油烧至180℃，放入鸭条煸干水分，待锅中油质变清亮后，放入郫县豆瓣、泡仔姜片、蒜片、花椒炒香后掺入鲜汤，再入食盐、酱油、料酒，改用小火烧至软熟后放入魔芋条烧至入味，最后放入蒜苗节、味精、水淀粉收浓汤汁，出锅装盘即成。

Sonw conjak jelly, a frosted or frozen version of conjak jelly, is soft and springy like sponge after being water-soaked. Braised with duck, this classic Sichuan cuisine has a special flavor.

Ingredients

400g duck; 50g snow conjak jelly; 15g baby leeks, segmented; 30g Pixian chili bean paste; 20g pickled ginger, sliced; 10g garlic, sliced; 2g Sichuan pepper; 1g salt; 5ml soy sauce; 20ml Shaoxinml Shaoxing cooking wine; 2g MSG, 1,000ml stock; 15g cornstarch batter; 100ml cooking oil

Features

reddish brown in color; soft in texture; savory and slightly spicy in taste

Preparation

1. Cut the duck into 6cm-long, 2cm-wide slivers, blanch, remove and drain.
2. Fully soak the conjak jelly in water until soft, and cut into 5cm-long, 1.8cm-thick slivers. Blanch, remove and soak in warm water.
3. Place a wok over the flame, add the oil and heat to 180℃. Stir fry the duck slivers to evaporate the water contents. Add the Pixian chili bean paste, pickled ginger, garlic and Sichuan pepper, and continue to stir fry till aromatic. Add the salt, soy sauce and cooking wine, turn down the flame and braise over a low flame till the duck is soft and the conjake jelly fully absorbs the flavors. Blend in the leeks, MSG and batter, wait till the sauce is thick, and transfer to a serving dish.

仔姜鲜兔

Rabbit with Tender Ginger

兔肉质地细嫩，具有高蛋白、低脂肪的优点，素有“保健肉”“美容肉”之称，深受女士青睐，兔肉又被视为“百味肉”，适合于各类味型。仔姜鲜兔选用具有增进食欲、健脾开胃之功的仔姜及辣椒等原料烹制而成，鲜香辣爽，是炎热夏日不可多得的一道开胃菜。

食材配方

兔肉1 000克　大蒜30克　仔姜100克　青尖椒100克　红尖椒50克　小米辣50克　青花椒10克　郫县豆瓣60克　姜片5克　白糖3克　食盐5克　胡椒粉1.5克　味精2克　料酒30毫升　淀粉20克　鲜汤1 000毫升　食用油1 500毫升（约耗150毫升）

成菜特点

色泽红亮、兔肉细嫩、咸鲜香辣。

制作工艺

1. 将兔肉斩成约1.5厘米见方的丁，加入食盐（2克）、胡椒粉（0.5克）、料酒（10毫升）及淀粉拌匀，入150℃的食用油中滑熟后捞出。
2. 仔姜切成长约6厘米、粗约0.3厘米的细丝；青尖椒、红尖椒、小米辣分别切成丁。
3. 锅置火上，下食用油烧至150℃时，放入大蒜、红尖椒、小米辣（25克）、青花椒、郫县豆瓣、姜片炒香后掺入鲜汤，接着入仔姜、白糖、食盐（3克）、胡椒粉（1克）、料酒（20毫升）烧沸出味，然后下兔肉烧2～3分钟，再入青尖椒、小米辣（25克）烧1分钟，最后下味精，出锅后装入凹盘中即成。

High in protein and low in fat, rabbit meat is called health meat or beauty meat, which is especially popular among women. It is also called hundred-flavor meat because of its suitability for use in different flavors. Appetizing tender ginger and chilies are used as ingredients and cooked with rabbit, presenting an appealing dish on hot summer day.

Ingredients

1,000g rabbit; 30g garlic; 100g tender ginger; 100g green chili peppers; 50g red chili peppers; 50g bird's eye chilies; 10g green Sichuan peppercorns; 60g Pixian chili bean paste; 5g ginger, sliced; 3g white sugar; 5g salt; 1.5g pepper; 2g MSG; 30ml Shaoxinml Shaoxing cooking wine; 20g corn starch; 1,000ml stock; 1,500ml cooking oil (about 150ml to be consumed)

Features

lustrous reddish color; savory and spicy in taste; tender rabbit

Preparation

1. Cut the rabbit into 1.5cm cubes, and mix well with 2g salt, 0.5g pepper, 10ml cooking wine and the cornstarch. Deep fry in 150℃ till cooked through and remove.
2. Cut the tender ginger into 6cm-long, 0.3cm-thick slivers. Cut the green chili peppers, red chili peppers, bird's eye chili peppers into small cubes.
3. Place a wok over the flame, add oil and heat to 150℃. Stir fry the garlic, red chilies, 25g bird's eye chilies, green Sichuan peppercorns, Pixian chili bean paste and ginger slices till fragrant. Add the stock, tender ginger, white sugar, 3g salt, 1g pepper and 20ml cooking wine, bring to a boil and blend in the rabbit. Braise for two to three minutes, add the green chilies and 25g bird's eye chilies, and braise for one more minute. Blend in the MSG, transfer to a deep dish and serve.

藿香具有芳香化湿、和胃止呕、祛湿解表的功效。除入药外，也常用来烹饪食物，藿香鲫鱼就是其中的代表，成菜具有藿香特殊的香味，夏季食用，可消暑、祛湿、理气。

食材配方

鲫鱼750克　藿香碎100克　食盐3克　姜片5克　葱段10克胡椒粉2克　料酒20毫升　郫县豆瓣20克　泡辣椒末20克　泡菜碎20克　泡姜米10克　蒜米10克　葱花15克　白糖5克　味精1克　鸡精2克　料酒20毫升　醪糟汁15毫升　酱油10毫升　醋10毫升　芝麻油10毫升　水淀粉20克　鲜汤500毫升　食用油100毫升

成菜特点

色泽红亮、鱼肉细嫩、咸鲜微辣、略带酸甜，霍香味浓郁。

制作工艺

1. 鲫鱼治净后，在鱼身两侧各剞数刀，用食盐(1克)、姜片、葱段、胡椒粉、料酒(10毫升)拌匀，码味10分钟。
2. 锅置火上，入食用油烧至120℃时，下郫县豆瓣、泡辣椒末、泡菜碎、泡姜米、蒜米、葱花炒香，然后掺入鲜汤，放入鲫鱼、食盐(2克)、胡椒粉(1克)、白糖、料酒(10毫升)、醪糟汁、酱油、醋，烧至鱼熟后捞出装盘。
3. 在锅中汤汁中放入味精、鸡精、芝麻油、水淀粉、藿香碎收汁至浓稠，出锅后淋在鱼身上即成。

Huoxiang, a fragrant herb often used in both Chinese medicine and as food ingredient, helps to remove body humidity, harmonize the stomach and relieve vomiting. Huxiang Carp, a typical dish using huoxiag herb, promotes sweat excretion and is suitable for the simmer.

Ingredients

750g crucian carp; 100g huoxiang herb, finely chopped; 3g salt; 5g ginger, sliced; 10g scallion, segmented; 2g pepper; 20ml Shaoxinml Shaoxing cooking wine; 20g Pixian chili bean paste; 20g pickled chilies, finely chopped; 20g pickles, finely chopped; 10g pickled ginger, finely chopped; 10g garlic, finely chopped; 15g scallion, finely chopped; 5g white sugar; 1g MSG; 10ml Shaoxinml Shaoxing cooking wine; 2g chicken essence; 15g fermented rice juice; 10ml soy sauce; 10ml vinegar; 10ml sesame oil; 20g cornstarch batter; 500ml stock; 100ml cooking oil

Features

reddish brown in color; tender fish with huoxiang fragrance; savory, slightly spicy, sweet and sour in taste

Preparation

1. Make several cuts into the two sides of the crucian carp, add 1g salt, ginger slices, scallion segments, pepper and 10ml cooking wine, mix well and leave to marinate for ten minutes;
2. Heat oil in a wok to 120℃, and stir fry the Pixian chili bean paste, pickled chilies, pickles, chopped ginger, garlic and scallion until fragrant. Add the stock, carp, 2g salt, 1g pepper, white sugar, 10ml cooking wine, fermented rice juice, soy sauce and vinegar, and braise till cooked through. Remove the fish from the wok and transfer to a serving plate.
3. Add MSG, chicken essence, sesame oil, batter and huoxiang to the wok, simmer until thick, and pour over the fish.

大蒜烧鳝段

Braised Eels with Garlic

大蒜烧鳝段是川中极具家常风味的传统名菜。鳝鱼具有高蛋白、低脂肪、无肌间刺、味道鲜美等优点，老少皆宜；大蒜具有抗菌杀毒、去腥压异的作用，两者搭配，相得益彰，不失为夏季很好的养生食物。

食材配方

去骨鳝鱼片300克　郫县豆瓣35克　独头大蒜100克　姜片5克　葱段20克　食盐1克　味精2克　酱油5毫升　料酒15毫升　水淀粉15克　鲜汤200毫升　食用油80毫升

成菜特点

色泽红亮、咸鲜微辣、质地滑嫩。

制作工艺

1. 鳝鱼洗净，切成长约8厘米的段。
2. 锅置火上，入食用油烧至180℃时，下鳝鱼段煸炒至卷缩、表面无水分后，加入郫县豆瓣、姜片、大蒜、葱段炒香，然后掺入鲜汤，下食盐、酱油、料酒烧至大蒜质地变软，最后入味精、水淀粉，待汁浓亮油时起锅装盘即成。

Braised Eels with Garlic is popular among old and young, for eels are high in protein, low in fat and delicate in taste with fewer bones. Garlic is anti-virus, anti-bacteria and helps to remove unpleasant smells.

Ingredients

300g eels; 35g Pixian chili bean paste; 100g garlic; 5g ginger, sliced; 20g scallion, cut into sections; 1g salt; 2g MSG; 5ml soy sauce; 15ml Shaoxinml Shaoxing cooking wine; 15g cornstarch batter; 200ml stock; 80ml cooking oil

Features

reddish brown in color; savory and slightly spicy in taste; soft and tender in texture

Preparation

1. Rinse the eels, and cut into 8cm lengths.

2. Heat oil in a wok to 180℃, and stir-fry the eels till the eel pieces curl up and the skin dries. Add Pixian chili bean paste, ginger, garlic and scallion, and stir-fry till aromatic. Add the stock, salt, soy sauce and Shaoxinml Shaoxing cooking wine, and braise till the garlic is soft. Blend in the MSG and batter, wait till the sauce is thick and lustrous, and transfer to a serving dish.

干煸鳝丝

Dry-Fried Eel Slivers

干煸鳝丝也是川菜的传统名菜之一。“干煸”是川菜特有的烹饪方式，是一种用旺火、热油，将食材快速煸炒至干香的方法。干煸类菜肴酥软干香，最宜佐酒。

食材配方

去骨鳝鱼片300克　芹菜35克　姜5克　蒜20克　郫县豆瓣30克　食盐1克　料酒15毫升　花椒粉1克　辣椒油10毫升　芝麻油3毫升味精1克　食用油70毫升

成菜特点

色泽棕红、麻辣咸鲜、酥软干香。

制作工艺

1. 将鳝鱼片斜切成头粗丝；姜、蒜分别切成长约2厘米的细丝；芹菜切成长约4厘米的段。
2. 锅置火上，入食用油烧至180℃时，放入鳝鱼丝煸炒至干香、吐油后下郫县豆瓣炒香，然后放入姜丝、蒜丝、料酒炒匀，待其出香后，下芹菜、食盐、芝麻油、辣椒油、花椒粉、味精，炒断生后出锅装盘即成。

This dish uses dry frying, a cooking method peculiar to Sichuan where ingredients are quickly stir fried in hot oil over a high flame until the ingredients are aromatic, tender and meltingly crispy, most suitable as accompaniment to alcohols.

Ingredients

300g eel slices; 35g celery; 5g ginger; 20g garlic;30g Pixian chili bean paste; 1g salt; 15ml Shaoxinml Shaoxing cooking wine; 1g ground Sichuan pepper; 10ml chili oil; 3ml sesame oil; 1g MSG; 70ml cooking oil

Features

reddish brown in color; spicy and savory in taste; tender and meltingly crispy in texture

Preparation

1. Cut the eel slices into course slivers; ginger and garlic into 2cm long thin slivers; celery into 4cm long sections.
2. Heat oil in a wok to 180℃, and stir fry eel slivers until dry and aromatic. Blend in Pixian chili bean paste, and continue to stir until fragrant. Add ginger, garlic and cooking wine, stir fry until fragrant, and add celery, salt, sesame oil, chili oil, ground Sichuan pepper and MSG. Continue to stir until the celery is just cooked, and transfer to a serving dish.

甲鱼又称鳖、团鱼、水鱼，肉质软糯、味道鲜美，可滋补强身，是上等的优质食材，适宜炖、烧、焖、煨等多种烹饪方法。此菜是将甲鱼与土豆一同红烧而成，营养搭配合理，口感丰富，是当今川菜中较为流行的一道菜品，很多川菜馆都有销售。

食材配方

甲鱼500克　小土豆400克　郫县豆瓣40克　香辣酱20克　姜片10克　葱段30克　食盐2克　味精2克　胡椒粉1克　醪糟汁15毫升　料酒40毫升　辣椒油20毫升　花椒油5毫升　芝麻油5毫升　鲜汤500毫升　食用油800毫升（约耗120毫升）

成菜特点

色泽红亮、质地软糯、麻辣香醇。

制作工艺

1. 甲鱼宰杀后洗净，入沸水中略煮后捞出，刮掉表面的白膜后洗净，并将其从背甲与底板处分开，去尽内脏，斩成约3厘米见方的块，入160℃的食用油中略炸后捞出备用。
2. 土豆去皮、洗净。
3. 锅置火上，入食用油烧至100℃时，放入郫县豆瓣、香辣酱、姜片、葱段炒香，再掺入鲜汤烧沸，去掉料渣，下甲鱼块、食盐、胡椒粉、醪糟汁、料酒，用中小火烧至软熟，再放入土豆烧至入味，最后下辣椒油、花椒油、芝麻油、味精炒匀，出锅装盘即成。

土豆烧甲鱼

Braised Shell Fish with Potatoes

Shell Fish, a soft shelled turtle found in China, is a quality ingredient in Chinese cuisine with its tender and glutinous texture, delicate taste and health benefits. It is suitable for stewing, braising, simmering and pressure simmering. In this dish, turtle is braised with potatoes, providing balanced nutrition and layered texture. It is a popular modern dish, sold in a number of Sichuan cuisine restaurants.

Ingredients

500g turtle; 400g potatoes; 40g Pixian chili bean paste; 20g chili pepper paste; 10g ginger, sliced; 30g scallion, segmented; 2g salt; 2g MSG; 1g pepper; 15ml fermented glutinous rice wine; 40ml Shaoxinml Shaoxing cooking wine; 20ml chili oil; 5ml Sichuan pepper oil; 5ml sesame oil; 500ml stock; 800ml cooking oil (about 120ml to be consumed)

Features

brown and lustrous in color; soft and glutinous in texture; savory and spicy in taste

Preparation

1. Kill the turtle. Boil the turtle briefly in water, remove, rinse, and separate back shell from bottom shell. Remove the entrails, rinse, chop into 3cm³ cubes and fry briefly in 160℃ oil.
2. Peel the potatoes, rinse, chop into 3cm-long diamonds and fry in 160℃ oil till the surface hardens.
3. Heat some oil in a wok to 100℃, add Pixian chili bean paste, chili pepper paste, ginger, scallion and stir-fry till aromatic. Add the stock and bring to a boil. Remove the scums, and slide in the turtle cubes. Add the salt, pepper, fermented glutinous rice wine, Shaoxinml Shaoxing cooking wine and simmer over a medium-low flame till the turtle cubes become soft and cooked through. Add the potatoes and braise still they have absorbed the flavors. Blend in chili oil, Sichuan pepper oil, sesame oil and MSG, mix well, transfer to a serving dish and sprinkle with coriander leaves.

家常海参

Home-Style Sea Cucumbers

家常海参是一道美味可口的川菜传统名菜，始于清代末年。海参为世界八大珍品之一，属于高蛋白、低脂肪、低胆固醇的优质食材，特别适合老年人和高血压、冠心病等疾病患者食用。

食材配方

水发海参300克　猪肉50克　黄豆芽100克　瓢儿白100克　姜米10克　蒜苗花40克
郫县豆瓣55克　食盐2克　酱油6毫升　料酒10毫升　味精2克　芝麻油3毫升
水淀粉20克　鲜汤150毫升　食用油200毫升

成菜特点

色泽红亮、咸鲜香辣、质地软糯。

制作工艺

1. 将水发海参放入热鲜汤中喂味2～3次；猪肉剁碎。
2. 锅置火上，入食用油烧热，先将猪肉碎入锅炒散，再下食盐（0.5克）、料酒（3毫升）炒香后备用。
3. 锅置火上，入食用油烧至150℃时，放入黄豆芽、食盐1克炒至断生，装入长盘的一端；瓢儿白焯水断生，捞出后装入长盘的另一端。
4. 锅置火上，入食用油烧至100℃时，下郫县豆瓣炒香至油呈红色，然后入姜米、蒜苗花炒香，掺入鲜汤，下海参、酱油、食盐（0.5克）、料酒（7毫升）、味精、芝麻油烧沸，最后用水淀粉收汁浓稠呈二流芡，出锅后盛入盘的中央即成。

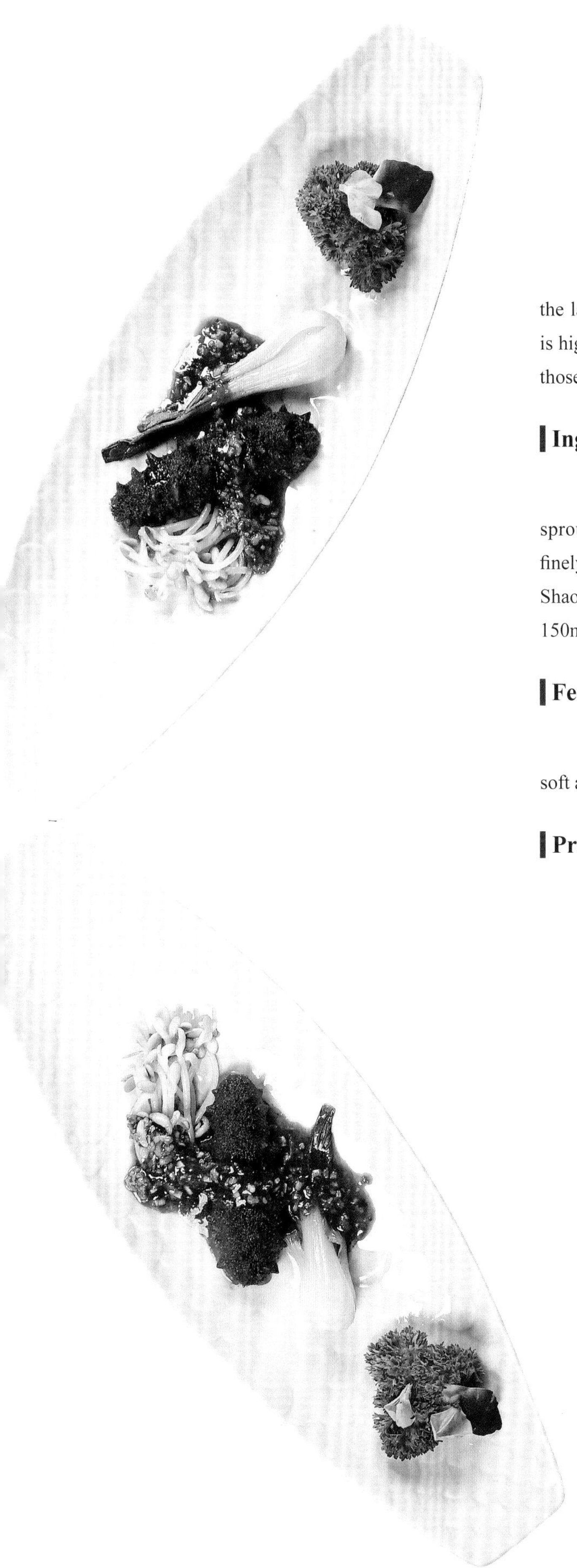

This dish, a traditional classic of Sichuan cuisine, was developed in the late Qing Dynasty. Sea cucumber, one of the Eight Rare Delicacies, is high in protein, low in late and cholesterol, suitable for the seniors and those suffering from high blood pressure and CHD.

Ingredients

300ml water-soaked sea cucumber; 50g pork; 100g soy bean sprouts; 100g boy choy; 10g ginger, finely chopped; 40g baby leeks, finely chopped; 55g Pixian chili bean paste; 2g salt; 6ml soy sauce; 10ml Shaoxing cooking wine; 2g MSG; 3ml sesame oil; 20g peastarch batter; 150ml stock; 200ml cooking oil

Features

reddish and lustrous in color; savory, aromatic and spicy in taste; soft and glutinous in texture

Preparation

1. Pour the stock into a pot, bring to a boil, turn off the heat, and soak the seas cucumbers in the hot stock to absorb the flavors. Repeat the above process for two more times. Finely chop the pork.
2. Heat oil in a wok, stir fry the pork mince pieces until they separate, blend in 0.5g salt and 3ml Shaoxing cooking wine, and continue to stir until fragrant.
3. Heat oil in a wok to 150℃, blend in the soybean sprouts and 1g salt, stirring until just cooked, and transfer to one side of a serving dish. Blanch the boy choy until just cooked, and transfer to the other side of the dish.
4. Heat oil in a wok to 100℃, add Pixian chili bean, and stir fry until fragrant. Blend in the chopped ginger and baby leeks, and continue to stir fry until fragrant. Pour in the stock, add sea cucumbers, soy sauce, 0.5g salt, 7ml Shaoxing cooking wine, MSG and sesame oil, bring to a boil, and add the peastarch batter. Stir until the sauce is thick, and transfer to the center of the dish.

麻婆豆腐以豆腐为主料烧制而成，是闻名世界的川菜传统名菜。此菜是因始创于清朝同治年间一个叫“陈麻婆”的女厨师而得名。

食材配方

豆腐400克　牛肉50克　蒜苗30克　郫县豆瓣40克　蒜米20克　豆豉6克
食盐11克（浸泡10克、牛肉臊子1克）　辣椒粉7克　花椒粉1克　酱油10毫升
料酒5毫升　鲜汤200毫升　水淀粉30克　食用油70毫升

成菜特点

麻、辣、咸、鲜、烫、酥、嫩面面俱到，色泽红亮、形整不烂、味道浓厚。

制作工艺

1. 豆腐切成大丁，放入加有食盐的沸水中浸泡待用；牛肉剁成末，先入油锅中炒散，再下食盐、料酒炒至酥香；蒜苗切成马耳朵形或长约3厘米的节；豆豉剁细。
2. 锅置火上，入食用油烧至100℃时，下郫县豆瓣、辣椒粉、蒜米、豆豉炒香，掺入鲜汤，再下酱油、豆腐、牛肉末烧至入味，然后入蒜苗、水淀粉、花椒粉收汁浓稠呈干二流芡，出锅后装入钵中即成。

As a world famous dish of traditional Sichuan cuisine, Mapo Tofu had its origins during Emperor Tongzhi's Reign of the Qing Dynasty. It was nicknamed Mapo (pock-faced Granny) Tofu by the diners for the fact that its inventor had pockmarks in her face.

Ingredients

400g tofu; 50g beef; 30g baby leeks; 40g Pixian chili bean paste; 20g garlic, finely chopped; 6g fermented soy beans; 11g salt (10g for soaking; 1g for beef topping); 7g ground chilies; 1g ground Sichuan pepper; 10ml soy sauce; 5ml Shaoxing cooking wine; 200ml stock; 30g peastarch batter; 70ml cooking oil

Features

lustrous reddish color; a combination of numbing, pungent, salty, savory, hot, salty, savory, crispy and tender tastes

Preparation

1. Cut the tofu into big cubes, and soak into salted boiling water. Mince the beef, stir fry until the pieces separate, add salt and cooking wine, and continue to stir until crispy and aromatic. Cut the baby leeks into 3cm-long horse-ear sections. Finely chop the fermented soy beans.
2. Heat oil in a wok to 100℃, stir fry Pixian chili bean paste, ground chilies, chopped garlic and fermented soy beans until fragrant, blend in the stock, soy sauce, tofu and beef mince, and braise until cooked through. Add the baby leeks, peastarch batter and ground Sichuan pepper, stir until the sauce is thick, and transfer to a serving dish.

麻婆豆腐
Mapo Tofu

红烧牛肉面

Noodles with Red-Braised Beef Topping

红烧牛肉面在南北各地有多种做法，其重点区别在于面臊的制作。川味红烧牛肉面的面臊，是采用郫县豆瓣等调料烧制而成，色泽红亮、鲜辣味美，具有浓郁地方风味。

食材配方

湿细面条150克　牛肉350克　水发竹笋150克　瓢儿白50克姜15克　葱20克　花椒2.5克　香料10克　郫县豆瓣100克　料酒15毫升　食盐6克　糖色80克　味精8克　牛肉汤500毫升　辣椒面20克　食用油150毫升

成菜特点

色泽红亮、面条滑爽、牛肉软糯、咸鲜香辣。

制作工艺

1. 将牛肉切成约3厘米见方的小块，入锅中焯水后备用；竹笋切成约2厘米见方的小块，焯水后备用；瓢儿白入沸水中焯熟。
2. 锅置火上，入食用油加热至100℃时，下郫县豆瓣、辣椒面炒至油红出香，然后加入牛肉汤、牛肉、竹笋、香料、姜、葱、花椒、料酒、食盐、糖色，用小火煨烧至熟软、入味、汁浓后成牛肉面臊（三人分量）。
3. 面条入锅煮熟后捞入碗中，放入牛肉面臊和瓢儿白即成。

This dish varies from place to place, and the key lies in the preparation of beef toppings. Sichuan-style toppings are braised beef with Pixian chili bean paste, presenting an appealing red color and savory and spicy tastes.

Ingredients

150g fresh noodles; 350g beef; 150ml water-soaked bamboo shoots; 50g bok choy; 15g ginger; 20g scallions; 2.5g Sichuan pepper; 10g spices; 100g Pixian chili bean paste; 15ml Shaoxing cooking wine; 6g sat; 80g caramel; 8g MSG; 500ml beef stock; 20g ground chilies; 150ml cooking oil

Features

reddish and lustrous in color; spicy and savory in taste; smooth noods and tender beef

Preparation

1. Cut the beef into 3cm cubes, and blanch. Cut the bamboo shoots into 2cm cubes and blanch.
2. Heat oil in a wok to 100℃, add Pixian chili bean paste and ground chilies, stirring until fragrant, and blend in the beef stock, beef, bamboo shorts, spices, ginger, scallions, Sichuan peppercorns, cooking wine, salt and caramel. Turn down the heat, simmer for the beef and bamboo shorts to absorb the flavors of the sauce, wait until the sauce the thick and turn off the heat. (three portions)
3. Boil the noodles until cooked through, transfer to a serving bowl, and top with the braise beef and blanched bok choy.

热拌米凉粉

Hot Rice Jelly Salad

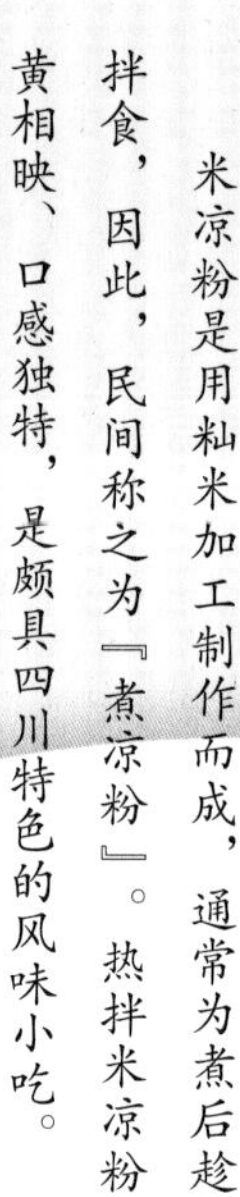

米凉粉是用籼米加工制作而成，通常为煮后趁热拌食，因此，民间称之为『煮凉粉』。热拌米凉粉红黄相映、口感独特，是颇具四川特色的风味小吃。

食材配方

米凉粉300克　芹菜碎50克　葱花20克　白糖2克　香菜碎10克　蒜泥15克
郫县豆瓣40克　味精2克　豆豉蓉10克　酱油6毫升　花椒粉1克　花椒油3毫升
芝麻油5毫升　食用油50毫升

成菜特点

色泽红亮、质地细腻、口感滑爽、咸鲜麻辣、米香浓郁。

制作工艺

1. 米凉粉切成约1.5厘米见方的丁，入沸水中焯水后捞出，沥干余水。
2. 锅置火上，入食用油烧至120℃时，下郫县豆瓣炒香，接着下豆豉蓉炒香，出锅后倒入碗中，加入蒜泥、白糖、味精、酱油、花椒粉、花椒油、芝麻油、芹菜碎、葱花、香菜碎调匀，最后放入米凉粉中拌匀装盘即成。

Rice jelly, as the name suggests, is made of indica rice. It is often called boiled jelly because it is often eaten hot directly after being boiled and seasoned. It is a snack peculiar to Sichuan.

Ingredients

300g rice jelly; 50g celery, finely chopped; 20g scallions, finely chopped; 10g coriander, finely chopped; 15g garlic, finely chopped; 40g Pixian chili bean pate; 10g fermented soy beans, finely chopped; 2g white sugar; 2g MSG; 6ml soy sauce; 1g ground Sichuan pepper; 3ml Sichuan pepper oil; 5ml sesame oil; 50ml cooking oil

Features

appealing reddish and yellow in color; tender and smooth in texture; savory and spicy in taste

Preparation

1. Cut the rice jelly into 1.5cm cubes blanch and drain.
2. Heat oil in a wok to 120℃, stir fry Pixian chili bean paste until fragrant, add the fermented soy beans, stirring until fragrant, and transfer to a bowl. Add garlic, white sugar, MSG, soy sauce, ground Sichuan pepper, Sichuan pepper oil, sesame oil, celery, scallion and coriander, mix well and add the rice jelly cubes. Blend well and transfer to a serving dish.

家常豆腐

Home-Style Tofu

家常豆腐又名熊掌豆腐，是将豆腐煎（或炸）至两面金黄后同猪肉一起烧制而成的一道家常菜。豆腐软嫩多汁、鲜香可口，常见于便餐和家庭。

食材配方

豆腐300克　猪肉100克　蒜苗20克　郫县豆瓣35克
食盐2克　酱油5毫升　味精1克　水淀粉15克　鲜汤200毫升
食用油1 000毫升（约耗100毫升）

成菜特点

色泽红亮、咸鲜香辣、软嫩可口。

制作工艺

1. 猪肉切成长约5厘米、宽约3. 5厘米、厚约0.2厘米的片；豆腐切成长约5厘米、宽约3厘米、厚约0.5厘米的片；蒜苗切成马耳朵形。
2. 锅置火上，入食用油烧至220℃时，将豆腐片入锅炸至表皮金黄、起硬膜后捞出。
3. 锅置火上，入食用油烧至160℃时，放入肉片炒香、出油，再下郫县豆瓣炒香上色，然后掺入鲜汤、食盐、酱油、豆腐，用小火烧至入味，最后加入蒜苗、味精、水淀粉，待收汁浓稠、亮油后起锅装盘即成。

Homely-Style Tofu, also known as Bear's Paw Tofu, is a homely-style dish of fried tofu until golden brown on both sides and cooked with pork. Tofu is soft, juicy and delicious. It is often found in casual meals and homes.

Ingredients

300g tofu; 100g pork; 20g baby leeks; 35g Pixian chili bean paste; 3g salt; 5ml soy sauce; 1g MSG; 15g peastarch batter; 200ml stock; 1,000ml cooking oil (about 100ml to be consumed)

Features

reddish lustrous color; savory and spicy taste; tender and appetizing tofu

Preparation

1. Cut the pork into 5cm-long, 3.5cm-wide and 0.2cm-thick slices. Cut the tofu into 5cm-long, 3cm-wide and 0.5cm-thick slices. Cut the baby leeks into horse's ear.
2. Heat oil in a wok to 220℃, deep fry the tofu slices until the surface becomes golden and hard.
3. Heat oil in a wok to 160℃, and stir fry the pork slices until fragrant. Blend in the Pixian chili bean paste, continue to stir until fragrant and the oil becomes reddish. Add the stock, salt, soy sauce and tofu, turn down the heat and simmer so that the tofu and pork can absorb the flavors of the sauce. Add baby leeks, MSG and peastarch batter, stir until the sauce becomes thick and lustrous, and transfer to a serving dish.

麻辣小龙虾

Mala Crayfish

小龙虾又称克氏原螯虾、红螯虾和淡水小龙虾，因价廉物美而广受人们欢迎。近年来，小龙虾在我国已成为重要的经济养殖品种，并进而形成了完整的产业链。小龙虾蛋白质含量高、肉质松软、肉味鲜美，适合制作多种菜品，麻辣小龙虾就是其代表性品种。

食材配方

小龙虾750克　干辣椒200克　花椒50克　大蒜100克　姜片15克
葱段20克　八角3克　草果3克　香叶2克　郫县豆瓣30克
火锅底料20克　食盐2克　料酒20毫升　白酒10毫升　味精5克
胡椒粉1克　芝麻油5毫升　鲜汤200毫升　火锅油100毫升
食用油1 000毫升（约耗50毫升）

成菜特点

色泽红亮、咸鲜麻辣、肉嫩多汁。

制作工艺

1. 小龙虾初加工后治净，入160℃的油锅中炸1分钟后捞出。
2. 锅置火上，入火锅油烧至120℃时，下干辣椒、花椒、大蒜、姜片、葱段、八角、香叶、草果、郫县豆瓣、火锅底料炒香，然后加入小龙虾、鲜汤、食盐、料酒、白酒、胡椒粉，用中火烧5分钟，待汤汁将干时，放入味精、芝麻油炒匀，出锅装盘即成。

Crayfish or crawfish, popular for its high-protein, tender and delicate meat, has become an important economic aquafarming species in China. It is used in a number of dishes, and Mala (hot and numbing) Crayfish is the most renowned.

Ingredients

750g crayfish; 200g dried chilies; 50g Sichuan peppercorns; 100g garlic; 15g ginger, sliced; 20g scallion, segmented; 3g star anise; 3g tsaoko amomum; 2g bay leaves; 30g Pixian chili bean pate; 20g hot pot soup base; 2g salt; 20ml Shaoxing cooking wine; 10ml liquor; 5g MSG; 1g pepper; 5ml sesame oil; 200ml stock; 100ml hot pot oil; 1,000ml cooking oil (about 50ml to be consumed)

Features

red and lustrous in color; spicy and savory in taste; tender and juicy meat

Preparation

1. Slaughter, devein and rinse the crayfish. Deep fry in 160℃ oil for one minute and remove.
2. Heat oil in a wok to 120℃, and add dried chilies, Sichuan peppercorns, garlic, ginger, scallions, star anise, bay leaves, tsaoko amomum, Pixian chili bean paste and hot pot soup base, stir frying until aromatic. Add the crayfish, stock, salt, cooking wine, liquor and pepper, braise over medium heat for about five minutes, and blend in MSG and sesame oil when the sauce almost dries up. Transfer to a serving dish.

郫县豆瓣传统品种与火锅系列

Pixian Chili Bean Paste and Hot Pot

从毛肚火锅开始，郫县豆瓣就与火锅结下了不解之缘。随着人们对火锅的热爱和市场需求的不断变化，四川火锅已从单一菜品发展成为一个系列品种，并出现了火锅的衍生品，大有与川式菜肴和面点小吃共成鼎足之势。但不可否认的是，不管火锅怎样发展，郫县豆瓣都一直与其相伴，始终都是其中不可或缺的重要角色。

Pixian chili bean paste has been an indispensible part of hot pot since the invention of Beef Tripe Hot Pot. Sichuan people's passion and enthusiasm for hot pot has pushed its development into a series of hotpot styles and categories. Hot pot is now an important component of Sichuan cuisine, alongside Sichuan dishes and snacks.

食材配方

肉蟹1 000克　干辣椒节50克　郫县豆瓣30克　花椒20克　姜片20克　葱段30克　蒜米30克　豆豉蓉10克　食盐5克　料酒20毫升　鸡精2克　味精2克　芝麻油5毫升　熟芝麻10克　酥花仁碎20克　火锅油500毫升　干淀粉20克　食用油1 000毫升（约耗50毫升）

成菜特点

色泽红亮、麻辣香醇、口感独特。

制作工艺

1. 肉蟹宰杀后治净，斩成块，沾上干淀粉，入180℃的食用油中炸至酥香后捞出备用。
2. 锅置火上，入火锅油烧至120℃时，下干辣椒节、花椒、葱段、姜片、蒜氺、郫县豆瓣、豆豉蓉炒至香味浓郁后，加入肉蟹、食盐、料酒、鸡精、味精、芝麻油、熟芝麻、酥花仁碎炒香，出锅后装入锅仔中上桌即成。
3. 食客吃完蟹肉后，可再放入素菜炒制，或掺入汤汁及添加火锅底料后点火涮食其他食材，主推食材为红苕粉、土豆、藕片、黄瓜等。

This spicy and savory dish is a typical dry hot pot where the main ingredients are stir fried and served with no soup. Other ingredients can be stir fried in the pot after the main ingredients have been consumed, or otherwise soup can be added to the pot where other ingredients can be boiled like the average hot pot.

Ingredients

1,000g crabs; 50g dried chilies, cut into sections; 20g Sichuan pepper; 20g ginger, sliced; 30g scallions, segmented; 30g garlic, finely chopped; 30g Pixian chili bean pate; 10g fermented soy beans, finely chopped; 5g salt; 20ml Shaoxing cooking wine; 2g chicken essence; 2g MSG; 5ml sesame oil; 10g roasted sesame seeds; 20g crispy peanuts, crushed; 1,000ml cooking oil (about 50ml to be consumed); 500ml hot pot oil; 20g peastarch

香辣蟹火锅

Spicy Crabs Hot Pot

香辣蟹火锅属于干锅系列的麻辣香锅，以麻、辣、鲜、香和混搭为特色，吃完螃蟹后，可再放入素菜炒制，或掺入汤汁及添加火锅底料后点火加温涮食其他食材。香辣蟹火锅口味多样，可将多种食材任意搭配。

Features

reddish brown in color; spicy and savory in taste

Preparation

1. Slaughter the crabs, rinse and cut into chunks. Coat the crab chunks with peastarch and deep fry in 180℃ oil until crispy and aromatic.
2. Heat oil in a wok to 120℃, and stir fry dried chilies, Sichuan peppercorns, scallion, ginger, garlic, Pixian chili paste and fermented soy beans until fragrant. Add the crabs, salt, cooking wine, ehicken essence, MSG, sesame oil, sesame seeds and crispy peanuts, stirring until aromatic. Transfer the contents of the wok to a small pot and serve.
3. When crabs have been consumed, vegetables can be stir fried in the small pot. Alternatively, soup and hot pot soup base can be added, and vegetables can be boiled in the pot until cooked through. Recommended vegetables include sweet potato noodles, potatoes, lotus roots and cucumber.

毛肚火锅

Beef Tripe Hot Pot

毛肚火锅是一道川菜传统名品，因主要食材为毛肚而得名。火锅制作的核心技术是其汤卤的调制，它决定着火锅的风味走向，是衡量火锅成败的关键。传统的毛肚火锅是麻辣卤汁，重用牛油、辣椒、花椒，脂香、辣香兼具，汤汁醇香而浓厚。

食材配方

①食材：毛肚400克　鸭肠400克　黄喉400克　鸭胗400克　午餐肉400克　牛肉400克　牛肝400克　猪腰400克　猪脑2个　鳝鱼400克　鱼肉400克　金针菇200克　香菇200克　黄豆芽200克　豆皮200克　鸭血500克　粉皮400克　白菜心200克　菠菜200克　青笋200克　鹌鹑蛋200克

②调料：郫县豆瓣150克　辣椒粉30克　干辣椒100克　干花椒15克　豆豉15克　花椒粉2克　姜片30克　葱段30克　芽菜20克　香料40克　食盐5克　料酒20毫升　白酒10毫升　白糖5克　味精5克　牛油750克　熟菜油750毫升　鲜汤1 500毫升

③油碟：芝麻油300毫升　食盐10克　味精10克　蒜泥200克

成菜特点

汤汁红亮、质感多样、麻辣鲜香。

制作工艺

1. 锅置火上，入熟菜油烧至120℃时，下郫县豆瓣、辣椒粉、香料、姜片、葱段炒香，接着下豆豉、芽菜炒香，掺入鲜汤烧沸后，加食盐、料酒、白酒、白糖、味精、花椒粉调成麻辣味的卤汁。
2. 锅置火上，入牛油烧化，放入干辣椒、花椒炒香后，倒入火锅卤汁熬至出味，再装入火锅盛器中上桌。
3. 将所有食材加工成一定的形状，装盘或装碗后上桌。
4. 将芝麻油、食盐、味精、蒜泥分别装入10个小碗中制成味碟，供食客将食材入火锅中涮熟后蘸食。

Beef Tripe Hot Pot is a traditional Sichuan cuisine famous for its main ingredients. The core technology of hot pot production is the preparation of soup and brine, which determines the flavor of hot pot and is the key to measure the success of hot pot. Traditional Beef Tripe Hot Pot is a spicy sauce, uses more beef butter, pepper, Sichuan pepper. The soup is mellow and strong.

Ingredients

Food Materials 400g beef tripe; 400g duck intestines; 400g beef arteries; 400g duck gizzards; 400g luncheon meat; 400g beef; 400g beef livers; 400g pork kidneys; 2 pork brains; 400g eels; 400g fish; 200g golden needle mushrooms; 200g shiitake mushrooms; 200g soy bean sprouts; 200g tofu sheets; 500g duck blood curd; 400g sweet potato noodles; 200g tender napa cabbage; 200g spinach; 200g asparagus lettuce; 200g quail eggs

Seasonings 150g Pixian chili bean paste; 30g ground chilies; 100g dried chilies; 15g dried Sichuan pepper; 15g fermented soy beans; 2g ground Sichuan pepper; 30g ginger, sliced; 30g scallions, segmented; 20g yacai (preserved mustard green, finely chopped); 40g spices; 5g salt; 20ml Shaoxing cooking wine; 10ml liquor; 5g white sugar; 5g MSG; 750g beef butter; 750ml cooked rapeseed oil; 1,500ml stock

Dipping Sauce 300ml sesame oil; 10g salt; 10g MSG; 200g garlic, crushed

Features

red and lustrous in color; spicy and savory in taste; various in texture

Preparation

1. Heat oil in a wok to 120℃, add Pixian chili bean pate, ground chilies, spices, ginger and scallions, stir frying until fragrant, blend in fermented soy beans and yacai, and continue to stir until fragrant. Pour in the stock, and blend in salt, cooking wine, liquor, white sugar, MSG and ground Sichuan pepper to make mala (hot and numbing) soup.
2. Heat beef oil in a wok until it melts, and stir fry dried chilies and Sichuan peppercorns until aromatic. Pour in the mala soup, bring to a boil, and simmer to bring out the aroma. Pour the contents of the wok into a hot pot container.
3. Cut the ingredients into different shapes as desired, and serve in dishes or bowls.
4. Combine the sesame oil, salt, MSG and garlic in a small bowl (Dividing the dipping sauce ingredients into ten portions and served in ten bowls), and serve as dipping sauce.

冷锅鱼火锅

Fish in Cold Pot

所谓冷锅，是指在厨房将菜品烹制成熟后倒入火锅盛器中，上桌时不点火，待食完锅中的主菜后，再点火烫涮其他食材的一种火锅形式。冷锅鱼火锅汤色红亮，汤面平静悠然、汤下热情似火，鱼片雪白，入口兼具麻、辣、鲜、香、嫩、滑、爽，回味悠长，犹如一曲以麻辣味为主旋律、层次丰富的交响乐，为冷锅中的上品。

食材配方

①食材：花鲢2 500克　肥牛200克　黄喉200克　草原毛肚200克　午餐肉200克　牛肉片200克　猪腰片200克　豆腐皮100克　红苕粉200克　香菇200克　鸡腿菇200克　藕片200克　生菜200克

②调料：郫县豆瓣200克　泡红辣椒末400克　泡姜末30克　榨菜片50克　干青花椒50克　八角1个　葱段20克　芝麻油10毫升　秘制香料粉3克　胡椒粉2克　味精5克　牛骨汤2 000毫升　食用油800毫升

③油碟：酥黄豆100克　大头菜粒100克　榨菜粒100克　蒜泥100克　香葱100克　香菜60克

成菜特点

色泽红亮、鱼肉细嫩、咸鲜麻辣、花椒香浓。

制作工艺

1. 花鲢治净，取净鱼肉片成厚片，鱼头、鱼尾、鱼骨斩成小块；其他食材分别装盘或装碗。
2. 锅置火上，入食用油烧至100℃时，下郫县豆瓣、泡红辣椒末、干青花椒炒制10分钟，再下八角炒20分钟，然后加入榨菜、泡姜、葱段炒2分钟，掺入牛骨汤烧沸5分钟至香味浓郁后，滤去料渣，放入秘制香料粉、胡椒粉、味精、芝麻油、鱼骨、鱼头、鱼尾，用小火煮至刚熟后下鱼片焖熟，然后倒入火锅盛器中上桌。
3. 将酥黄豆、大头菜粒、榨菜粒、蒜泥、香葱、香菜分别放入10个味碟中，食用时舀一勺锅里的原汁加入其中。
4. 先食用鱼肉，然后再点火烫涮其他食材。

The so-called cold pot refers to the fact that the main ingredients are cooked through when served and no flame or heating is needed. However, it is still a form of hot pot because the heat will be turned on to boil other ingredients after the main precooked ingredients have been consumed. Fish in Cold Pot, the most renowned dish in the category of cold pot, features dark red soup what looks tranquil on the surface but spicy and hot underneath. The snowy fish slices soaked in the soup tastes numbing, hot, savory, tender, smooth and aromatic with a lingering fragrance.

Ingredients

Food Materials 2,500g silver spotted carp; 200g streaky beef; 200g beef arteries; 200g beef tripe; 200g luncheon meat; 200g beef slices; 200g pork kidney slices; 100g tofu sheets; 200g sweet potato noodles; 200g shiitake mushrooms; 200g chicken feet mushrooms; 200g lotus roots slices; 200g lettuce

Seasonings 200g Pixian chili bean paste; 400g pickled red chilies, finely chopped; 30g pickled ginger, finely chopped; 50g zhacai (preserved mustard tubers) slices; 50g dried green Sichuan peppercorns; 1 star anise; 20g scallions, segmented; 10ml sesame oil; secret-recipe spices; 2g pepper; 5g MSG; 2,000ml beef bone soup; 800ml cooking oil

Dipping Sauce 100g crispy soy beans; 100g preserved kohlrabi; 100g zhacai, finely chopped; 100g garlic, crushed; 100g scallion; 60g coriander

Features

dark red in color; tender fish; spicy and savory tastes with a strong aroma of green Sichuan peppercorns

Preparation

1. Rinse the carp, remove the bones and slice. Chop the head, tail and bones into small chunks. Transfer the rest of the ingredients into serving plates or bowls.
2. Heat oil in a wok to 100℃, stir fry Pixian chili bean paste, pickled red chilies and green Sichuan peppercorns for ten minutes, add the star anise, and stir fry for another twenty minutes. Blend in the zhaicai, pickled ginger and scallions, stir frying for two minutes before pouring in the beef stock. Bring to a boil, and continue to boil for five minutes to bring out the fragrance. Skim, add the secret-recipe spices, pepper, MSG, sesame oil, fish bones, heads and tails, and simmer until just cooked. Add the fish slices, cover and simmer till cooked through. Remove from the heat pour into a hot pot container, and serve.
3. Divide among ten bowls the crispy soy beans, preserved kohlrabi, zhaicai, garlic, scallion and coriander. Add one scoop of fish soup into the bowl when eating.
4. Turn on the heat and boil other ingredients after consuming the fish in the pot.

肥肠鸡火锅

Chicken and Pork Intestines Hot Pot

肥肠鸡火锅是当今四川兴起的一道创新川菜，是将肥肠与土鸡结合烹制而成，味道麻辣鲜香，价廉物美，深受『好吃嘴』的喜爱，在成都彭州一带尤为流行。

食材配方

①食材：土鸡1只（约2 000克）　肥肠1 000克　青尖椒200克　红尖椒200克　青笋200克

②调料：干辣椒节50克　青花椒20克　郫县豆瓣100克　火锅底料200克　辣椒粉50克　姜片15克　葱段25克　八角3克　草果3克　白蔻2克　香叶2克　食盐5克　胡椒粉3克　白糖5克　料酒20毫升　白酒20毫升　味精3克　鸡精3克　青花椒油5毫升　食用油750毫升　鲜汤1 500毫升

③油碟：蒜泥100克　芹菜碎50克　香菜碎50克　榨菜末50克

成菜特点

色泽红亮、鸡肉嫩滑、肥肠软糯、味厚麻辣。

制作工艺

1. 肥肠洗净后焯水，入锅煮至软熟后捞出，切成滚刀块；土鸡斩成约4厘米见方的块，焯水后备用；青尖椒、红尖椒分别切成马耳朵形；青笋切成滚刀块。
2. 将蒜泥、芹菜碎、香菜碎、榨菜末分别装入10个小碗中制成味碟。
3. 锅置火上，入食用油烧至120℃时，下干辣椒节、青花椒、郫县豆瓣、辣椒粉、姜片、葱段、八角、草果、白蔻、香叶、火锅底料一同炒香，然后掺入鲜汤，放入鸡块、食盐、胡椒粉、白糖、料酒、白酒，用小火烧制1小时后，放入肥肠继续烧制30分钟至入味，之后放入青尖椒、红尖椒、青笋块、味精、鸡精、青花椒油烧沸，再倒入火锅盛器中上桌。
4. 食用时，在味碟碗中舀一勺锅中的原汁，吃完锅中的食材后，可再加热烫涮其他食材。

A newly invented hot pot of Sichuan with pork intestines and free-range chicken as the main ingredients, this dish enjoys immense popularity among food lovers, especially in Pengzhou and Chengdu.

Ingredients

Food Materials 1 free-range chicken (about 2,000g); 1,000g pork intestines; 200g green chili peppers; 200g red chili peppers, 200g asparagus lettuce

Seasonings 50g dried chilies, cut into sections; 20g green Sichuan peppercorns; 100g Pixian chili bean paste; 200g hot pot soup base; 50g ground chilies; 15g ginger, sliced, 25g scallions, segmented; 3g star anise; 3g tsaoko amomum;;2g round cardamom; 2g bay leaves; 5g salt; 3g pepper; 5g white sugar; 20ml Shaoxing cooking wine; 20ml liquor; 3g MSG; 3g chicken essence; 5ml green Sichuan pepper oil; 750ml cooking oil; 1,500ml stock

Dipping Sauce 100g garlic, crushed; 50g celery, finely chopped; 50g coriander, finely chopped; 20g zhacai (preserved mustard tuber), finely chopped

Features

lustrous red in color; soft and glutinous chicken; springy pork intestines; spicy and savory in taste

Preparation

1. Rinse the pork intestines, blanch and boil until cooked through. Remove and cut ito rolling chunks. Chop the chicken into 4cm cubes and blanch. Cut the green and red chili peppers into horse-ear slices. Cut the asparagus lettuce into rolling chunks.
2. Divide the garlic, celery, coriander and zhacai into ten dippin sauce bowls.
3. Heat oil in a wok to 120℃, and stir fry dired chilies, green Sichuan peppercorns, Pixian chili bean pate,ground chilies, ginger, scallions, star anises, tsaoko adamom, round cardamum, bay leaves and hot pot soup base until fragrant. Add the stock, chicken, salt, pepper, white sugar, cooking wine and liquor, simmer over low heat for an hour, and add the pork intestines. Braise for another thirty minutes, add green and red chili peppers, asparagus lettuce, MSG, chicken essence and green Sichuan pepper oil, bring to a boil, and pour into a hot pot container.
4. Add one scoop of hot pot soup into the dipping sauce bowl when eating. Reheat the hot pot and boil other ingredients after consuming the chicken and intestines.

串串香

串串香是火锅的另外一种呈现形式，又称麻辣烫、小火锅，起源于四川成都，是四川地区的特色品种，因为是将食材用竹签穿成串后再放入锅中烫涮食用而得名。如今，串串香以其鲜明的特色和魅力而遍布于中国众多城市，充分体现了广大民众对美食最宽厚的包容。

食材配方

①食材：排骨30克　牛肉片30克　五花肉30克　黄喉30克　脆皮肠30克腊肉片30克　牛板筋30克　鸡肉条30克　鸭肉条30克　鸡翅30克　鸡胗30克　鸡心30克　鲫鱼1尾（约100克）　鳝鱼片30克　泥鳅30克　海虾30克　水发鱿鱼30克　冻豆腐30克　鹌鹑蛋30克　金针菜30克　香菇30克　玉兰片30克　竹笋30克　魔芋30克　豆腐皮30克　水发木耳30克　白菜30克　菠菜30克　青笋30克　土豆片30克　莲藕片30克　鱼腥草30克　鲜黄花30克　空心菜30克

②调料：郫县豆瓣150克　辣椒粉100克　干辣椒节40克　花椒20克　豆豉10克　姜片20克　葱段30克　香料20克　食盐10克　味精5克　料酒20毫升　白酒20毫升　冰糖5克　醪糟汁30毫升　食用油300毫升　熟菜油300毫升　鲜汤2 000毫升

③味碟：食盐10克　辣椒粉100克　花椒粉10克　味精5克　熟芝麻50克　酥花生碎50克

成菜特点

用料多样，麻、辣、烫、鲜兼具。

制作工艺

1. 所有食材经清洗、刀工处理后分别用竹签穿成串。

2. 锅置火上，入食用油烧至120℃时，下干辣椒节炸至棕红色后放入花椒炸香，连油带料倒入大碗中成煳辣汁。
3. 锅置火上，入熟菜油烧至120℃时，下郫县豆瓣、辣椒粉、豆豉、姜片、葱段、香料炒香，然后加入鲜汤、白酒、煳辣汁，用中小火熬至香味浓郁后，加入食盐、味精、料酒、冰糖、醪糟汁调成卤汁。
4. 用食盐、辣椒粉、花椒粉、味精、熟芝麻、酥花生碎制成味碟；将食材放入卤汁中烫涮成熟后取出，蘸上味碟中的调料食用。

Chuan Chuan Xiang, another form of hot pot peculiar to Sichuan which had its origins in Chengdu, is so named because ingredients are strung along bamboo sticks (called chuan chuan in Chinese) and boiled in hot pot. It is now popular in a number of cities in China, a reflection of the vigor of grassroot foods.

Ingredients

Food Materials 30g pork ribs; 30g beef; 30g pork belly; 30g beef arteries; 30g crispy sausage; 30g cured pork, sliced; 30g beef tendon; 30g chicken; 30g duck, cut into stirps; 30g chicken wings; 30g chicken gizzards; 30g chicken hearts; 1 carp (about 100g); 30g eels; 30g loaches; 30g shrimps; 30g water-soaked squid; 30g frozen tofu; 30g quail eggs; 30g golden needle mushrooms; 30g tsutake mushrooms; 30g slieced bamboo shorts; 30g bamboo shoots; 30g conjak jelly; 30g tofu sheets; 30g water-soaked wood ear fungus; 30g napa cabbage; 30g spinach; 30g asparagus lettuce; 30g potatoes, sliced; 30g lotus roots, sliced; 30g fish mint; 30g fresh daylilies; 30g water spinach

Seasonings: 150g Pixian chili bean paste; 100g ground chiliesl; 40g dried chilies, segmented; 20g Sichuan pepper; 10g fermented soy beans; 20g ginger, sliced; 30g scallions, segmented; 20g spices; 10g salt; 5g MSG; 20ml Shaoxing cooking wine; 20ml liquor; 5g rock sugar; 30ml fermented glutinous rice juice; 300ml cooking oil; 300ml cooked rapeseed oil; 2,000ml stock

Dipping Sauce: 10g salt; 100g ground chilies; 10g ground Sichuan pepper; 5g MSG; 50g roasted sesame seeds; 50g crspy peanuts, crushed

Features

a variety of ingredients; numbing, spicy, hot and savory tastes

Preparation

1. String all the ingredients after proper processing.
2. Heat oil in a wok to 120℃, and deep fry dried chilies until dark red. Add Sichuan peppercorns, deep fry until fragrant to make the spicy oil, and pour into a big bowl.
3. Heat the cooked rapeseed oil in a wok to 120℃, and stir fry Pixian chili bean paste, ground chilies, fermented soy beans, ginger, scallions and spices until fragrant. Add the stock, liquor, and the spicy oil, simmer over medium-low heat until aromatic, and blend in salt, MSG, cooking wine, rock sugar and fermented rice juice to make the hot pot soup.
4. Combine the salt, ground chilies, ground Sichuan pepper, MSG, sesame seeds and peanuts to make the dipping sauce. Boil the ingredients in the hot pot soup until cooked through, remove from the strings, and dip into the dipping sauce before eating.

郫县豆瓣的新产品与川菜创新

Pixian Chili Bean Paste Chain Products and Sichuan Cuisine Innovation

进入21世纪后，随着人们生活节奏的不断加快，大众对方便、快捷的饮食方式，也提出了越来越高的要求。为了满足广大消费者的新需求，郫县豆瓣在继承传统的基础上不断创新，陆续研制出诸多以郫县豆瓣为核心的新型复合调味品，如豆瓣蘸水、豆瓣蘸粉、豆瓣牛肉酱、豆瓣蘸料及豆瓣芝士烤肉酱等，并将其融会贯通于川菜的创新研发之中。

Since the beginning of the 21st century, the increasing pace of life has made convenient and instant foods a trend. Pixian chili bean paste has been developing new products such as chili bean paste dipping sauce, chili bean paste dipping powder, beef chili bean sauce, chili bean spices and cheese chili bean barbecue sauce. All these products have been used in the innovation of Sichuan cuisine.

过江豆花与豆瓣蘸水

Silken Tofu with Chili Bean Dipping Sauce

豆瓣蘸水是基于“郫县豆瓣+”定位而首创的郫县豆瓣即食产品，是用郫县豆瓣、辣椒、花椒、姜、大蒜、豆豉、白糖、鸡油、食用油等炒制而成。该产品辣而不燥且有酱香，开袋即食，方便快捷，可用于蘸食火锅、豆花、夹馍、拌面、拌饭、拌凉菜等。这里的过江豆花就是用豆瓣蘸水作为蘸碟来调味。

食材配方

豆花500克　豆瓣蘸水100克　葱花20克

成菜特点

豆花洁白细嫩、味道麻辣浓厚。

制作工艺

1. 锅置火上，注水烧热，将豆花入锅煮至热透后装入碗中。
2. 将豆瓣蘸水分装入数个味碟中，加入葱花成豆瓣蘸碟。食用时，将豆花捞入豆瓣蘸碟中蘸味食用。

Chili Bean Dipping Sauce is the first instant food attempt by Pixian chili bean paste. This instant food has a peculiar aroma, and can be used as dipping sauce for hot pot and silken tofu, fillings for mantou sandwich, and seasonings for noodles and salads.

Ingredients

500g silken tofu; 100g chili bean dipping sauce; 20g scallions, finely chopped

Features

snowy and tender tofu; spicy and savory in taste

Preparation

1. Heat water in a wok, add the silken tofu and boil till it's hot thoroughly.
2. Combine chili bean dipping sauce and scallions in a saucer. Dip the tofu into the saucer before eating.

柠檬茶香煎鲈鱼与豆瓣蘸粉

Lemon-Flavor Tea Bass & Chili Bean Dipping Powder

豆瓣蘸粉是采用现代食品加工技术，将“鹃城牌”一级郫县豆瓣磨成粉，再配以辣椒等原料所研发的即食产品，可蘸食以烧烤、火锅、卤菜、串串、凉菜、香煎热菜等方式烹制的食物。柠檬茶香煎鲈鱼即是采用该特色调料烹制的一款特色菜品。

食材配方

鲈鱼750克　白玉菇80克　蟹味菇80克　生菜50克　绿茶叶10克　豆瓣蘸粉15克　黄柠檬片10克　姜葱水5毫升　料酒2毫升　干白葡萄酒5毫升　干柠檬草5克　白胡椒粉1克　食盐4克　味精1克　色拉油500毫升（约耗油80毫升）

成菜特点

色泽美观、咸鲜微辣，柠檬及茶香味浓郁。

制作工艺

1. 鲈鱼宰杀后洗净，取净鱼肉切成长约5厘米，宽约3厘米的块；生菜、白玉菇、蟹味菇洗净；柠檬草用沸水泡发备用；绿茶用90℃水温冲泡后滤出茶水及茶叶备用。
2. 锅置旺火上，入色拉油烧至180℃时，放入茶叶炸至酥香后盛入碗中。
3. 鱼肉加柠檬片、胡椒粉、食盐（2克）、料酒、姜葱水、茶叶水、柠檬草码味15分钟。
4. 平底锅置中小火上，入色拉油（50毫升）加热至150℃时，先将鱼肉放入锅中略煎片刻，再烹入干白葡萄酒，待煎至鱼肉表面呈浅黄色后装入盘中。
5. 锅置旺火上，入色拉油（20毫升），用旺火加热至200℃时，将白玉菇、蟹味菇略微挤去多余的水分，放入锅内炒至断生，再加入食盐（2克）、味精调好味后装入盘中。
6. 将烤箱预热至200℃，湿度调节至30%，放入鱼肉烤制7分钟后取出。
7. 在盘中依次摆入生菜及炸制好的茶叶垫底，再放上白玉菇和蟹味菇，然后摆上鱼肉，撒上豆瓣蘸粉即成。

Chili bean dipping powder, which is a mixture of chili and ground first-class Pixian chili bean paste made by Juancheng by using modern food processing technology, is an instant seasoning for barbecues, hot pot, luzhu, chuan chuan, cold dishes and pan-fried hot dishes. This dish is an invention using this seasoning.

Ingredients

750g bass; 80g white beech mushrooms; 80g brown beech mushrooms; 50g lettuce; 10g green tea; 15g chili bean dipping powder; 10g lemon, sliced; 5ml scallion-ginger juice; 2ml Shaoxing cooking wine; 5ml white wine (dry); 5g dried lemon grass; 1g pepper; 4g salt; 1g MSG; 500ml salad oil (about 80ml to be consumed)

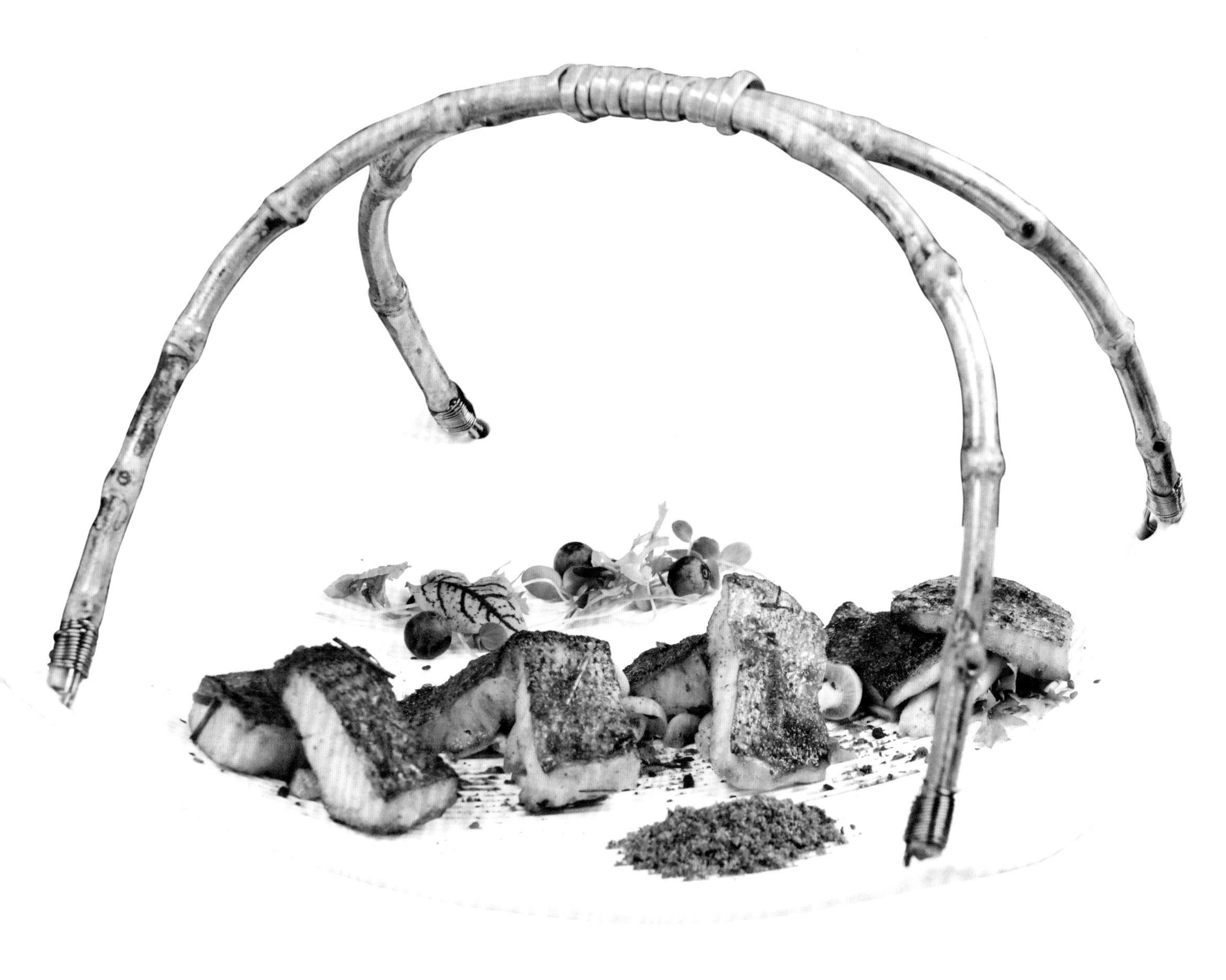

Features

appealing colors; savory and slightly spicy fish with lemon and tea aromas

Preparation

1. Slaughter the bass, rinse and remove the bones. Cut into 5cm-long, 3cm-wide chunks. Rinse the lettuce, white and brown beech mushrooms. Soak the dried lemon grass in boiling water. Make green tea with 90℃ water, remove the tea leaves from the water and drain. Save the tea water for later use.
2. Heat oil in a wok to 180℃, deep fry tea leaves until fragrant and crispy, remove and transfer to a bowl.
3. Combine the bass, lemon slices, pepper, 2g salt, cooking wine, ginger-scallion juice, tea water and lemon grass, blend well and leave to marinate for fifteen minutes.
4. Heat 50ml salad oil in a plan to 150℃, fry the fish briefly, add the white wine, and wait till the surface of the fish becomes light brown. Remove the fish from the pan and transfer to a plate.
5. Heat 20ml salad oil in a wok to 200℃, squeeze the mushrooms to remove extra water, and stir fry until just cooked. Add 2g salt and MSG to season, and transfer to a dish.
6. Preheat the oven to 200℃, set the humidity to 30%, and roast the fish for seven minutes.
7. Place the lettuce and tea leaves on the bottom of a serving plate, stack the mushrooms on top and then the fish. Sprinkle with chili bean dipping powder and serve.

芝士豆瓣烤鸡翅与豆瓣芝士烤肉酱

Roast Chicken Wings with Cheese Chili Bean Sauce & Cheese Chili Bean Barbecue Sauce

芝士豆瓣烤肉酱是采用郫县豆瓣、芝士、黄油、奶油、辛香料、洋葱、白糖等原料加工而成，口感柔和、清爽，酱酯香、芝士香及奶油香兼而有之，回味悠长，最适宜烤鸡翅、羊排和制作川味比萨。

食材配方

鸡中翅500克　芝士豆瓣烤肉酱45克　狗牙生菜5克　三色堇3朵　圣女果10克

成菜特点

色泽红亮、外酥内嫩、咸鲜麻辣，具有芝士的特殊香味。

制作工艺

1. 鸡翅洗净后沥干余水，用芝士豆瓣烤肉酱拌匀腌制30分钟。
2. 将鸡翅平铺在烤盘上，放入万能蒸烤箱中用中高火烤制5分钟后取出，摆放在盘中，点缀上狗牙生菜、三色堇及圣女果即成。

Cheese chili bean barbecue sauce is made with Pixian chili bean pate, cheese, butter, cream, spices, onions and white sugar. It has a strong milk fragrance and an appetizing and delicate taste, most suitable for chicken wings, lamb chops and Sichuan-style pizzas.

Ingredients

500g medium chicken wings; 45g cheese chili bean barbecue sauce; 5g lettuce; 3 pansies; 10g cherry tomatoes

Features

lustrous brown in color; spicy and savory in taste; chicken wings crispy on the outside but tender on the inside with a strong cheese aroma

Preparation

1. Rinse the chicken wings, drain and blend well with cheese chili bean barbecue sauce. Leave to marinate for thirty minutes.
2. Lay the chicken wings flat on the tray, roast in an oven with medium-high heat for five minutes, and remove. Transfer the chicken wings to a serving plate, and garnish with lettuce, pansies and tomatoes.

烤串与豆瓣蘸料

Barbecue Strings & Chili Bean Spices

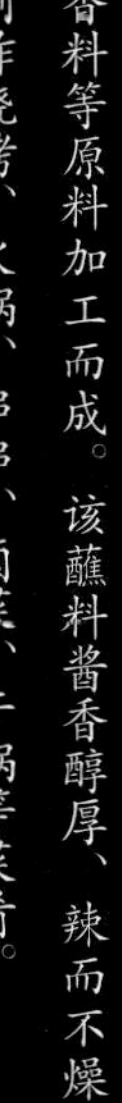

豆瓣蘸料是用郫县豆瓣、干辣椒、芝麻、大豆、花生、辛香料等原料加工而成。该蘸料酱香醇厚、辣而不燥，可用于制作烧烤、火锅、串串、卤菜、干锅等菜肴。

食材配方

①食材：羊肉50克　羊腰50克　牛肉50克　牛板筋50克　猪排骨50克　五花肉50克　兔腰50克　鸡肉50克　鸡胗50克　鸡心50克　鱿鱼须50克　墨鱼仔50克　大虾50克　鱼丸50克　豆腐干50克　韭菜50克　土豆50克　玉米50克　小馒头50克

②调料：豆瓣蘸料200克

成菜特点

辣而不燥，酱香浓郁。

制作工艺

1. 所有食材经清洗及刀工处理后分别用竹签穿成串，制成烤串生坯。
2. 将烤串生坯分批分次放在烤炉上烤制，边烤边撒上豆瓣蘸料，烤至成熟后装盘即成。

Chili bean spices, taking dried chilies, Pixian chili bean paste, sesame seeds, soy beans, peanuts and spices as the main ingredients, features spicy and savory tastes, suitable for barbecues, hot pot, chuan chuan, luzhu and dry pot.

Ingredients

Food Materials 50g lamb; 50g lamb kidneys; 50g beef; 50g beef tendon; 50g pork chops; 50g pork belly; 50g rabbit kidneys; 50g chicken; 50g chicken gizzards; 50g chicken heart; 50g squid; 50g baby cuttlefish; 50g prawns; 50g fish balls; 40g dried tofu; 50g chives; 50g potatoes; 50g corns; 50g minibuns

Seasonings: 200g chili bean spices

Features

spicy but moist; aromatic and savory

Preparation

1. String the ingredients after proper processing.
2. Grill the strung ingredients until cooked through, turning from time to time and sprinkling the chili bean spices over the ingredients during the process.

川式三明治与豆瓣牛肉酱

Sichuan-Style Sandwich & Beef Chili Bean Sauce

豆瓣牛肉酱是用牛肉、香菇、郫县豆瓣、豆豉、辣椒、大蒜、洋葱、香辛料等原料加工而成，开袋即食，既可用于拌饭、拌面、拌凉菜，也可用于馒头、面包、三明治等食物的即食调味。

食材配方

吐司面包4片　西式火腿100克　芝士片50克　番茄100克　生菜50克
豆瓣牛肉酱60克　三色堇数朵

成菜特点

层次分明、色彩明快、吐司酥香、味道略辣。

制作工艺

1. 先切去吐司面包四周的硬边，再切成约6厘米见方的小片，入烤箱烤至酥脆后取出。
2. 将西式火腿、芝士片、番茄、生菜分别切成与吐司面包片同样大小的片。
3. 取一片吐司面包，先抹上一层豆瓣牛肉酱，再依次放上西式火腿、芝士片、番茄片、生菜，每放一种食材均要在上面抹上一些豆瓣牛肉酱，然后放上一片吐司面包压实，最后再对切成三角形装入盘中，点缀上生菜、三色堇即成。

Beef chili bean sauce, a ready to eat sauce made of beef, shiitake mushrooms, Pixian chili bean paste, chilies, garlic, onions and other spices, can be used to as noodle and rice toppings, salad seasonings, sandwich fillings and side dishes for mantou and bread.

Ingredients

4 pieces of toast; 100g western ham; 50g cheese slices; 100g tomatoes; 50g lettuce; 60g beef chili bean sauce; 3 pansies

Features

layered and colorful presentation; crispy toast; slightly spicy taste

Preparation

1. Cut off the hard crust of the toast. Cut and trim the toast into 6cm squares, place in an oven and bake till crispy.
2. Cut and trim the ham, cheese slices, tomatoes and lettuce the same size as the toast.
3. Smear half of the toast pieces with beef chili bean sauce, and stack in order ham, cheese, tomatoes and lettuce, smearing each layer with the sauce. Cover with the other half of the toast pieces, press and cut diagonally into triangles. Transfer to a serving dish and garnish with lettuce and pansies.

郫县豆瓣宴
Pixian Chili Bean Paste Banquet

郫县豆瓣宴，是『郫县豆瓣风情宴』的简称，它是以『川菜之魂』的郫县豆瓣为主题及核心调料，并以『古蜀之源』的郫都农耕文明及移民文化风情为基调，再辅以特色食材与川菜烹调技艺研制而成的创意宴席。

该宴席的创制始于二〇一四年，是由成都蜀都川菜产业投资发展有限公司、四川省郫县豆瓣股份有限公司率先提出，而且得到了郫县餐饮同业公会的积极响应，并由该同业公会牵头，逐步开展了郫县豆瓣宴研发的相关筹备工作。二〇一五年，郫都区举办『郫县豆瓣宴』比赛，二〇一六年，郫都区不仅举办了『二〇一六川菜文化旅游节暨第四届郫县豆瓣博览会郫县豆瓣宴制作大赛』，而且还与四川旅游学院川菜发展研究中心合作，对已有的郫县豆瓣宴进行改良、提升，从而创制出郫县豆瓣宴的传统版与时尚版，这里选择了郫县豆瓣宴改良版与时尚版的部分菜品介绍如下。

Pixian Chili Bean Paste Banquet, an innovation attempt that features Pixian chili bean paste as the core seasoning, is developed on the basis of the agriculture and immigration culture of Pidu, and uses local specialty ingredients and Sichuan cooking techniques.

The Banquet was first launched in 2014 by Chengdu Shudu Sichuan Cuisine Industry Investment and Development Co., Ltd. and Development Corporation and Sichuan Pixian Chili Bean Paste Incorporated Company. The initiative was warmly received by Pidu Catering Association, which then started the preparation of the Banquet research and development. In 2015, Pidu District held the Pixian Chili Bean Paste Banquet Competition t. In 2016, in addition to the 2016 Sichuan Culinary Cultural and Tourism Festival, the Fourth Pixian Chili Bean Paste Expo and Pixian Chili Bean Paste Banquet Preparation Competition, Pidu District also worked with the Sichuan Cuisine Development and Research Center of Sichuan Tourism University in an effort to improve and upgrade the banquet, integrating both the Chinese and western health principles. New dishes were developed, and the Banquet was divided into two versions: traditional and fashionable. Here presented are only parts of the dishes from both versions.

寻味探源·川味冷碟

Taste Origin: Sichuan Flavor Cold Dishes

本款菜品的设计思路来源于“寻川菜之味，探蜀人之源”，均以郫县豆瓣为调料，同时结合胡豆瓣、辣椒等制作郫县豆瓣的食材研发而成，食材丰富多样，味型搭配合理，菜品陈列精致。

The dishes were inspired by the passion to explore the origins of Sichuan flavors. They are based on Pixian chili bean paste and the main ingredients for the paste (broad beans and chilies), pursuing varied ingredients, multiple flavors and exquisite presentation.

郫县豆瓣卤汁鸡

Pixian Chili Bean Paste Flavored Chicken

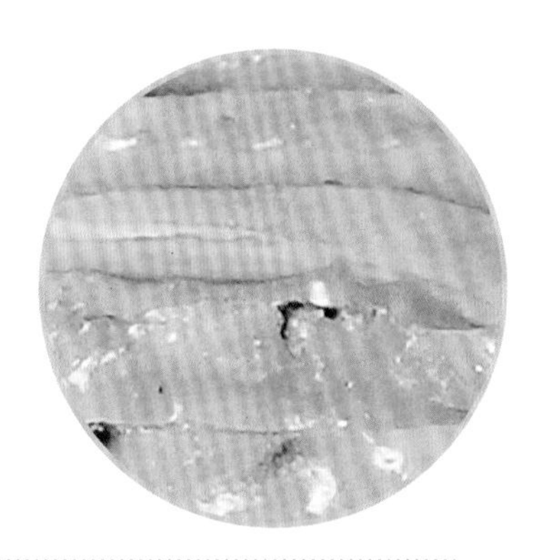

食材配方

土鸡500克　大葱50克　郫县豆瓣100克　香料10克　姜米20克　葱花30克　料酒20毫升　胡椒粉1克　白糖100克　味精2克　芝麻油5毫升　花椒油3毫升　辣椒油20毫升　熟芝麻5克　食用油100毫升

成菜特点

色泽红亮、细嫩紧实、咸甜辣香、豆香浓郁。

制作工艺

1. 土鸡煮熟后晾凉，切成粗丝；大葱斜切成马耳朵形。
2. 锅置火上，入食用油烧至100℃时，下郫县豆瓣、香料、姜米、葱花炒香，先掺入煮鸡的汤烧沸，再加入胡椒粉、料酒，用小火熬制10分钟后过滤、取汁，然后加入白糖、味精、芝麻油、花椒油调制成豆瓣卤汁。
3. 将大葱放入盘中垫底，用鸡肉盖面，先淋入豆瓣卤汁，再放入辣椒油、熟芝麻即成。

Ingredients

500g free-range chicken; 50g scallion; 100g Pixian chili bean paste; 10g spices; 20g ginger, finely chopped; 30g scallion, finely chopped; 20ml Shaoxing cooking wine; 1g pepper; 100g white sugar; 2g MSG; 5ml sesame oil; 3ml Sichuan pepper oil; 20ml chili oil; 5g roasted sesame seeds; 100ml cooking oil

Features

lustrous and reddish brown in color; savory, sweet and spicy in taste; tender and tight chicken with a strong aroma from Pixian chili bean paste

Preparation

1. Boil the chicken until cooked through, remove, drain and cut into coarse slivers. Save the chicken stock for later use. Cut the scallions into horse-ear sections;
2. Heat oil in a wok to 100℃, stir fry Pixian chili bean paste, spices, ginger and chopped scallions until fragrant, add the chicken stock and bring to a boil. Add pepper and cooking wine, turn down the heat and simmer over low heat for ten minutes. Remove from the heat, filter out the dregs and save the sauce. Combine the sauce with white sugar, MSG, sesame oil and Sichuan pepper oil, stirring well to make the seasoning sauce.
3. Stack the scallion on a serving plate, top with the chicken and drizzle with the seasoning sauce. Sprinkle with chili oil and sesame seeds.

五香豆瓣鱼条

Five Spice Fish Strips in Chili Bean Sauce

食材配方

虹鳟鱼肉500克　郫县豆瓣30克　姜米5克　蒜米8克　葱花15克　姜片5克　葱段10克　食盐1克
料酒20毫升　醪糟汁20毫升　糖色20克　五香粉2克　味精2克　芝麻油5毫升
食用油1 500毫升（约耗80毫升）

成菜特点

色泽棕红、鱼肉香酥、味道咸鲜，具有浓郁的五香和豆瓣风味。

制作工艺

1. 鱼肉切成长约8厘米，粗约1厘米的条，加入姜片、葱段、食盐、料酒10克拌匀，码味10分钟，再分两次放入油锅中炸至色泽金黄，表皮起硬膜后捞出备用。
2. 锅置火上，入食用油烧至100℃时，先下郫县豆瓣炒香，再掺入鲜汤，用小火熬制5分钟后滤去料渣，取豆瓣汁水留用。
3. 锅中再下油，烧至100℃时，放入姜米、蒜米、葱花炒香后，倒入豆瓣汁水及料酒（10毫升）、糖色、醪糟汁、五香粉烧沸，然后放入鱼条收汁至浓味，最后放入味精、芝麻油推匀，出锅倒入大盘中，凉后装碟即成。

Ingredients

500g rainbow trout; 30g Pixian chili bean paste; 5g ginger, finely chopped; 8g garlic, finely chopped; 15g scallions, finely chopped; 5g ginger, sliced; 10g scallions, segmented; 1g salt; 20ml Shaoxing cooking wine; 20ml fermented glutinous rice juice; 20g caramel; 2g five-spice powder; 2g MSG; 5ml sesame oil; 1,500ml cooking oil (about 80ml to be consumed)

Features

reddish brown in color; savory and crispy fish with strong aromas from five-spice powder and Pixian chili bean paste

Preparation

1. Cut the trout into 8cm-long and 1cm-thick strips, mix with ginger slices, scallion segments, salt and 10ml Shaoxing cooking wine, stir well and leave to marinate for ten minutes. Deep fry twice until the surface becomes golden and hard.
2. Heat oil in a wok to 100℃, stir fry Pixian chili bean paste until fragrant, pour in the stock and simmer over low heat for five minutes. Skim and save the sauce for later use.
3. Heat oil in a wok to 100℃, and stir fry chopped ginger, garlic and scallion until fragrant. Blend in the chili bean sauce, 10ml Shaoxing cooking wine, caramel, fermented rice juice and five-spice powder, bring to a boil, and add fish strips. Simmer to thicken the sauce, add MSG and sesame oil, stir well and pour into a large plate. Leave to cool, and transfer to a serving dish.

祈愿天府·长寿回锅肉

Good Wishes for Sichuan: Longevity Twice-Cooked Pork

此菜是在传统回锅肉基础上的创新之作，其中加入了郫都特产的太和豆豉和韭菜（谐音九，寓意长久、长寿），为了增加菜品的营养价值和提升口感，还为此搭配了四川名小吃——“蛋烘糕”的外皮，既松软可口，又减轻了猪肉的油腻感。在装盘时，还精心设计了小陶猪摆件与小碟盛装的郫县豆瓣，与主盘中的菜品和谐统一，相映成趣。

食材配方

带皮猪二刀肉500克　青韭菜200克　蒜苗100克　干姬松茸20克　面粉500克　鸡蛋液330毫升　酵母6.5克　泡打粉5克　清水550毫升　白糖180克　郫县豆瓣650克　甜面酱20克　白糖7克　酱油7毫升　味精2克　食用油70毫升

成菜特点

色泽红亮、香气浓郁、咸鲜微辣、质软略甜、肥而不腻。

制作工艺

1. 带皮猪二刀肉入锅煮至断生，切成长约6厘米，厚约0.15厘米的片；青韭菜、蒜苗切成段；干姬松茸用热水泡软后切片，再入油锅中炸至酥香。
2. 面粉、鸡蛋液、酵母、泡打粉、水、白糖入盆搅匀，静置30分钟，入不粘锅摊成小圆饼，对切成4块，装入圆盘的上端。
3. 锅置火上，入食用油烧至150℃时，下猪肉片炒香并呈“灯盏窝”形后，入郫县豆瓣、甜面酱、酱油、白糖炒香上色，最后放入青韭菜段、蒜苗段、姬松茸炒断生，起锅后装入圆盘的下端即成。

This dish is an innovated version of the traditional Twice-Cooked Pork with Taihe fermented soy beans and Chinese chives added. Taihe fermented soy beans are a local specialty, and chives, called Jiucai in Chinese which is a homonym of eternity, has the connotation of longevity. To enhance the nutrition and texture of the dish, puffy pancakes are served with the dish to reduce the greasiness of the pork. Porcelain pigs and small saucers containing Pixian chili bean paste are also used to garnish the dish when serving.

Ingredients

500g Erdao (second-cut) pork with skin; 200g chives; 100g baby leeks; 20g dried Brazilian mushrooms; 500g wheat flour; 330ml beaten eggs; 6.5g yeast; 5g baking powder; 550ml water; 180g white sugar; 650g Pixian chili bean paste; 20g sweet wheat flour paste; 7g white sugar; 7ml soy sauce; 2g MSG; 70ml cooking oil

Features

reddish brown in color; savory, slightly spicy and sweet in taste; fatty but not greasy

Preparation

1. Boil the pork until just cooked, and cut into 6cm-long, 0.15cm-thick slices. Cut the chives and baby leeks into sections; Soak the Brazilian mushrooms in hot water until soft, cut into slices and deep fry until fragrant and crispy.
2. Combine in a basin the wheat flour, beaten eggs, yeast, baking powder, water and white sugar, stir well and leave to stand for thirty minutes. Pan fry the batter to make pancakes. Cut the pancake into four even portions, and transfer to the upper end of a serving dish.
3. Heat oil in a wok to 150℃, and stir fry the pork slices until they curl up. Blend in the Pixian chili bean paste, sweet wheat flour paste, soy sauce and white sugar, stir well, and add the chives, leeks and mushrooms, stirring until just cooked. Transfer to the other end of the serving dish.

太极和合·太极蚕豆羹

Tachi Harmony: Taichi Broad Bean Soup

此菜采用酿造郫县豆瓣的主料之一——蚕豆瓣制作而成，造型采用绿、黄双色蚕豆羹组拼成太极图案，传递出和谐与融合的文化寓意。郫都当地有许多以『和』为意的场镇，如太和村、团结镇、友爱镇等，此菜的创意，即蕴含了对和谐社会、政通仁和的良好祝愿。

食材配方

嫩胡豆200克　老南瓜200克　橄榄油30毫升　黄油30克　食盐4克　白胡椒粉0.5克　法香碎0.5克　鲜百里香碎0.5克　莳萝碎0.5克　白糖100克　水淀粉30克　鲜汤200毫升　清水200毫升

成菜特点

色形美观、甜香爽口、咸鲜入味。

制作工艺

1. 蚕豆入锅煮熟后捞出，捣成蚕豆泥；老南瓜去皮，切成大片入锅蒸熟，再捣成南瓜泥。
2. 锅置火上，入橄榄油、黄油加热至融化，下蚕豆泥炒香，再掺入鲜汤烧沸，然后加入食盐、白胡椒粉、法香碎、莳萝碎、鲜百里香碎搅匀制成咸羹。
3. 锅置火上，入清水烧沸，下南瓜泥、白糖烧沸后，用水淀粉勾芡，调制成甜羹。
4. 将咸羹和甜羹分别装入盛器中的太极形模具内造型，取出模具即成。

This dish uses broad beans, the main ingredients of Pixian chili bean paste, to create a taichi pattern as an embodiment of the traditional philosophy of harmony and unity. There are a lot of townships in Pidu District whose names suggest harmony, unity or love (pronounced as “he” in Chinese), such as Taihe, Tuanjie and You’ai, etc.

Ingredients

200g tender broad beans; 200g pumpkins; 30ml olive oil; 30g butter; 4g salt; 0.5g pepper; 0.5g parsley, finely chopped; 0.5g

thyme, finely chopped; 0.5g dill, finely chopped; 100g white sugar; 30g cornstarch batter; 200ml stock; 200ml water

Features

beautiful presentation; fresh, fragrant and appetizing tastes; a combination of both sweet and salty soups

Preparation

1. Boil the broad beans until cooked through, remove and mash. Peel the pumpkins, steam until cooked through and mash.
2. Heat olive oil and butter in a wok, stir fry mashed broad beans, and add the stock. Bring to a boil, blend in salt, pepper, parsley, thyme and ill, stirring well to make the salty soup.
3. Heat water in a wok, bring to a boil, and add mashed pumpkins and white sugar. Bring to a boil, add the batter, and stir well to make the sweet soup.
4. Transfer the salty and sweet soup into a container with taichi mode, and remove the mode

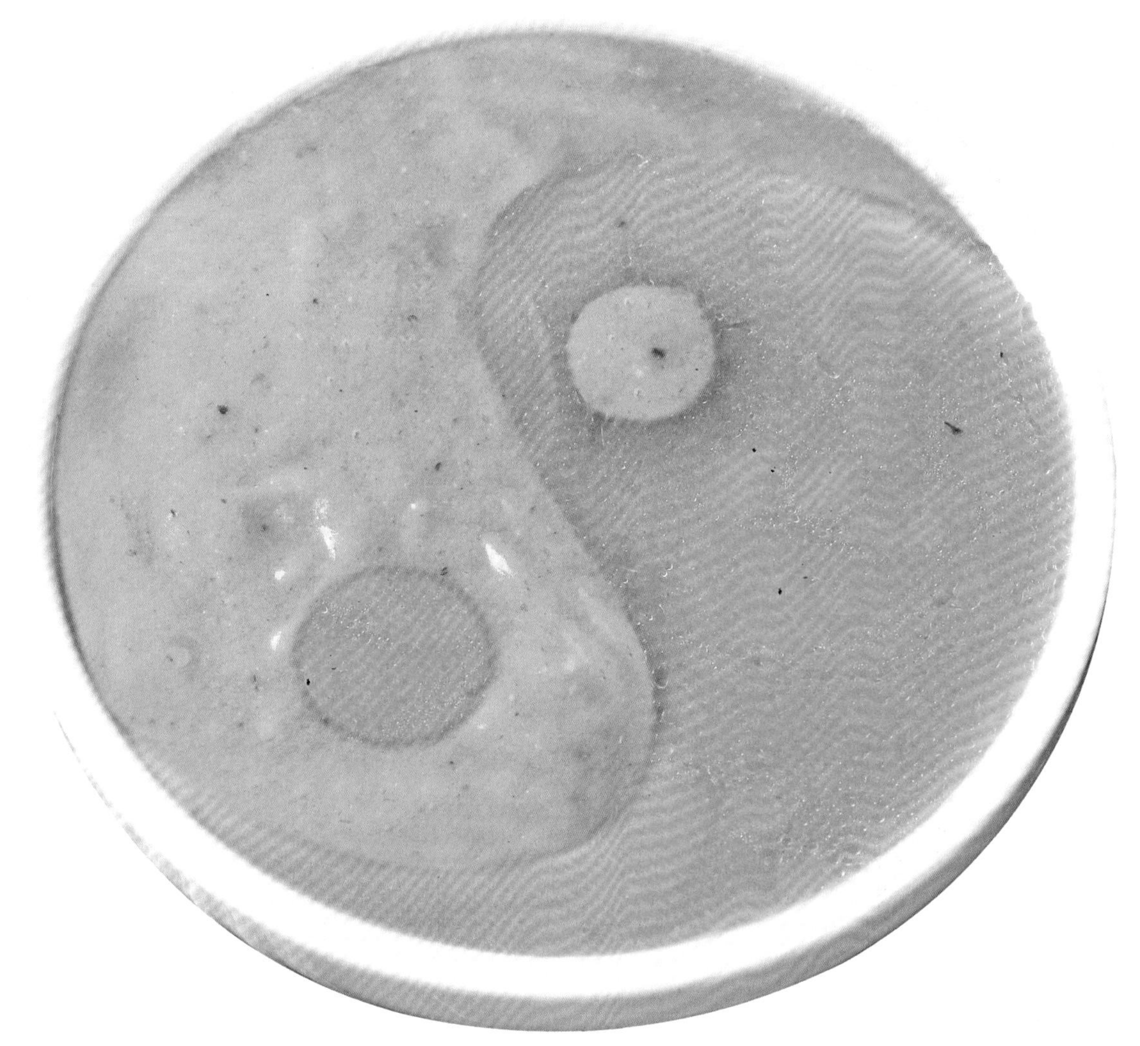

稻香郫都·蚂蚁杂粮饭

Capital of Grains: Ants-Dotted Steamed Coarse Grains

苕菜又称“巢菜”，四川地区以其入馔，已有近千年历史。苕菜鲜品可作为蔬菜食用，清香细嫩；干品可入羹、粥，甘滑清香。选用苕菜干品与大米、玉米碎做成的这款粗粮米饭，因干苕菜碎在米饭中形似蚂蚁而得名。

食材配方

大米500克　玉米碎100克　干苕菜100克　清水750毫升

成菜特点

色彩分明、香味浓郁、特色显著。

制作工艺

1. 大米、玉米碎洗净，入干苕菜拌匀，置盛器中加清水上笼蒸熟。
2. 将蚂蚁杂粮饭盛入碗中，食用时配以清炒时蔬和泡菜尤佳。

Shaocai is a local vegetable also called chaocai. Sichuan people have been using shaocai in their diet for nearly a thousand years. It can be eaten fresh or dry. Fresh shaocai is fragrant and tender while dried shaocai can be added to soups or porridge. Here in this dish, dry shaocai and corn are added to steamed rice, and the dotted shaocai looks like ants, hence the name.

Ingredients

500g rice; 100g corn grains; 100g dry shaocai; 750ml water

Features

colorful grains with a lingering smell; dotted shaocai like ants

Preparation

1. Rinse the rice and corn grains. Add shaocai, and stir well. Transfer to a container, add water and steam until cooked through.
2. Transfer to a serving bowl, and serve with stir fried vegetable or pickles.

五香豆瓣牛排

Five Spice Steak in Chili Bean Sauce

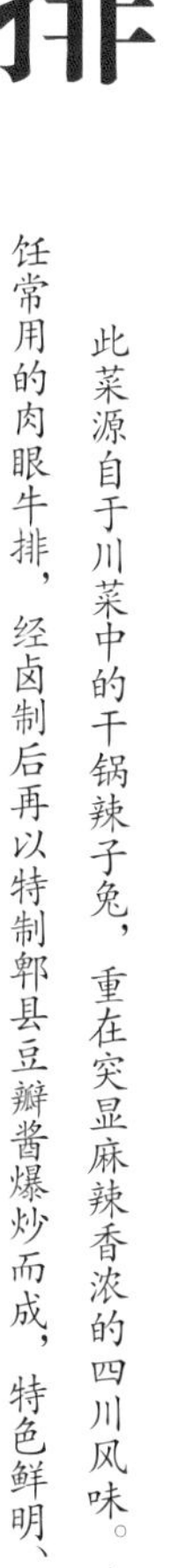

此菜源自于川菜中的干锅辣子兔，重在突显麻辣香浓的四川风味。此菜选取西餐烹饪常用的肉眼牛排，经卤制后再以特制郫县豆瓣酱爆炒而成，特色鲜明、风味独特。

食材配方

肉眼牛排200克　苹果50克　时蔬50克　五香卤水1 000毫升　干辣椒30克　花椒20克　郫县豆瓣20克　姜米10克　蒜米20克　酥花生仁100克　熟芝麻20克　橄榄油40毫升

成菜特点

色泽棕红、外表干香、肉质滋润、麻辣突出、酱香浓郁。

制作工艺

1. 肉眼牛排切片，入五香卤水中用小火煮至微沸，再煮30分钟至入味，切成大丁。
2. 锅置火上，入橄榄油烧至120℃时，放入干辣椒、花椒爆出香味后，再加郫县豆瓣炒香，下牛肉丁炒至入味，下姜米、蒜米、酥花生仁、熟芝麻翻炒匀后装盘。

This dish is in inspired by a classic in Sichuan cuisine, Spicy Rabbit in Dry Pot which features spicy and savory tastes. Here fillet steak is first stewed in seasoned broth and then quick stir fried.

Ingredients

200g rib eye steak; 50g apples; 50g seasonal vegetables; 1,000ml five spice broth; 30g dried chilies; 20g Sichuan pepper; 20g Pixian chili bean pate; 10g ginger, finely chopped; 20g garlic, finely chopped; 100g crispy peanuts; 20g roasted sesame seeds; 40ml olive oil

Features

reddish brown in color; spicy beef crispy on the outside with a strong aroma of chili bean paste

Preparation

1. Cut the beef steaks, add to five spice broth, bring to a boil over low heat and simmer for thirty minutes. Remove the beef from the broth and cut into big cubes.
2. Heat oil in a wok to 120℃, stir fry dried chilies and Sichuan peppercorns over high flame till fragrant. Blend in the Pixian chili bean paste, and stir fry till fragrant. Add the beef cubes, and continue to stir for the beef to absorb the flavors. Add the ginger, garlic, peanuts and sesame seeds, stir to bring out the aroma, and transfer to a serving dish.

豆瓣酱烤五花肉

Roasted Pork Belly with Chili Bean Sauce

此菜源自于川菜最为经典的“回锅肉”，在原有基础上，本菜结合西式暗火烧烤的方式，并借鉴西餐鸡尾酒会中CANAPE的制作形式，以豆瓣酱烧烤而成回锅肉，同时配以小黄瓜和小番茄，在不失咸鲜微辣、干香滋润的同时，又可清香解腻，风味十分独特。

食材配方

带皮猪五花肉200克　苹果50克　新鲜时蔬30克　水果黄瓜30克
郫县豆瓣40克　花椒1克　姜片10克　姜米10克　蒜米5克　葱段10克
葱花15克　辣椒粉20克　花椒粉3克　孜然粉1克　味精1克
干红葡萄酒1毫升　焦糖100克　君度酒50毫升　食用油50毫升

成菜特点

色泽棕红、微辣干香、咸鲜微甜、酱香滋润。

制作工艺

1. 先将郫县豆瓣入食用油中炒香，再加姜米、蒜米、葱花炒匀，之后加辣椒粉、花椒粉、孜然粉和味精拌匀，制成豆瓣烤肉酱。
2. 带皮猪五花肉放入沸水中，加姜片、葱段、花椒煮沸，用小火煮约40分钟，关火浸泡至凉后捞出，切成厚约1厘米的片，用豆瓣烤肉酱拌匀腌制备用。
3. 苹果切成圈，用焦糖煎香，再用君度酒点燃烧出味，制成焦糖苹果。
4. 先将腌制好的肉片放入200℃的烤炉中烤15分钟后翻面，再烤10分钟后取出，切成约2. 5厘米见方的块，与黄瓜块、焦糖苹果块一同用牙签串上装盘即成。

This dish combines the cooking method of Twice-Cooked Pork, roasting in western cuisine and CANAPE in western cocktail parties. Roasted pork belly is served with cucumber and tomatoes, achieving a balance between savory and spicy pork and fragrant and fresh vegetables.

Ingredients

200g pork belly with skin; 50g apples; 30g seasonal vegetables; 30g fruit cucumber; 40g Pixian chili bean paste; 1g Sichuan peppercorns; 10g ginger, sliced; 10g ginger, finely chopped; 5g garlic, finely chopped; 10g scallion, segmented; 15g scallion, finely chopped; 20g ground chilies; 3g ground Sichuan pepper; 1g cumin; 1g MSG; 1ml dry wine; 100g caramel; 50ml cointreau; 50ml cooking wine

Features

reddish brown in color; grilled pork crispy on the outside and juicy on the inside with a strong aroma of chili bean sauce; savory, slightly spicy and sweet in taste

Preparation

1. Stir fry the Pixian chili bean paste until fragrant, add chopped ginger, garlic and scallion, stir well, and blend in ground chilies, ground Sichuan pepper, cumin and MSG to make the barbecue sauce.
2. Add the pork to boiling water, blend in sliced ginger, segmented scallion and Sichuan peppercorns, bring to a boil, turn down the heart and simmer for forty minutes. Remove from the heat, and leave to cool. Remove the pork from the water, cut into 1cm-thick slices, and mix well with the barbecue sauce.
3. Slice the apples, pan fry with the caramel, burn with cointreau to make caramel apples.
4. Roast the pork belly in 200℃ oven for fifteen minutes, turn the slices, and roast for another ten minutes. Remove from the oven, and cut into 2.5cm chunks. String the pork with cucumber and apples with tooth picks, and serve.

豆瓣调理软欧包

Chili Bean Sauce Flavored European Bread

软欧面包是非常受欢迎的面包品种，低油低塘、口感柔软，比较符合现代消费者的需求。这道面点将青椒、红椒、培根切丁后加豆瓣酱炒香，待冷却后再加入到面团中，经醒发、造型、烘烤，色泽金黄，口感丰富。

食材配方

高筋粉600克　白糖120克　淡奶油30克　食盐7克　鸡蛋液35毫升　奶粉18克　酵母9克　改良剂3克　牛奶120毫升　清水150毫升　黄油80克　青椒150克　红椒150克　培根100克　郫县豆瓣57克

成菜特点

色泽金黄、松软滋润、酱香浓郁，鲜香、麦香、奶香兼具。

制作工艺

1. 青椒、红椒、培根切丁备用。
2. 锅置火上，入黄油（50克）烧至融化，先下郫县豆瓣炒香，再放入青椒丁、红椒丁和培根丁炒香，出锅后在常温下冷却成馅料。
3. 将高筋粉、白糖、淡奶油、食盐、鸡蛋液、奶粉、酵母、改良剂、牛奶、清水、黄油（30克）依次加入揉成面包面团，并持续搅拌至面筋充分扩展阶段，然后在面团中拌入冷却好的馅料，并将拌入馅料的面团分割为每个25克的小面团，搓圆醒发后，以200℃/160℃的温度烘烤15分钟左右，待其冷却后装入面包篮，撒上糖粉装饰即成。

European bread is becoming increasingly popular with its soft texture and low fat and sugar contents. Here in this dish, vegetable are dices, stir fried with chili bean paste, and then mixed with bread dough. Bread made in this way is golden in color, presenting a multiple of textures.

Ingredients

600g bread flour; 120g white sugar; 30g whipping cream; 7 g salt; 35ml beaten eggs; 18g milk powder; 9g yeast; 3g improver; 120ml milk; 150ml water; 30g butter; 150g green chili peppers; 150g red chili peppers; 100g bacon; 57g Pixian chili bean paste

Features

golden in color; soft, fresh bread with mixed fragrance from milk, wheat and chili bean paste

Preparation

1. Cut the green and red chili peppers into cubes.
2. Heat a wok over heat, add 50g butter to melt, blend in Pixian chili bean paste, and stir fry till fragrant. Add green and red chili peppers, continue to stir until aromatic, remove from the heat and leave to cool.
3. Mix the bread flour, white sugar, whipping cream, salt, eggs, milk powder, yeast, improver, milk, water and 30g butter to make dough. Continue to stir until the gluten fully expands. Add the filling made in the second step, blend well and divide into dough pieces of 25g each. Knead and roll into buns, and leave to expand. Bake in a 200℃/160℃ oven for fifteen minutes, remove from the oven, and leave to cool. Transfer to a basket, and sprinkle with icing.

第四章 CHAPTER FOUR

美馔佳肴 活色生香 Delicacies

郫都区气候温润、土地肥沃、物产丰富，农耕文化、移民文化与水文化在这里和谐交融，再加上智慧的郫都人喜食擅烹，从而为我们贡献了众多的美馔佳肴，极大地丰富了当地民众和外来消费者的饮食需求。本章仅遴选出其中的部分名特菜肴及面点小吃共60个品种进行介绍，与第三章中以郫县豆瓣为调料的40个品种，共同构成了郫都饮食文化活色生香的独特画卷。其入选标准，主要基于三个原则：第一，具有较为鲜明的郫都地方特色，包括烹饪技法、风味特色等；第二，主要采用本地名特食材制作而成，具有从田间到餐桌，乃至一、二、三产业联动的作用，能够助推乡村振兴；第三，具有一定的文化和旅游表现属性，具备传统与创新兼顾，大气与地气兼有，代表性与常见性兼备等要素，能够促进文旅融合，满足人们美好生活的需要。

Pidu District has mild climate and fertile land, blessing the locals with abundant produce. Development of agriculture and aquaculture as well as immigration waves has made it possible for Pidu people to use ingredients from different places in their cookery innovation, thus creating a large number of delicate dishes to cater for the pallet of local and travelling diners. In this chapter, sixty dishes are listed. Together with the forty dishes introduced in chapter III, they present a profile of the colorful culinary picture of Pidu District. These dishes are chosen according to the following three criteria. First, they are representative of local cooking techniques and flavors. Secondly, they use as the main ingredients local produce, which helps with the integrated growth of the primary, secondary, and tertiary industries and rural rejuvenation efforts. Lastly but not least, they have high cultural and tourism values, combing traditional with innovation, universality with locality.

第一节 冷 菜

Section 1 COLD DISHES

特色卤耳骨

Specialty Spiced Pork Ear Bones

食材配方

猪耳骨1 000克　秘制卤水2 500毫升

成菜特点

色泽酱红、肉质鲜嫩、口感柔软、咸鲜略辣、香味浓郁。

制作工艺

1. 猪耳骨洗净，焯水后用清水浸泡1小时。
2. 卤锅置火上，倒入秘制卤水烧沸，放入猪耳骨卤至软熟后捞出，晾凉后装盘即成。

评鉴

猪耳骨又称猪眉骨，是猪耳朵下面一块带有部分肉质的骨头，食用部分为依附在猪耳骨上的肉。此菜是郫都区著名的“黄老二牛脑壳店”老板黄建研制的特色卤菜，闻着香，入口鲜嫩柔软、唇齿留香，令无数“好吃嘴”慕名而来，吃后依然念念不忘。

Ingredients

1,000g pork ear bones; 2,500ml secret-recipe broth

Features

dark brown in color; tender and soft in texture; savory, aromatic and slightly spicy in taste

Preparation

1. Rinse pork ear bones, blanch, remove to water and leave to soak for one hour.
2. Heat a pot over a flame, pour in the secret-recipe seasoned broth, and bring to a boil. Add the pork ear bones, boil till cooked through and soft, and remove to a serving dish.

Reviews

Pork ear bones, also called pork eyebrow bones, refer to the bones and meat under pork ears. It is the meat rather than bones that the diners would eat. Developed by the boss and chef of Huang Laoer Beef Head, a humble restaurant in Pidu District, the dish attracts a number of food lovers coming to the restaurant.

平乐腌猪头

Pingle Cured Pork Heads

食材配方

去骨新鲜猪头肉5 000克　腌料（朝天椒50克　花椒50克　葱段500克　芹菜段500克　姜片300克　料酒500毫升　醪糟500克　胡椒粉20克　冰糖50克　食盐210克　酱油200毫升　八角25克　丁香10克　山柰20克　桂皮15克　白蔻15克　小茴香40克　甘草10克　排草20克）

秘制卤水3 000毫升

成菜特点

色泽酱红、质地软糯、咸鲜香浓。

制作工艺

1. 猪头肉洗净，对剖成两块。
2. 将腌料均匀地涂抹在猪头肉上，放入2～3℃的保鲜柜（或冰箱）腌制48小时（每隔12小时将猪头肉翻动一次）。
3. 将腌制好的猪头肉放进烘房烘制5小时左右，待肉质干香后取出，再放入秘制卤水中卤制1～1.5小时，至肉质软熟后捞出，晾凉后切片装盘即成。

评鉴

此菜是郫都区馨苑印象休闲庄在“花园场腌肉”基础上改良而来的一道特色菜品，不但秉承了花园场腌肉制作的传统技艺，而且还在原有基础上结合秘制卤煮进行了大胆改良，在横山平乐寺一带专点销售，产品供不应求，故命名为“平乐腌猪头”。

Ingredients

5,000g pork heads (deboned); marinade (50g sky-pointing chilies; 50g Sichuan pepper; 500g scallion, segmented; 500g celery, segmented; 300g ginger, sliced; 500ml Shaoxing cooking wine; 500ml fermented glutinous rice juice; 20g pepper; 50g rock sugar; 210g salt; 200ml soy sauce; 25g star anise; 10g cloves; 20g sand ginger; 15g cinnamon; 15g round cardamom; 40g fennel; 10g liquorice; 20g nephrolepis); 3,000ml secret-recipe broth

Features

dark brown in color; soft and glutinous in texture; savory and delicate in taste

Preparation

1. Rinse the pork head and halve.
2. Smear the halved pork heads with the marinade, and leave in a 2℃-3℃ fridge to marinate for forty-eight hours, turning every twelve hours.
3. Transfer the pork heads to a drying chamber, and leave to dry for five hours until the pork is aromatic. Simmer the pork heads in secret-recipe broth for one to one and a half hours until cooked through and soft. Remove from the broth, and leave to cool. Cut into slices, and transfer to a serving dish.

Reviews

This dish is an upgraded version of the Huayuanchang Cured Pork, using traditional curing techniques with a stewing process added. It is called Pingle Cured Pork Heads because it is sold around Pingle Temple and enjoys immense popularity.

椒香缠丝兔

Sichuan Pepper Flavor Cord Wrapped Rabbit

食材配方

兔子1只（约750克）　食盐7.5克　秘制香料15克
白酱油15毫升　甜酱30克　青花椒120克
红花椒10克　豆豉20克　五香粉3克　卤水2 000毫升

成菜特点

色泽酱红、质地紧实、椒香味浓。

制作工艺

1. 兔子洗净后沥干余水，加入食盐、秘制香料、白酱油、甜酱拌匀，腌制8小时，再取出自然晾干水分，然后用五香粉、青花椒、红花椒、豆豉均匀涂抹兔子全身，再继续腌制1小时后，用麻绳缠绕成圆筒状，放入烟熏室烟熏后取出。
2. 卤水锅置火上，放入烟熏过的兔子卤制1小时后捞出晾凉，斩成丁后装盘即成。

评鉴

此菜是在原缠丝兔的基础上演变而来，采用的是两次腌制，再经烟熏后卤而成，成菜椒麻味突出，香味悠长，佐酒尤佳。

Ingredients

a rabbit (about 750g); 7.5g salt; 15g secret-recipe spice; 15ml light soy sauce; 30g sweet flour paste; 120g green Sichuan peppercorns; 10g Sichuan pepper; 20g fermented soy beans; 3g five-spice powder; 2,000ml seasoned broth

Features

dark brown in color; tight in texture; savory and numbing in taste

Preparation

1. Mix rabbit well with salt, secret-recipe spice, light soy sauce and sweet flour paste. Leave to marinate for eight hours. Drain and smear with five-spice powder, green Sichuan pepper, Sichuan peppercorns and fermented soy sauce. Leave to marinate for another hour. Wrap with linen cord into a cylinder, and smoke.
2. Heat the broth, and simmer the rabbit for an hour. Remove the rabbit from the broth, leave to cool, dice and transfer to a serving dish.

Reviews

This dish is an upgraded version of Cord Wrapped Rabbit, where rabbit is marinated twice before being smoked and boiled in seasoned broth.

五香猪肝
Five Spice Pork Livers

食材配方

新鲜猪肝1 500克　花椒70克　食盐50克　秘制香料60克
白酒75毫升　熟芝麻20克　熟菜籽油50毫升

成菜特点

干香滋润、入口化渣、花椒味浓。

制作工艺

1. 将新鲜猪肝切成粗条，用花椒、食盐、秘制香料、白酒拌匀腌制4小时。
2. 将腌制好的猪肝均匀地铺在烤网上，入熏烤炉熏烤，待基本烤干猪肝中的水分后晾凉、洗净。
3. 锅置火上，将猪肝放入煮10分钟后捞出，刷上一层熟菜籽油，撒上熟芝麻，切片后装盘即成。

评鉴

此菜是郫都区余记腌卤店的特色菜品，曾荣获成都市第一届美食节“成都市名小吃”“成都名菜”称号。如果蘸取少许“老白干”食用，更是越嚼越香、回味无穷。

Ingredients

1,500g pork livers; 70g Sichuan pepper; 50g salt; 60g secret-recipe spices; 75ml liquor; 20g roasted sesame seeds; 50ml cooked rapeseed oil

Features

aromatic and savory livers with a strong Sichuan pepper fragrance, melting in the mouth

Preparation

1. Cut pork livers into thick strips, add Sichuan peppercorns, salt, secret-recipe spices and liquor, stir well and leave to marinate for four hours.
2. Place the pork livers on a grill, and transfer to a smoking stove to smoke. Till they are almost dry, remove from the cooker, and leave to cool and rinse.
3. Heat a wok over a flame, add water and the pork livers, and boil for ten minutes. Remove the livers from the wok, smear with rapeseed oil and sprinkle with sesame seeds. Cut into slices and transfer to a serving dish.

Reviews

This dish is a specialty of Yu's Luzhu Business in Pidu District. It has won honorable titles such as Renowned Snack of Chengdu and Famous Dish of Chengdu. Dipping into laobaigan liquor before eating gives the pork livers an extra peculiar flavor and a lingering aroma in the mouth.

花园场腌肉
Huayuanchang Cured Pork

食材配方

猪肉2 500克　食盐1 000克（约耗150克）　花椒150克　香料150克　醪糟200克

成菜特点

肉皮棕黄、质地香软、味道咸鲜、烟香浓郁。

制作工艺

1. 将猪肉切成每根重约600克的长条。
2. 锅置火上，入食盐、花椒、香料，用小火炒至食盐变色后关掉火源，将猪肉条分别放入抹匀，再放入容器中加醪糟拌匀，密封腌制72小时，捞出晾干水分（约2～3天）。

3. 将晾干水分的猪肉挂入熏房，用锯末烟熏4小时后取出，洗净后放入笼中蒸熟，取出切为薄片装盘即成。

评鉴

受汉代扬雄美食思想的影响，郫都区西部的花园场、横山平乐、三元场一带的人们向来崇尚美食生活，民间美食众多，“花园场腌肉”就是其中的代表作之一，至今已有一百多年的历史。此菜耗时费工，大多在春节前制作，是当地民众春节期间招待贵宾的必上菜品。

Ingredients

2,500g pork; 1,000g salt (about 150g to be consumed); 150g Sichuan pepper; 150g spices; 200ml fermented glutinous rice juice

Features

brown in color; soft in texture; salty and savory in taste with a strong smoky flavor

Preparation

1. Cut pork into strips of about 600g each.
2. Heat a wok over a flame, dry roast salt, Sichuan peppercorns and spices until the colro of the salt changes. Turn off the heat, add the pork strips, and blend well so that each pork strip is evenly coated with the salt mixture. Remove the pork strips to a container, add fermented rice juice, mix well and seal the container. Leave to marinate for seventy-two hours, remove and air dry (two or three days needed).
3. Hang the pork strips in a smoking chamber, and smoke with saw powder for four hours. Remove from the chamber, rinse and steam till cooked through. Cut into thin slices and transfer to a serving plate.

Reviews

Influenced by the dietary philosophy of Yang Xiong, a famous philosopher and literary figure in the Han Dynasty, people in the west of Pidu District such as Huayuanchang have a tradition of pursuing culinary delicacies, and a lot of folk gourmet foods can be found in this area, among which Huayuanchang Cure Pork is one of the most renowned. This dish has a history of over one hundred years and is often prepared before the Spring Festival. It takes a long time and complex processes to make, and is a must to serve distinguished guests during the Spring Festival.

糙糖酥

Crispy Sugar Cubes

食材配方

猪肥肉500克　鸡蛋液150毫升　红苕淀粉150克　红糖250克　熟白芝麻100克　菜籽油1 500毫升（约耗50毫升）

成菜特点

色泽酱红、外酥内空、肥而不腻、香甜可口。

制作工艺

1. 猪肥肉切成约1.5厘米见方的丁；鸡蛋液与红苕淀粉搅匀成全蛋淀粉。
2. 锅置火上，入菜籽油烧至150℃时，将肥肉与全蛋淀粉拌匀，分散放入油锅中炸至定型后捞出，待油温回升至180℃时，再将肉丁入锅复炸至外酥内空、色呈金黄时捞出沥去余油。
3. 锅中留油少许，放入红糖炒化后关掉火源，先倒入炸酥的肥肉丁搅拌均匀，再撒上熟白芝麻拌匀，出锅晾凉后装盘即成。

评鉴

糙糖酥在郫都俗称“干盘子”，有较长历史，是川西坝子田席中的必上菜品。目前在郫都区的一些乡镇和街道都有销售，代表店家有廖氏糙糖酥、唐昌花田人家休闲庄、郫都区红星饭店等。

Ingredients

500g pork fat; 150ml beaten eggs; 150g sweet potato starch; 250g brown sugar; 100g roasted sesame seeds; 1,500ml rapeseed oil (about 50ml to be consumed)

Features

reddish brown in color; crispy in texture; sweet and appetizing in taste; fatty but not greasy

Preparation

1. Cut the pork fat into 1.5cm cubes. Mix the beaten eggs with sweet potato starch, and blend well.
2. Heat oil in a wok to 150℃, coat the pork cubes with the above-made mixture, and deep fry to harden the surface. Reheat the oil to 180℃, deep fry the pork cubes for a second time until the fat melts and the cubes becomes golden and crispy on the outside, remove and drain.
3. Heat oil in a wok, add the brown sugar, stir until the sugar melts, and turn off the heat. Pour in the pork cubes, stir well and blend in the sesame seeds. Remove from the wok, leave to cool and transfer to a serving dish.

Reviews

Crispy Sugar Cubes, also called Gan Pan Zi (dry plates if literally translated), is a time-honored dish in Pidu District. It is a must in rural feasts, and can be found in local stores and restaurants, among which the most famous are Liao's Crispy Cubes, Tangchang Floral Farm Restaurant and Red Star Restaurant of Pidu District.

太和牛肉

Taihe Beef

食材配方

牛腱肉500克　姜片10克　葱段15克　大蒜15克　料酒20毫升　肉桂3克　丁香2克　八角3克　草果3克　茴香2克　花椒1克　食盐50克　酱油50毫升　红糖20克　豆瓣蘸粉10克　鸡汤2 000毫升　食用油300毫升

成菜特点

色泽红亮、质地软糯、咸鲜浓香、回味悠长。

制作工艺

1. 牛腱肉洗净，切成大块焯水后备用。
2. 锅置火上，入食用油烧至150℃时，放入姜片、葱段、大蒜、料酒炒香，接着下肉桂、丁香、八角、草果、茴香、花椒、酱油、红糖、食盐、鸡汤、牛腱肉，先用大火煮20～30分钟，再改用小火煮至牛腱肉软熟、入味时捞出，晾凉后切片装盘，配上豆瓣蘸粉上桌即成。

评鉴

太和牛肉始创于清朝末年，因源于郫都区太和场（今郫都区团结街道）而得名。卤制后的太和牛肉表面红润、质地紧密、纹理细腻，为宴席和馈赠亲朋好友之上品，远近闻名，并于2019年入选“成都市郫都区十大名小吃”之列。

Ingredients

500g beef; 10g ginger, sliced; 15g scallion, segmented; 15g garlic; 20ml Shaoxing cooking wine; 3g cinnamon; 2g cloves; 3g star anise; 3g tsaoko amomum; 2g fennel; 1g Sichuan pepper; 50g salt; 50ml soy sauce; 20g brown sugar; 10g dipping sauce; 2,000ml chicken stock; 300ml cooking oil

Features

reddish brown in color; soft in texture; savory in taste with a lingering arom

Preparation

1. Rinse the beef, cut into chunks and blanch.
2. Heat oil in a wok to 150℃, and stir fry ginger, scallion, garlic and cooking wine until fragrant. Add the cinnamon, star anise, tsaoko amomum, fennel, Sichuan pepper, soy sauce, brown sugar, salt, chicken stock and beef, boil over high heat for twenty to thirty minutes, turn down the heat and continue to boil until the beef is soft and cooked through. Remove the beef from the wok and leave to cool. Cut into slices, transfer to a serving dish and serve with the sauce.

Reviews

This dish is so named because it was invented in Taihe Village in the late Qing Dynasty. The beef is renowned for tight texture, savory taste and appealing colors, often served on banquets or presented as a gift. It was honored as Top 10 Snacks of Pidu District of Chengdu.

火鞭牛肉
Firecracker Beef

食材配方

黄牛腿肉5 000克　花椒50克　食盐220克　姜片50克　葱段200克　醪糟汁200毫升　料酒300毫升　八角25克　山柰15克　桂皮18克　丁香10克　小茴香36克　白蔻15克　甘草10克

成菜特点

色泽红亮、质地松软、入口化渣、咸鲜香浓。

制作工艺

1. 牛肉顺筋改刀为粗约2.5厘米见方的长条，依次加入花椒、食盐、姜片、葱段、醪糟汁、料酒、八角、山柰、桂皮、丁香、小茴、白蔻、甘草，拌匀后装盆，入冰箱腌制24小时（温度控制在2～3℃，中途翻动一次）。

2. 将牛肉从冰箱中取出，在气温为16℃左右的环境中风干8小时，使牛肉脱水30%～35%，再将牛肉放入熏炉，用香樟叶、柏树枝熏制12小时，取出晾凉，入蒸箱蒸制40分钟后取出，切条后装盘即成。

评鉴

火鞭牛肉是在太和牛肉基础上的改良菜品，因牛肉改刀、晾晒后形似鞭炮而得名。此菜选用三年以上新鲜黄牛腿肉经腌制、风干、烟熏、蒸制而成，松软化渣、肉含烟香，曾荣获“成都名菜”称号。

Ingredients

5,000g beef (leg cut); 50g Sichua pepper; 220g salt; 50g ginger, sliced; 200g scallion, segmented; 200ml fermented glutinous rice juice; 300ml Shaoxing cooking wine; 25g star anise; 15g sand ginger; 18g cinnamon; 10g cloves; 36g fennel; 15g round cardamom; 10g liquorice

Features

reddish brown in color; soft and meltingly crispy in texture; savory and aromatic in taste

Preparation

1. Cut the beef along the tendons into long strips about 2.5cm in width and thickness. Add Sichuan peppercorns, salt, ginger slices, segmented scallion, fermtend rice juice, cooking wine, star anise, sand ginger, cinnamon, clove, fennel, round cardamom and liquorice, stir well and transfer to a basin. Put the basin into 2℃ to 3℃ fridge, and marinate for twenty-four hours, turning once during the process.
2. Remove the beef from the fridge, and leave in room temperature (about 16℃) to air dry for eight hours. When 30% to 35% of the water contents evaporate, place the beef into a smoking stove and smoke with camphor tree leaves and cypress branches for twelve hours. Remove from the stove, leave to cool and steam in a steamer for forty minutes. Slice and transfer to a serving dish.

Reviews

Firecracker Beef, an upgraded version of Taihe Beef with an honorable title of Renowned Dish of Chengdu, is so named because the beef strips, when air drying, look like firecrackers. The dish uses beef leg from cattle of three years old or more, which is marinated, air dried, smoked and finally steamed.

凉拌大肉

Large Pork Slices in Chili Sauce

食材配方

猪头肉400克　姜片10克　葱段15克　花椒0.5克　料酒15毫升　食盐1克　味精2克　白糖1克　酱油10毫升　花椒粉1克　辣椒油50毫升

成菜特点

色泽红亮、质地脆糯、麻辣咸鲜、回味略甜。

制作工艺

1. 锅置火上，放入猪头肉、清水、姜片、葱段、花椒、料酒煮熟，捞出晾凉后片成长约8厘米、宽约5厘米、厚约0.15厘米的大片装盘备用。
2. 将食盐、味精、白糖、酱油、花椒粉、辣椒油调成麻辣味汁，浇淋在盘中的猪头肉片上即成。

评鉴

此处的“大肉”，既是指猪肉本身，也是指肉片形大质薄，说明制作技艺高超。郫都区的三道堰镇、郫筒街道等地，有多家餐馆都在经营凉拌大肉。如戴大肉、陈大肉、幺哥大肉等。

Ingredients

400g pork head; 10g ginger, sliced; 15g scallion, segmented; 0.5g Sichuan pepper; 15ml Shaoxing cooking wine; 1g salt; 2g MSG; 1g white sugar; 10ml soy sauce; 1g ground Sichuan pepper; 50ml chili oil

Features

lustrous red in color; crispy and glutinous in texture; spicy, savory and slightly sweet in taste

Preparation

1. Heat a wok over the flame, and add pork head, water, ginger slices, segmented scallion, Sichuan peppercorns and cooking wine. Boil till the pork head is cooked through, remove, cool and cut into big slices of 8cm in length, 5cm in width and 0.15cm in thickness. Transfer to a serving dish.
2. Combine salt, MSG, sugar, soy sauce, ground Sichuan pepper and chili oil in a bowl, stir well and pour over the pork.

Reviews

In this dish, pork head is cut into large thin slices, a showcase of the chef's exquisite cutting techniques. This dish is sold in a number of catering businesses of Sandaoyan Town and Pitong Township of Pidu District. Famous restaurants include Dai's Large Pork Slices, Chen's Large Pork Slices and Yaoge's Large Pork Slices.

丽妹肺片

Limei's Beef Offal Slices in Chili Sauce

食材配方

牛腱肉80克　牛筋80克　牛舌80克　牛肚80克　牛心80克芹菜碎20克　酥花生碎20克　香菜碎10克　熟芝麻3克　食盐1克　酱油10毫升　味精2克　花椒粉1克　芝麻油3毫升　辣椒油60毫升　白卤水2 000毫升

成菜特点

色泽红亮、麻辣鲜香、肉质软韧。

制作工艺

1. 牛腱肉、牛筋、牛舌、牛肚、牛心分别焯水后放入白卤水中卤至软熟，捞出后晾凉，再切成长约6厘米、宽约3厘米、厚约0.2厘米的片。

2. 用卤水（60毫升）、食盐、味精、酱油、花椒粉、芝麻油、辣椒油调成味汁。
3. 将牛肉、牛筋、牛舌、牛肚、牛心混合后装盘，淋入味汁，撒上酥花仁碎、芹菜碎、熟芝麻、香菜碎即成。

评鉴

在四川成都，有一款家喻户晓、人人皆知的传统名菜叫作夫妻肺片，而在郫都区，则有丽妹肺片与之遥相呼应。这道丽妹肺片深受当地人喜爱，曾获得县、市级多个名小吃称号。

Ingredients

80g beef shank; 80g beef tendon; 80g beef tongue; 80g beef tripe; 80g beef heart; 20g celery, finely chopped; 20g crispy peanuts, finely chopped; 10g coriander, finely chopped; 3g roasted sesame seeds; 1g salt; 10ml soy sauce; 2g MSG; 1g ground Sichuan pepper; 3ml sesame oil; 60ml chili oil; 2,000ml seasoned white broth

Features

lustrous red in color; spicy, fresh and aromatic in taste; soft and chewy in texture

Preparation

1. Blanch beef shank, tendon, tongue, tripe and heart, and boil in seasoned white broth until cooked through. Remove from the broth, leave to cool and cut into slices of 6cm in length, 3cm in width and 0.2cm in thickness.
2. Combine 60ml broth, salt, MSG, soy sauce, ground Sichuan pepper, sesame oil and chili oil in a bowl, stirring well to make the seasoning sauce.
3. Mix beef shank, tendon, tongue, tripe and heart, transfer to a serving dish, drizzle with the seasoning sauce, and sprinkle with crispy peanuts, celery, sesame seeds and coriander.

Reviews

In Chengdu, there is a household dish Fuqi Feipian (Fuqi Beef Offal Slices), while in Pidu District the most famous beef offal slices would be Limei's. It enjoys immense popularity among local people, and has been entitled Famous Snack on county and city levels.

食材配方

土公鸡1只（约2 000克）　姜20克　葱30克　食盐50克　胡椒粉8克　秘制香料10克　醪糟10克
清水4 000毫升　椒麻味碟1个（约60毫升）　红油味碟1个（约60毫升）　姜汁味碟1个（约60毫升）

成菜特点

外皮金黄、肉质紧实、香味浓郁、色泽淡雅、味感多样。

制作工艺

1. 土公鸡洗净，用醪糟均匀地涂抹全身；姜（10克）、葱（15克）拍破后塞进鸡肚，腌制15分钟。
2. 锅置火上，放入清水、土公鸡、食盐、胡椒粉、小茴香、秘制香料、姜（10克）、葱（15克）烧沸，去掉浮沫，改用中火煮10分钟，再用小火煮20分钟，关掉火源焖15分钟，捞出后置通风处自然冷却。
3. 将冷却后的鸡肉斩为长条后装盘，配上椒麻味碟、红油味碟和姜汁味碟上桌即成。

杨鸡肉

Yang's Chicken

评鉴

杨鸡肉既是店招，又是菜名，因创始者姓杨，招牌菜品以凉拌鸡肉为主打而得名。郫都区是“中国农家乐发源地”，最初的农家乐多用凉拌鸡款待游客，杨鸡肉便是其中的佼佼者，有椒麻味汁、红油味汁、姜汁味汁三种味型供食客任意选用。该菜曾荣获“2011年度成都市十佳风味菜品”“2021年成都名菜”等称号。

Ingredients

1 free-range rooster (about 2,000g); 20g ginger; 30g scallion; 50g salt; 5ml soy sauce; 5g MSG; 8g pepper; 10g secret-recipe spice; 10ml fermented glutinous rice juice; 4,000ml water; 1 saucer of jiaoma paste (about 60ml); 1 saucer of chili oil (about 60ml); 1 saucer of ginger juice (about 60ml)

Features

savory chicken firm in texture and golden on the skin

Preparation

1. Rinse the rooster, smear with fermented rice juice, stuff with 10g ginger and 15g scallion, and leave to marinate for fifteen minutes.
2. Heat a wok over a flame, add water, rooster, salt, pepper, fennel, secret-recipe spice, 10g ginger and 15g scallion, bring to a boil and skim. Turn down the heat, simmer for ten minutes over medium heat and twenty minutes over low heat. Turn off the heat, leave the chicken in the pot for another fifteen minutes before removing from the pot and cooling in well-ventilated places.
3. Cut the chicken into long strips, transfer to a serving platter and serve with the three saucers.

Reviews

Yang's Chicken is the name for both a dish and the restaurant that serves the dish as its specialty. Pidu District is the starting place of farm restaurants of China, and most early farm restaurants served Chicken in Chili Sauce, among which Yang's Chicken is one the best. The dish has appealing colors and varied textures, winning honorable titles such as “Top10 Dishes of Chengdu, 2011” and “Renowned Dish of Chengdu, 2021”. It offers three flavors of sauce for the diners to choose from: pepper salt, chili oil and ginger juice.

食材配方

土鸭1只（约1 800克）　花椒50克　干辣椒50克　冰糖10克　鸡精30克　味精20克
食盐130克　香料水2 000毫升

成菜特点

色泽棕红、形如琵琶、肉质紧实、咸鲜香浓、肥而不腻。

制作工艺

1. 土鸭治净，入香料水中浸泡两小时后捞出，用食盐涂抹全身至腌制入味。
2. 将腌制好的鸭子从腹部剖开，压断鸭胸骨和背骨，撑开鸭腹中部，置于通风处晾晒至干。
3. 将风干后的鸭子入冷水中浸泡2～3小时后捞出，放入卤水锅中，入花椒、干辣椒、冰糖、鸡精、味精，用旺火烧沸，转中火卤制20分钟，再改用小火卤制40分钟后捞出，置通风处晾凉后斩成块装盘即成。

评鉴

唐昌板鸭是唐昌的著名特产，已有上百年历史，其经久不衰的秘诀，就在于香料水的调制。制作唐昌板鸭所使用的香料水，是由八角、茴香、桂皮、山柰等30余种香料熬制而成，故而香味浓郁。该菜先后荣获“天府食品博览会优秀名小吃”“成都十大家乡菜”“成都名菜”等荣誉称号。

唐昌板鸭
Tangchang Flat Duck

Ingredients

1 free-range duck (about 1,800g); 50g Sichuan pepper; 50g dried chilies; 10g rock sugar, 30g chicken essence; 20g MSG; 130g salt; 2,000ml spiced water

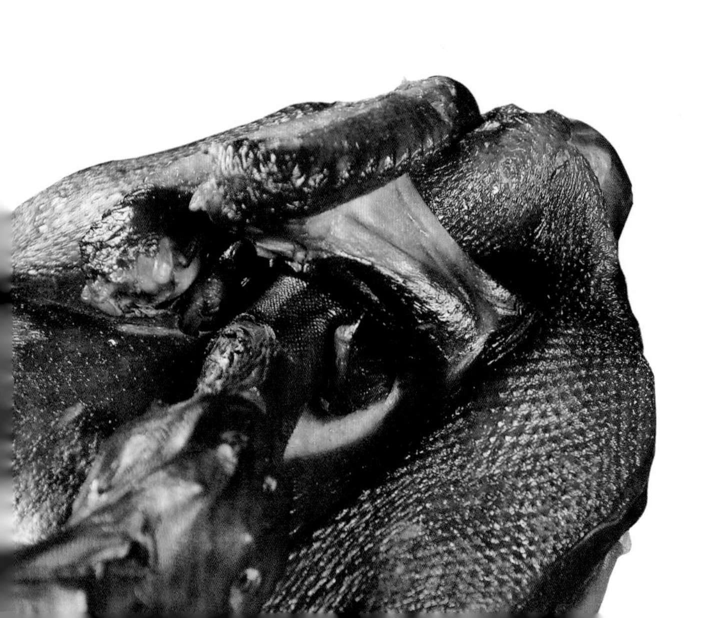

Features

reddish brown in color; tight and melting in texture; savory and aromatic in taste; fatty but not greasy

Preparation

1. Rinse the duck, soak in spiced water for two hours, remove and smear with salt. Leave to marinate.
2. Cut the duck open from the belly, break its chest and back bones, and brace out its belly. Leave to air dry in well-ventilated places.
3. Soak the duck in cold water for two to three hours, remove to a pot with seasoned broth, and add Sichuan peppercorns, dried chilies, rock sugar, chicken essence and MSG. Bring to a boil over high heat, and continue to simmer for twenty minutes over medium heat and forty minutes over low heat. Remove the duck from the pot, and leave to cool in well-ventilated places. Chop the duck into chunks and transfer to a serving dish.

Reviews

Tangchang Flat Duck, a local specialty of Tangchang, has a history of over a hundred years. The secret to its success lies in the aromatic spiced water, and more than thirty spices are used including star anise, fennel, cinnamon and sand ginger, etc. The dish has won a number of honorable titles such as Excellent Snack on Food Expo of Chengdu, Top10 Hometown Dish of Chengdu and Renowned Dish of Chengdu, etc.

花椒鸭

Sichuan Pepper Duck

食材配方

鸭子1只（约1 500克） 食盐20克 秘制香料20克 白酱油30克 卤水2 000毫升

成菜特点

色泽酱红、肉质紧实、咸麻兼具。

制作工艺

1. 鸭子治净后沥干余水，加入食盐、秘制香料、白酱油拌匀，腌制8小时。

2.卤水锅置火上，放入腌制好的鸭子卤制1. 5小时后捞出，晾凉后斩成条装盘即成。

评鉴

花椒鸭是四川近年来较为流行的一道菜品，花椒香味特别浓郁，且能去腥解腻，是一道很有特色的佐酒美肴。这里介绍的这款花椒鸭，是郫都区范氏花椒名鸭店的代表菜品，其制作工艺除了用花椒腌制外，还采用了卤水卤制，风味更是别具一格。

Ingredients

1 duck (about 1,500g); 20g salt; 20g secret-recipe spice; 30ml light soy sauce; 2,000ml seasoned broth

Features

reddish brown in color; tight in texture; savory in taste with a strong Sichuan pepper aroma

Preparation

1. Rinse the duck, drain, and mix well with salt, secret-recipe spice and light soy sauce. Leave to marinate for eight hours.
2. Heat a pot with seasoned broth, add the duck and simmer for one and a half hours. Remove from the pot and leave to cool. Cut into strips and transfer to a serving dish.

Reviews

A recently popular dish in Sichuan, Sichuan Pepper Duck has a strong Sichuan pepper flavor and goes best with Chinese liquor. The cooking method used here comes from Fan's Sichuan Pepper Duck Restaurant in Pidu District, where the duck is first seasoned with Sichuan pepper and then boiled in seasoned broth.

周鹅肉油烫鹅

Zhou's Oil Scorched Goose

食材配方

仔鹅1只（约2 500克） 饴糖30克 八角3克 小茴香1克 山柰4克 草果（拍破）5克 桂皮5克 丁香1.5克 良姜8克 荜拨2克 花椒4克 姜片20克 葱段25克 食盐15克 料酒20毫升 卤水3 000毫升 菜籽油2 000毫升（约耗60毫升）

成菜特点

色泽红亮、质软化渣、咸鲜略甜。

制作工艺

1. 鹅治净，用食盐、花椒、姜片、葱段、料酒拌匀，腌制3小时，入沸水中焯水5分钟捞出备用。
2. 八角、小茴香、山柰、草果、桂皮、丁香、良姜、荜拨用纱布包上成香料包。
3. 卤水锅置火上，入鹅及香料包，用旺火烧沸，再改用小火卤制20分钟，关掉火源焖30分钟后出锅、晾凉；再将鹅身均匀涂抹上饴糖，自然晾放1个小时。
4. 锅置火上，入菜籽油烧至180℃时，将鹅放入炸至色金黄后捞出，改刀后装盘即成。

评鉴

此菜是因周姓创制人自营的油烫鹅而得名。该菜选用约2500克重的仔鹅烹制而成，肉质细嫩、香味扑鼻，深受大众喜爱，曾荣获“成都名菜”“成都市郫都区十大小吃”等称号。

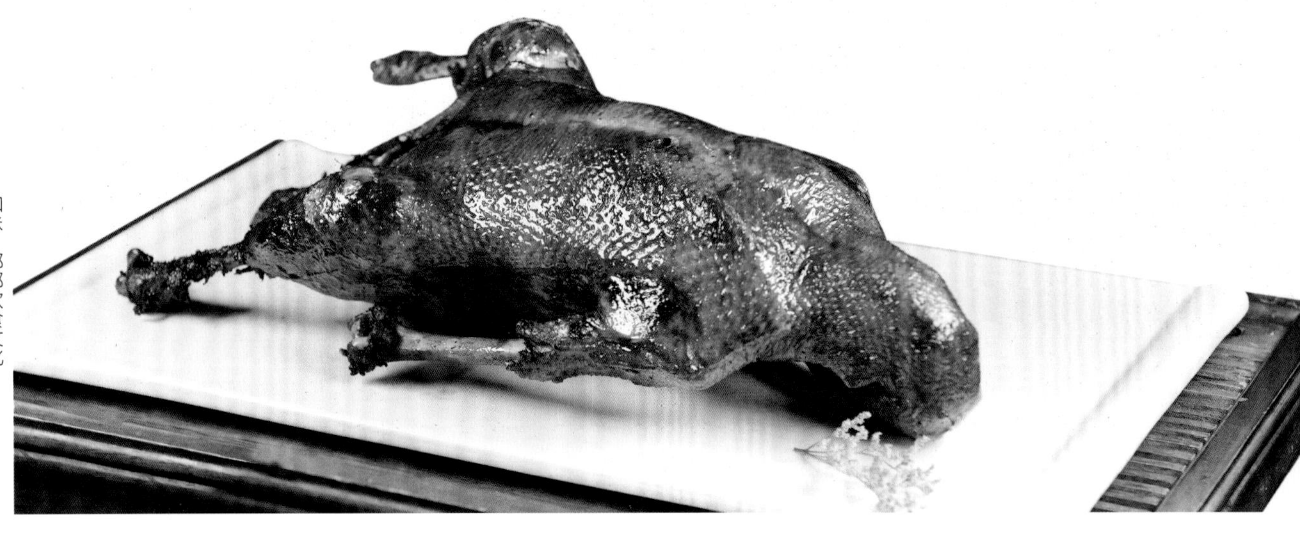

Ingredients

1 goose (bout 2,500g); 30g maltose; 3g star anise; 1g fennel; 4g sand ginger; 5g tsaoko amomum (crushed); 5g cinnamon; 1.5g clove; 8g galangal; 2g long pepper; 4g Sichuan pepper; 20g ginger, sliced; 25g scallion, segmented; 15g salt; 20ml Shaoxing cooking wine; 3,000ml seasoned broth; 2,000ml rapeseed oil (about 60ml to be consumed)

Features

reddish brown in color; soft and melting textures; savory and slightly sweet in taste

Preparation

1. Rinse the goose, and mix well with salt, Sichuan peppercorns, sliced ginger and cooking wine. Leave to marinate for three hours, and blanch in boiling water for five minutes.
2. Wrap up star anise, sand ginger, fennel, tsaoko amomum, cinnamon, clove, galangal and long pepper with cheese cloth to make a spice bag.
3. Heat seasoned broth in a pot, add goose and the spice bag, bring to a boil and turn down the heat. Simmer over low heat for twenty minutes, turn off the heat and leave the duck in the pot for another thirty minutes. Remove the duck from the pot and leave to cool. Smear the goose with maltose, and leave to marinate and drain for one hour.
4. Heat oil in a wok to 180℃, and deep fry the goose until golden. Cut and serve.

Reviews

The dish is name after its inventor Zhou, and has won such titles as "Renowned Dish of Chengdu" and "Top10 Snack of Pidu District" with its tender and savory goose. The geese used in this dish are small ones of about 2,500g in weight.

郫都腌熏排骨

Pidu Cured and Smoked Pork Ribs

食材配方

猪排骨5 000克　食盐1 500克（约耗150克）　八角80克　小茴香80克　香叶30克　桂皮50克　花椒30克　姜片100克　葱段150克　醪糟200克

成菜特点

色泽棕红、咸鲜干香、熏香味浓。

制作工艺

1. 猪排骨洗净，均匀地涂抹上姜片、葱段、醪糟，晾干水分后装入盆中。
2. 锅置火上烧热，入食盐、八角、小茴香、香叶、桂皮、花椒炒香，至色变微黄后，倒入装有排骨的盆中拌匀，腌制24小时，然后取出晾干水分，放入烘房，用柏枝、锯末熏制4小时。
3. 将熏制后的排骨用温水洗净，入笼蒸40分钟后取出，斩成长约8厘米的段装盘即成。

评鉴

此菜源于四川腊肉的传统制作方式，经腌制、熏制、蒸制而成，工序多、耗时长，在郫都乡下，以前只在逢年过节时才舍得拿出来款待宾客，向来都被视为家宴中的佐酒大菜。

Ingredients

5,000g pork ribs; 1,500g salt (about 150g to be consumed); 80g star anise; 80g fennel; 30g bay leaves; 50g cinnamon; 30g Sichuan pepper; 100g ginger, sliced; 150g scallion, segmented; 200ml fermented glutinous rice juice

Features

reddish brown in color; salty and savory in taste with a strong smoky flavor

Preparation

1. Rinse the pork ribs, blend well with ginger slices, scallion and fermented rice juice, drain and transfer to a basin.
2. Heat a wok over a flame, and dry roast salt, star anise, fennel, bay leaves, cinnamon and Sichuan peppercorns until the salt becomes light brown. Pour the salt mixture into the basin, mix well and leave to marinate for twenty-four hours. Remove the pork ribs from the basin, drain and transfer to a smoking chamber. Smoke the pork ribs with cypress branches and saw powder.
3. Rinse the ribs with warm water, and then steam in a steamer for forty minutes. Remove from the steamer, chop into 8cm sections, and transfer to a serving dish.

Reviews

This dish uses Sichuanese traditional curing method that including three processes, marinating, smoking and steaming. As it takes a long time to prepare, cured pork is only served for distinguished guests on special festive occasions. The dish goes best with Chinese liquor.

郫都萝卜干

Pidu Dried Radish Pickles

食材配方

春不老萝卜50千克　花椒粉200克　食盐1 500克　鸡精150克　味精100克　白糖200克　熟油辣椒1 500克　熟芝麻100克

成菜特点

色泽红亮、质地爽脆、麻辣回甜。

制作工艺

1. 春不老萝卜洗净，切成长约8厘米、粗约0.6厘米见方的长条，用食盐拌匀后，置容器内腌渍24小时左右，再放入专用袋中，用特制压制机压干水分成萝卜干初坯。
2. 将萝卜干初坯倒入盆内，放入熟油辣椒、花椒粉、鸡精、味精、白糖、熟芝麻拌匀即成。

评鉴

郫都萝卜干是当地人餐桌上不可或缺的名小菜和儿时记忆，被评为“郫县大众喜爱菜品”。此菜在郫都区各大乡镇均有销售，最有特色的是唐昌镇、友爱镇和三道堰镇。

Ingredients

50kg spring radish; 200g ground Sichuan pepper; 1,500g salt; 150g chicken essence; 100g MSG; 200g sugar; 1,500ml chili oil; 100g roasted sesame seeds

Features

reddish brown in color; crispy and crunchy in texture; spicy and slightly sweet in taste

Preparation

1. Rinse the radish, cut into strips of 8cm in length and 0.6cm in width, mix well with salt and transfer to a container. Leave to marinate for twenty-four hours and remove to specially-made bags. Press the radish strips dry using a special pressing machine.
2. Pour the radish into a basin, add chili oil, ground Sichuan pepper, chicken essence, MSG, sugar and sesame seeds, stir well and transfer to a serving dish.

Reviews

This dish is made of the spring radish of Pidu, which is a necessary side dish and enjoys immense popularity, available in every town and township in Pidu District. The most renowned are made in Tangchagn Town, You'ai Town and Sandaoyan Town.

罐罐肉
Guanguan Pork

食材配方

猪肘1 000克　雪豆250克　海带皮150克　姜片30克　葱节30克　花椒0.5克　料酒20毫升　食盐10克　葱花20克　清水1 500毫升

成菜特点

色泽奶白、质地软烂、咸鲜清香。

制作工艺

1. 猪肘切成约3厘米见方的块，焯水后备用；雪豆用清水浸泡两小时。
2. 将猪肘、雪豆放入陶罐中，加入清水、姜片、葱节、花椒、料酒，用中火烧沸，再改用小火炖2～3小时，食用前放入食盐、葱花即成。

评鉴

所谓罐罐肉，就用土陶罐煨制而成的肉。雨水时节，郫都民间盛行家中晚辈应精心烹制罐罐肉孝敬长辈的习俗，这一习俗秉承了“敬老孝亲”的优秀文化传统。此菜咸鲜清淡，软烂适口，最宜年长者食用，还可根据不同季节变换不同的辅料食材，如青蒿、菌类、红花藕等，使营养更加均衡。

Ingredients

1,000g pork knuckle; 250g snowy kidney beans; 150g kelp; 30g ginger, sliced; 30g scallion, segmented; 0.5g Sichuan pepper; 20ml Shaoxing cooking wine; 10g salt; 20g scallion, finely chopped; 1,500ml water

Features

milky white in color; soft and tender in texture; savory and fragrant in taste

Preparation

1. Cut the pork knuckle into 3cm cubes and blanch. Soak the kidney beans in water for two hours.
2. Place the pork knuckle cubes and kidney beans into an earthen pot, add water, ginger slices, scallion segments, Sichuan peppercorns and cooking wine, bring to a boil and simmer over low heat for two to three hours. Add salt and chopped scallion.

Reviews

Guanguan Pork, or Pork in Earthen Pot, is a stew that the young traditionally make for the seniors in the family during the rainy seasons to show their respect for the old. The dish is delicate, soft and savory, most suitable for the elderly. Seasonal vegetables can be added to improve and balance the nutritional values, such as wormwood, funguses, lotus roots, etc.

苕菜狮子头

Lion’s Head Meatballs with Shaocai

食材配方

去皮猪肥膘肉375克　猪前腿瘦肉375克　鲜苕菜300克　马蹄40克　鲜笋40克　金钩15克　鲜豌豆75克　食盐15克　白糖2克　胡椒粉3克　姜片10克　葱节15克　花椒2克　花雕酒50毫升　玉米淀粉45克　鸡蛋清75毫升　姜葱水20毫升　鲜汤1 000毫升　水淀粉15克　化鸡油30毫升　食用油1 000毫升（约耗30毫升）

成菜特点

色泽粉绿、苕菜清鲜、猪肉鲜嫩、味道咸鲜。

制作工艺

1. 去皮猪肥膘肉切成约0.5厘米见方的丁；猪前腿瘦肉、马蹄加工成绿豆大小的粒；鲜豌豆、鲜笋焯水后均切成绿豆大小的粒；金钩用水浸泡后剁细；鲜苕菜焯水后冲凉，沥干余水后切成碎末。
2. 将肥膘丁和瘦肉粒放入盆中，加食盐（8克）、胡椒粉（1克）及姜葱水搅打上劲，然后加入鸡蛋清搅拌均匀，再入金钩、鲜笋、马蹄、豌豆粒拌匀，最后加入玉米淀粉拌匀，做成每个重约150克的肉丸成狮子头生坯。
3. 锅置火上，入食用油烧至180℃时，将狮子头生坯入锅炸至定型及色为浅黄时捞出。
4. 砂锅置火上，放入鲜汤、花椒、姜片、葱节、食盐7克、白糖、胡椒粉（2克）、花雕酒、狮子头烧沸，然后改用小火㸆1小时。
5. 锅置火上，放入化鸡油烧至150℃时，入苕菜末煸炒，再掺入烧狮子头的原汤烧沸，然后放入狮子头，用小火煨至苕菜入味后，捞出狮子头装盘；将锅中原汁用水淀粉勾成薄芡，淋在狮子头上即成。

评鉴

这道苕菜狮子头，是在清炖狮子头的基础上，融入郫都当地特产的苕菜创制而成。苕菜又名“巢菜”，纤维较粗，质地虽不太细腻，但味道清香，有清热利湿、和血散瘀之功效。此菜在西御园乡村酒店推出后，一直是到店客人的必点菜品，曾荣获“成都名菜”称号。必须强调的是，在食用苕菜后4小时内不宜喝酒。

Ingredients

375g pork fat; 375g lean pork (front feet cut); 300g shaocai (local vegetable); 40g water chestnut; 40g fresh bamboo shoots; 15g dried shrimps; 75g green peas; 15g salt; 2g white sugar; 3g pepper; 10g ginger, sliced; 15g scallion, segmented; 2g Sichuan pepper; 50ml Huadiao wine; 45g cornstarch; 75ml egg white; 20ml ginger and scallion juice; 1,000ml stock; 15g cornstarch batter; 30ml chicken oil; 1,000ml cooking oil (about 30ml to be consumed)

Features

pink green in color; savory and delicate in taste; tender shaocai and pork

Preparation

1. Cut the pork fat into 0.5cm cubes. Cut lean pork and water chestnuts into mung bean size grains. Blanch the green peas and bamboo shorts, and cut into mung bean size grains. Soak the shrimps in water, and chop finely. Blanch shaocai, rinse till cool, drain and chop finely.
2. Combine in a basin the lean and fat pork, add 8g salt, 1g pepper, ginger and scallion juice, whisk well, add egg white and blend well. Add shrimps, bamboo shoots, water chestnuts and peas, mix well, and blend in cornstarch. Make meatballs of about 150g each.
3. Heat oil in a wok to 180℃, and deep fry the meatballs until their surface hardens and becomes light brown.
4. Place a wok over the flame, add the stock, Sichuan peppercorns, ginger slices, scallion segments, 7g salt, white sugar, 2g pepper, Huadiao wine and meatballs, bring to a boil, turn down the heat and simmer over low heat for an hour.
5. Heat the chicken oil in a wok to 150℃, stir fry the shaocai, and then add the gravy of the meatballs. Bring to a boil, add the meatballs, and simmer over low heat for the shaocai to absorb the flavors. Transfer the meatballs to a serving dish. Add cornstarch batter to the wok to thicken the sauce, and pour over the meatballs.

Reviews

Braised Lion's Head Meatballs with Shaocai is an innovated version of the famous Lion's Head Meatballs in Consumme with shaocai added. Shaocai is a local specialty with course fibers and fragrant smells. It helps to remove inner heat and humidity, replenish the blood and relieve blood stasis. After being launched in Xiyu Garden Rural Restaurant, the dish soon becomes a must for guests, and is honored as Renowned Dish of Chengdu. One thing worth noting is that it is advisable to drink alcoholic drinks at least four hours after taking in shaocai.

醪糟肉
Fermented Rice Flavored Pork

食材配方

带皮猪五花肉5 000克　辣椒粉100克　花椒粉20克　胡椒粉10克　五香粉50克　食盐100克　酱油1 000毫升　醪糟500克

成菜特点

色泽深红、肥瘦兼备、味道咸鲜、糟香浓郁。

制作工艺

1. 带皮猪五花肉切成重约1 000克的块；炒锅置火上烧热，入辣椒粉、花椒粉、胡椒粉、五香粉、食盐，用中火炒香后放凉待用。
2. 先将醪糟均匀地涂抹在每块肉上，再均匀地涂抹上炒好的调料，装入大盆内腌制两天，然后倒入酱油腌制两天，再翻面腌制两天，最后将其取出，放在室外风吹晾制40天。
3. 将晾干后的肉洗净，放入笼中蒸熟后取出，切为薄片装盘即成。

评鉴

此菜是郫都区郫筒镇的一款特色美食，肉皮暗红，瘦肉深红，肥肉透明，糟香浓郁，深受大众喜爱。除醪糟肉外，采用同样的方法还可制作醪糟鸡等。

Ingredients

5,000g pork belly with skin; 100g ground chilies; 20g ground Sichuan pepper; 10g pepper; 50g five-spice powder; 100g salt; 1,000ml soy sauce; 500g fermented glutinous rice

Features

dark brown in color; fresh and savory in taste with a strong fermented rice flavor

Preparation

1. Cut pork belly into chunks, about 1,000g each. Heat a wok over the flame, add ground chilies, ground Sichuan pepper, pepper, five-spice powder and slat, dry roast over medium heat until fragrant, and leave to cool.
2. Rub the pork with the fermented rice wine and the salt mixture. Transfer to a big basin and leave to marinate for two days. Add soy sauce to the basin and leave to marinate for two days. Turn the pork over and marinate for another two day. Remove the pork from the basin, and air dry outdoors for forty days.
3. Rinse the pork and steam in a steamer until fully cooked. Remove the pork from the steamer, cut into thin slices and transfer to a serving dish.

Reviews

As a specialty of Pitong Town in Pidu District, this popular dish features translucent fat and dark brown lean pork with a strong fermented rice flavor. The same cooking method can be used to make Fermented Rice Flavored Chicken.

食材配方

猪前肘1个（约1 750克） 自制五香粉5克 食盐20克 姜片15克 葱段20克 料酒20毫升 干辣椒面味碟1个（约100克） 食用油2 000毫升（约耗50毫升）

成菜特点

色泽棕红、咸鲜微辣、皮脆酥香、肥肉不腻、瘦肉不柴。

制作工艺

1. 猪前肘洗净，加入自制五香粉、食盐、姜片、葱段、料酒拌匀，腌制72小时，入笼蒸至软熟后取出晾干水分。
2. 锅置火上，入食用油烧至130～150℃时，将猪肘入锅炸至酥香后捞出，待油温回升至200℃时，再次将猪肘入锅复炸至外皮棕红、肉出焦香后捞出，改刀后装盘，配上干辣椒面味碟上桌即成。

评鉴

猪肘有皮，肥瘦兼备，口感丰富。猪肘的烹饪方法多种多样，此菜是其中极具特色的一种。此菜用自制香料进行腌制，再经蒸制、炸制成菜，皮酥肉糯，略带焦香。

盛兴飘香肘

Shengxing Aromatic Pork Knuckle

Ingredients

1 front pork knuckle (about 1,750g); 5g self-made five-spice powder; 20g salt; 15g ginger, sliced; 20g scallion, segmented; 20ml Shaoxing cooking wine; 1 saucer of ground chilies (about 100g); 2,000ml cooking wine (about 50ml to be consumed)

Features

dark brown in color; savory and slightly spicy in taste; juicy knuckle with crispy and aromatic skin

Preparation

1. Rinse the front pork knuckle, and mix well with five-spice powder, salt, ginger slices, scallion segments and cooking wine. Leave to marinate for seventy-two hours and steam in a steamer until soft and cooked through. Remove the knuckle from the steamer and drain.

2. Heat oil in a wok to 130℃ ~ 150℃, deep fry the pork knuckle until fragrant and crispy, and remove. Reheat the oil to 200℃, deep fry the pork knuckle for a second time until the skin becomes dark brown and slightly scorched, and remove. Slice the knuckle, transfer to a serving dish and serve with the saucer of ground chilies.

Reviews

This dish presents a multitude of textures through a series of cooking methods: marinating with self-made spice, steaming and deep frying. The meat is juicy and the skin is crispy and glutinous with a scorched flavor.

青蒿肘子
Wormwood Flavored Pork Knuckle

食材配方

猪肘1个（约1 000克） 青蒿300克 姜片20克 葱段30克鸡精3克 味精2克 食盐15克 花椒0.5克 清水1 500毫升 郫县豆瓣油碟1个（100毫升）

成菜特点

色泽青绿、质地软糯、肥而不腻、味道咸鲜、微带麻辣。

制作工艺

1. 猪肘治净，焯水后备用。
2. 砂罐置火上，入清水、猪肘、姜片、葱段、花椒烧沸，改用小火炖两个小时，再入青蒿炖10分钟，最后下鸡精、味精、食盐调味，出锅装入盛器中，与郫县豆瓣油碟一同上桌即成。

评鉴

将带有野菜香的青蒿与猪肘搭配，成菜色泽碧绿、肥而不腻、清淡爽口，且有清热解暑、提高免疫力等作用，特别适合夏季食用。

Ingredients

1 pork knuckle (about 1,000g); 300g sweet wormwood; 20g ginger, sliced; 30g scallion, segmented; 3g chicken essence; 2g salt; 15g salt; 0.5g Sichuan pepper; 1,500ml water; 1 saucer of Pixian chili bean paste oil (100ml)

Features

pleasant green in color; soft and glutinous in texture; fatty and not greasy; savaory, delicate and slightly spicy in taste

Preparation

1. Rinse and blanch the pork knuckle.
2. Combine in an earthen pot the water, knuckle, ginger, scallion and Sichuan pepper, bring to a boil over high flame, turn down and simmer over low heat for two hours. Add the wormwood, and simmer for another ten minutes. Blend in the chicken essence, MSG and salt to season. Transfer to a serving container, and serve with the dipping saucer.

Reviews

The combination of verdant and fragrant wormwood with pork knuckle presents an appetizing, delicate dish that is fatty but not greasy. It is suitable as summer diet, for it helps to remove body heat and boost immunity. The dish can be eaten directly or served with chili bean dipping sauce.

鲜仔姜双脆

Double Crispiness (Crispy Pork Nose Tendon and Pork Palate) with Tender Ginger

食材配方

鲜猪鼻筋100克　鲜猪天堂200克　鲜仔姜300克　小米辣150克
小青椒50克　红油豆瓣20克　蒜泥5克　食盐2克　味精2克　鸡精2克
白糖3克　醋3毫升　藤椒油5毫升　鲜汤100毫升　菜籽油180毫升

成菜特点

色泽自然、脆嫩爽口、咸鲜麻辣。

制作工艺

1. 鲜猪鼻筋洗净后焯水；鲜猪天堂焯水后切成长约10厘米、粗约0.4厘米的丝；鲜仔姜切丝；小米辣、小青椒切成圈。
2. 锅置火上，入菜籽油烧至120℃时，下红油豆瓣炒香，再下鲜仔姜、小米辣圈、蒜泥炒熟，然后入猪鼻筋、猪天堂炒匀，在注入鲜汤继续翻炒90秒后，下小青椒圈、白糖、醋、味精、鸡精、藤椒油炒匀，起锅装盘即成。

评鉴

川菜传统名菜“火爆双脆”的主料为猪肚和鸡肫，火候考究、脆嫩爽口。 所谓“双脆”，是泛指两种脆性食材，并无固定要求，此菜别出心裁地选择了较为冷门的猪鼻筋和猪天堂，在加入仔姜、小米辣爆炒后，质地脆嫩Q弹、有嚼劲，又很入味，适宜佐酒。

Ingredients

100g pork nose tendon; 200g pork palate; 300g tender ginger; 150g bird’s eye chilies; 50g small green chili peppers; 20g chili oil bean paste; 5g garlic, crushed; 2g salt; 2g MSG; 2g chicken essence; 3g sugar; 3ml vinegar; 5ml green Sichuan pepper oil; 100ml stock; 180ml rapeseed oil

Features

spicy and savory in taste; crispy and crunchy in texture

Preparation

1. Rinse pork tendon and blanch. Blanch pork palate, and cut into 10cm-long and 0.4cm-thick slivers. Cut the tender ginger into slivers. Cut the bird's eye chilies and small green peppers into rings.
2. Heat oil in a wok to 120℃, and stir fry chili oil bean paste until fragrant. Add tender ginger, bird's eye chilies and garlic, stir fry till cooked through, and blend in pork tendon and palate. Stir well, and blend in the stock, stirring for ninety seconds before adding small green pepper rings, sugar, vinegar, MSG, chicken essence and green Sichuan pepper oil. Blend well and transfer to a serving plate.

Reviews

Traditional ingredients for Double Crispiness are pork tripe and chicken gizzard while this dish uses pork nose tendon and pork palate, which are quick stir fried with tender ginger and bird's eye chilies. The dish is springy, chewy and savory, a good accompaniment for Chinese liquor.

食材配方

猪蹄500克　小芸豆300克　食盐10克　姜片30克　花椒5克　花椒粉0.2克　葱段30克　葱花15克　料酒10毫升　酱油10毫升　醋3毫升　芝麻油3毫升　辣椒油10毫升　清水1 500毫升

成菜特点

汤色乳白、肉质软烂、咸鲜香浓、微带麻辣。

芙蓉蹄花

Hibiscus Pork Feet Stew

制作工艺

1. 小芸豆浸泡、发胀后洗净；猪蹄去尽残毛后洗净，焯水备用。
2. 用酱油、醋、芝麻油、辣椒油、花椒粉、葱花（5克）制成蘸料味碟。
3. 砂锅置火上，注入清水，下蹄花、芸豆、姜片、葱段、花椒、料酒烧沸，加盖后用小火炖3小时至猪蹄软烂，食用前放入食盐、葱花（10克），与蘸料味碟一同上桌即成。

评鉴

此菜是郫都人几十年来乐此不疲的经典老味道，汤白味醇，肉质柔软，皮肉轻拨即下，蘸上味碟，就着红油泡菜，既饱口福，又解油腻，如果仅用一个字来概括，那就是“爽”。

Ingredients

500g pork feet; 300g white kidney beans; 10g salt; 30g ginger; 5g Sichuan pepper; 0.2 ground Sichuan pepper; 30g scallion, segmented; 15g scallion, finely chopped; 10ml Shaoxing cooking wine; 10ml soy sauce; 3ml vinegar; 3ml sesame oil; 10ml chili oil; 1,500ml water

Features

milkly white soup; soft and glutinous pork feet; savory and slightly spicy in taste

Preparation

1. Soak the kidney beans in water, and rinse; Remove the hair of the pork feet, rinse and blanch.
2. Combine in a small saucer the soy sauce, vinegar, sesame oil, chili oil, ground Sichuan pepper and 5g chopped scallion to making the dipping sauce.
3. Place an earthen pot over the flame, add water, pork feet, kidney beans, sliced ginger, segmented scallion, Sichuan peppercorns and cooking wine, bring to a boil and cover. Turn down the heat, and simmer over a low flame for three hours until the pork feet are soft and fully cooked. Add salt, sprinkle with 10g chopped scallion, and serve with the dipping sauce.

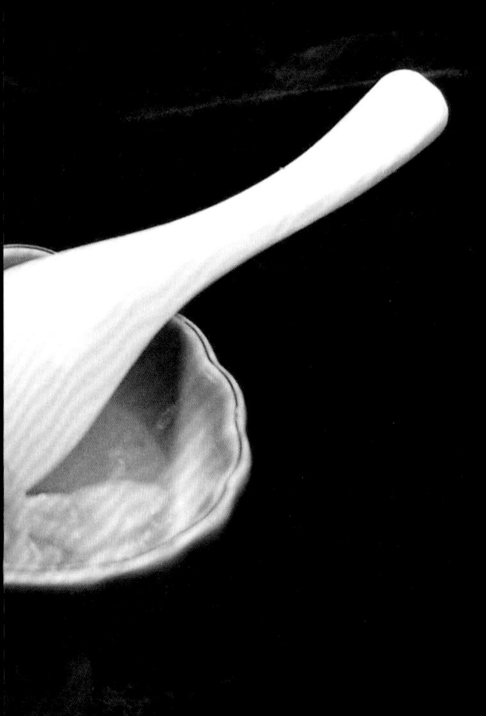

Reviews

As a traditional classic in Pidu District, this dish feature milky white soup and soft and glutinous pork feet. The skin and meat is easily tore off the bones, dipped into the saucer, and then enjoyed with chili oil pickles, which helps to relieve the greasiness.

郫筒酒醉牛腩

Braised Beef with Pitong Wine

食材配方

牛腩500克　山药150克　水发竹笋150克　香菜5克　郫筒酒100毫升　郫县豆瓣50克　姜片10克　葱段15克　八角3克　花椒1克　食盐2克　酱油5毫升　白糖1克　胡椒粉1克　味精3克　鲜汤1 500毫升　水淀粉15克　食用油100毫升

成菜特点

色泽红亮、质地软糯、咸鲜微辣、酒香浓郁。

制作工艺

1. 牛腩切为约4厘米见方的块，入冷水锅中焯水后捞出洗净；山药切成滚刀块；水发竹笋切成滚刀块后焯水。
2. 锅置火上，入食用油烧至120℃时，下郫县豆瓣、姜片、葱段、八角、花椒炒香，再加入鲜汤、牛肉、郫筒酒、胡椒粉烧两小时，之后下水发竹笋、山药烧半小时，然后下食盐、酱油、白糖、味精烧沸，最后用水淀粉勾成薄芡，装入碗中，撒上香菜即成。

评鉴

郫筒酒是郫都特产，历史悠久、色如琥珀、入口微甜、清香弥久，具有清热化痰、健脾养生、美容抗衰等功效，用郫筒酒烧制牛肉，与料酒同功，能去腥增香，因其酒精浓度更高，所以烹调出来的菜品味道更加醇厚、浓香。

Ingredients

500g beef; 150g Chinese yam; 150ml water-soaked bamboo shoots; 5g coriander; 100ml Pitong Wine; 50g Pixian chili bean paste; 10g ginger, sliced; 15g scallion, segmented; 3g star anise; 1g Sichuan pepper; 2g salt; 5ml soy sauce; 1g white sugar; 1g pepper; 3g MSG; 1,500ml stock; 15g pea starch batter; 100ml cooking oil

Features

reddish and lustrous in color; soft and glutinous in texture; savory and slightly spicy in taste with a strong wine aroma

Preparation

1. Cut the beef into 4cm cubes, add to cold water, and blanch. Cut the Chinese yam into rolling chunks. Cut the bamboo shoots into rolling chunks, and blanch.
2. Heat oil in a wok to 120℃, and blend in Pixian chili beanpaste, sliced ginger, segmented scallion, star anise and Sichuan peppercorns, stirring until fragrance. Add the stock, beef, Pitong Wine and pepper, and braise for two hours. Add the bamboo shoots and yam, and braise for half an hour. Add salt, soy sauce, white sugar and MSG, bring to a boil, and add the batter. Transfer to a serving dish and sprinkle with coriander.

Reviews

When braised with beef, Pitong Wine helps to remove the unpleasant smell and enhance the flavors better than Shaoxing cooking wine, for it has higher alcoholic contents.

火焰雪花牛肉

Streaky Beef over Flames

食材配方

雪花牛肉300克　西兰花200克　黑椒汁100毫升　姜20克　葱20克
蒜20克　香菜20克　清水100毫升　白兰地100毫升

成菜特点

色泽深红、肉质鲜嫩、口味独特。

制作工艺

1. 将姜、葱、蒜、香菜、清水放入搅拌器中搅打成蔬菜汁。
2. 雪花牛肉切成长约8厘米、宽约4厘米、厚约0.6厘米的片，用黑椒汁、蔬菜汁拌匀腌制1小时。
3. 将雨花石放进烤箱中加热后取出，放入垫有锡纸的砂锅中。
4. 将腌制好的雪花牛肉在平锅中煎熟，再沿着锅边平铺在雨花石上；西兰花入水焯熟后摆入砂锅中央，上桌时浇上白兰地，点火即成。

评鉴

此菜是在西餐“黑椒牛排”的基础上经改进、创新而成，烹制好后上桌点火再食，可起到烘托就餐气氛的作用。

Ingredients

300g streaky beef; 200g broccoli; 100ml black pepper sauce; 20g ginger; 20g scallion; 20g garlic; 20g coriander; 100ml water; 100ml brandy

Features

dark red in color; tender and juicy in texture

Preparation

1. Combine the ginger, scallion, garlic, coriander and water in a blender to make juice.
2. Cut the beef into 8cm-long, 4cm-wide and 0.6cm-thick slices. Add black pepper sauce and juice, mix well and leave to marinate for one hour.
3. Heat the rain-flower pebbles in an oven, remove and transfer to an earthen pot with tin paper.
4. Pan fry the beef until cooked through, and lay flat on the pebbles. Pour the brandy over the beef and light the fire, while serving.

Reviews

This dish is an innovated version of the traditional western-style black pepper steak. Lighting up the beef while serving is a delightful surprise and entertainment for the diners.

粉蒸牛肉

Steamed Beef with Rice Flour

食材配方

牛肉200克　姜4克　葱10克　花椒1克　郫县豆瓣25克　酱油2毫升　食盐0.5克
腐乳汁10毫升　醪糟汁8毫升　料酒5毫升　胡椒粉0.2克　鲜汤20毫升　菜籽油20毫升
味精1克　蒸肉米粉50克　辣椒油30毫升　花椒粉0.5克　蒜泥15克　香菜碎10克

成菜特点

色泽棕红、质地软嫩、咸鲜麻辣、香味浓郁。

制作工艺

1. 牛肉切为薄片；姜切末；花椒与葱混合剁碎成粗椒麻；郫县豆瓣剁细。
2. 牛肉片入盆，加郫县豆瓣、食盐、腐乳汁、醪糟汁、料酒、姜末、粗椒麻、胡椒粉、味精、鲜汤、菜籽油拌匀，再加入蒸肉米粉拌匀。
3. 将裹有米粉的牛肉片装入小竹蒸笼中蒸制30分钟后取出，食用前放入辣椒油、花椒粉、蒜泥、香菜碎即成。

评鉴

粉蒸牛肉是川菜的经典菜品，因采用川菜的特色烹调方法——“粉蒸”制作而得名，在川中包括郫都在内的许多地区都极为常见，多出现在家庭、大众便餐和田席等场所的餐桌上。

Ingredients

200g beef; 4g ginger 10g scallion; 1g Sichuan peppercorns; 25g Pixian chili bean pate; 2ml soy sauce; 0.5g salt; 10ml fermented tofu juice; 8ml fermented glutinous rice juice; 5ml Shaoxing cooking wine; 0.2g pepper; 20ml stock; 20ml rapeseed oil; 1g MSG; 50g rice flour for steaming; 30ml chili oil; 0.5g ground Sichuan pepper; 15g garlic, crushed; 10g coriander, finely chopped

Features

golden brown in color; soft and glutinous in texture; savory and spicy in taste

Preparation

1. Thinly slice the beef. Finely chop the ginger. Combine Sichuan peppercorns with scallion, and chop. Finely chop Pixian chili bean paste.
2. Combine in a basin the beef slices with Pixian chili bean past, salt, fermented tofu juice, cooking wine, chopped ginger, chopped mixture of Sichuan peppercorns and scallion, pepper, MSG, stock and rapeseed oil, blend well, add the rice flour, and stir well.
3. Place beef slices coated with rice flour into a small bamboo steamer, and steam for thirty minutes. Remove the beef from the steamer, and add chili oil, ground Sichuan pepper, garlic and coriander.

Reviews

As a classic of Sichuan cuisine, this dish uses a peculiar cooking method often used in families or restaurants of Pidu District, where ingredients are steamed with seasoned glutinous rice flour.

鱼香牦牛掌

Fish-Flavored Yak Paws

食材配方

牦牛掌1只（约1 500克） 瓢儿白200克 香菇100克 洋葱150克 姜片10克 葱段15克 料酒30毫升 红曲米水100毫升 泡辣椒末50克 郫县豆瓣40克 泡豇豆粒50克 泡姜米20克 蒜米15克 葱花20克 食盐2克 酱油3毫升 白糖3克 味精2克 醋10毫升 鲜汤200毫升 水淀粉25克 清水3 000毫升 食用油80毫升

成菜特点

色泽红亮、质地软糯、酸甜适口、咸鲜略辣。

制作工艺

1. 牦牛掌洗净后焯水，再放入锅中，加姜片、葱段、料酒、红曲米水、清水烧沸，用小火煨制6小时至软熟后捞出备用。
2. 洋葱切成粗丝，焯水后装在盘中垫底；瓢儿白、香菇焯水后摆放在盘子两侧，再放上煨制好的牦牛掌。
3. 锅置火上，入食用油烧至120℃时，下泡辣椒末、郫县豆瓣、泡豇豆粒、泡姜米、蒜米炒香，再掺入鲜汤、食盐、酱油、白糖、味精、醋，用水淀粉收汁浓稠后入葱花和匀，起锅浇淋在牦牛掌上即成。

评鉴

牦牛掌由皮、筋、骨及部分胶质脂肪组成，富含胶原蛋白，适宜烧、炖、焖等，常见菜品有红烧牦牛蹄、黄焖牦牛蹄等。此菜采用软熘的烹调方法，配搭富有特色的鱼香味汁，个性鲜明、特色突出，被评为“郫都区十大川菜菜品”。

Ingredients

1 yak paw (about 1,500g); 200g bok choy; 100g shiitake mushrooms; 150g onions; 10g ginger, sliced; 15g scallion, segmented; 30ml Shaoxing cooking wine; 100ml fermented red rice juice; 50g pickled chilies, finely chopped; 40g Pixian chili bean paste; 50g pickled yard long beans, chopped; 20g pickled ginger, finely chopped; 15g garlic, finely chopped; 20g scallion, finely chopped; 2g salt; 3ml soy sauce; 3g white sugar; 2g MSG; 10ml vinegar; 200ml stock; 25g cornstarch batter; 3,000ml water; 80ml cooking oil.

Features

reddish brown in color; soft and glutinous in texture; savory, sweet, sour and slightly spicy in taste

Preparation

1. Rinse and blanch the yak paws. Combine the paw, ginger, scallion, cooking wine, fermented red rice juice and water in a pot, bring a boil, turn down the heat and simmer for six hours over low heat until the paw is soft and cooked through. Remove the paw from the pot.
2. Cut the onion into thick slivers, blanch and transfer to a serving dish as the base. Blanch the bok choy and shiitake mushrooms, remove and lay along the rims of the serving dish. Place the yak paw on top.
3. Heat oil in a wok to 120℃, and stir fry pickled chilies, Pixian chili bean paste, pickled yard long beans, pickled ginger and garlic until fragrant. Add the stock, salt, soy sauce, white sugar, MSG, vinegar and batter, stirring until the sauce is thick, add the scallion, blend well, and pour over the yak paw.

Reviews

Yak paws consist of skin, tendon, bones and colloid fat, rich in collagen and suitable for braising, stewing and pressure-simmering. Popular dishes include Brown Braised Yak Paws and Pressure Simmered Yak Paws. This dish uses the cooking method of Soft Braising and is served with a peculiar fish fragrance sauce, making it one of the Top10 dishes of Pidu District.

火爆牛肝

Fire Exploded Beef Livers

食材配方

牛肝200克　甜椒100克　青椒50克　郫县豆瓣25克　泡辣椒20克　蒜片10克
姜米5克　葱节10克　食盐1克　白糖1克　味精2克　料酒6毫升
花椒油3毫升　干淀粉10克　水淀粉8克　鲜汤20毫升　食用油80毫升

成菜特点

色泽自然、质地滑嫩、爽口化渣、咸鲜醇厚。

制作工艺

1. 牛肝切成薄片，加入食盐（0.5克）、料酒（3毫升）、干淀粉拌匀；甜椒切成菱形块；泡辣椒、青椒切成短节。
2. 用食盐（0.5克）、白糖、味精、料酒（3毫升）、花椒油、水淀粉、鲜汤调成芡汁。
3. 锅置火上，入食用油烧至180℃时，下牛肝炒断生，再下郫县豆瓣炒至出香，然后下甜椒、青椒、泡辣椒、蒜片、姜米、葱节炒熟，最后烹入芡汁，收汁亮油后出锅装盘即成。

评鉴

此菜是川菜的传统特色菜品，制作极为讲究火候，成菜时间短，犹如火中取宝，不及则生，稍过则老，必须在短短几十秒钟成就牛肝最完美的质感和味道。

Ingredients

200g beef livers; 100g bell peppers; 50g green peppers; 25g Pixian chili bean paste; 20g pickled chilies; 10g garlic, sliced; 5g ginger, finely chopped; 10g scallion, segmented; 1g salt; 1g white sugar; 2g MSG; 6ml Shaoxing cooking wine; 3ml Sichuan pepper oil; 10g pea starch; 8g cornstarch batter; 20ml stock; 80ml cooking oil

Features

pleasant and natural in color; soft, tender and melting in texture; savory and delicate in taste

Preparation

1. Thinly slice the livers, and mix well with 0.5g salt, 3ml cooking wine and pea starch. Cut the bell peppers into diamond slices. Cut the pickled chilies and green peppers into sections.

2. Combine 0.5g salt, white sugar, MSG, 3ml cooking wine, Sichuan pepper oil, batter and stock in a bowl, stirring well to make the seasoning sauce.
3. Heat oil in a wok to 180℃, stir fry the beef livers until just cooked. Blend in the Pixian chili bean paste, and stir until fragrant. Add the bell peppers, green peppers, pickled chilies, garlic, ginger and scallion, stir frying until cooked through. Pour in the seasoning sauce, stir till the sauce is thick and the oil is clear, and transfer to a serving dish.

Reviews

As a traditional and popular dish of Sichuan cuisine, this dish focuses on the use of heat or fire and the control of time. Ingredients are stir fried over a high flame for a relatively short period of time to achieve the best texture and flavors of beef livers.

一鸡三吃

Chicken Prepared in Three Ways

吃法一：炒鸡杂

Course I: Stir Fried Chicken Offal

食材配方

鸡肝100克　鸡胗60克　鸡肠100克　鸡血100克　芹菜150克　泡辣椒20克　郫县豆瓣30克　食盐2克　酱油5毫升　白糖1克　醋2毫升　味精2克　胡椒粉0.5克　料酒15毫升　水淀粉25克　鲜汤15毫升　食用油100毫升

成菜特点

质地脆嫩、咸鲜微辣、泡椒味浓郁。

制作工艺

1. 鸡血入锅焯熟后切成长约3厘米、宽约2厘米、厚约1厘米的小块；鸡肠入锅焯熟后斩成长约8厘米的段；鸡肝切片；鸡胗剞“十”字花刀，再切成三刀一断的片；芹菜切成长约4厘米的节；泡辣椒切成马耳朵形。
2. 鸡肝、鸡胗入碗，加食盐（1克）、料酒（7毫升）、水淀粉（10克）码味上浆。
3. 用食盐（1克）、酱油、白糖、醋、味精、胡椒粉、料酒（8毫升）、水淀粉（15克）及鲜汤调成芡汁。
4. 锅置火上，入食用油烧至180℃时，先放入鸡肝、鸡胗炒撒籽，再下鸡肠、鸡血炒断生，然后下郫县豆瓣、泡辣椒、芹菜节炒香，最后烹入调好的芡汁，待收汁亮油后起锅装盘即成。

评鉴

所谓“一鸡三吃”，就是将一只整鸡用三种不同的烹调方法制作成三道菜品，最初流行于郫都及成都周边地区的农家乐，大多选用农家自养的土鸡，论斤计价，现点、现杀、现做。此菜采用红烧、凉拌和炒制三种方式烹制而成，是郫都农家乐“一鸡三吃”非常经典的呈现方式。

Ingredients

100g chicken livers; 60g chicken gizzards; 100g chicken intestines; 100g chicken blood; 150g celery; 20g pickled chilies; 30g Pixian chili bean paste; 2g salt; 5ml soy sauce; 1g white sugar; 2ml vinegar; 2g MSG; 0.5g pepper; 15ml Shaoxing cooking wine; 25g cornstarch batter; 15ml stock; 100ml cooking oil

Features

crunchy and tender chicken offal; savory and slightly spicy in taste with a strong pickled chili flavor

Preparation

1. Blanch the chicken blood till cooked through, and cut into 3cm-long, 2cm-wide and 1cm-thick pieces. Blanch the chicken intestines till cooked through and cut into 8cm-long sections. Slice the chicken livers. Make cross cuts into the chicken gizzards, and cut into slices. Cut the celery into 4cm segments, and the pickled chilies into horse-ear sections.
2. Combine the chicken livers and gizzard in a bowl. Add 1g salt, 7ml cooking wine and 10g batter, mix well and leave to marinate.
3. Mix 1g salt, soy sauce, white sugar, vinegar, MSG, pepper, 8g cooking wine, 15g batter and stock in a bowl to make the seasoning sauce.
4. Heat oil in a wok to 180℃, and stir fry the chicken liver and gizzard pieces so that they do to cling and stick to each other. Add the chicken intestines and blood, and stir fry till just cooked. Blend in the Pixian chili bean paste, pickled chilies and celery, and stir to bring out the fragrance. Pour in the seasoning sauce, simmer until the sauce is thick and the oil is clear, remove from the heat and transfer to a serving dish.

Reviews

A rooster is picked and slaughtered on the spot and then cooked and served in three different ways, which is a typical practice in farm restaurants of Pidu District and other rural areas of Chengdu. The free-range rooster is usually priced based on the weight. In this dish, chicken is braised, boiled and seasoned, stir fried, the most classical cooking methods in farm restaurants.

吃法二：土豆烧鸡

Course II: Braised Chicken with Potatoes

食材配方

公鸡半只（约1 000克） 土豆500克 青尖椒100克 郫县豆瓣50克 姜片10克 葱段15克 花椒0.5克 八角2克 桂皮2克 食盐3克 胡椒粉1克 白糖1克 醋3毫升 味精3克 鲜汤1 000毫升 食用油100毫升

成菜特点

肉质软糯、咸鲜香辣。

制作工艺

1. 鸡肉斩成约5厘米左右大的块；土豆切成滚刀块；青尖椒切成长约4厘米的段。
2. 锅置火上，入食用油烧至180℃时，先下鸡块煸炒至吐油，再下郫县豆瓣、花椒、姜片、葱段、八角、桂皮煸炒至出香，然后掺入鲜汤、食盐、胡椒粉、白糖、醋、土豆、青尖椒烧至成熟，出锅前加入味精和匀即成。

Ingredients

half a rooster (about 1,000g); 500g potatoes; 100g green peppers; 50g Pixian chili bean paste; 10g ginger, sliced; 15g scallion, segmented; 0.5g Sichuan pepper; 2g star anise; 2g cassia; 3g salt; 1g pepper; 1g white sugar; 3ml vinegar; 3g MSG; 1,000ml stock; 100ml cooking oil

Features

soft and glutinous chicken; savory and spicy in taste

Preparation

1. Cut the chicken into 5cm cubes, potatoes into chunks and green peppers into 4cm sections.
2. Heat oil in a wok to 180℃, stir fry the chicken until chicken fat bubbles up, and then blend in the Pixian chili bean paste, Sichuan peppercorns, ginger, scallion, star anise and cassia, stirring to bring out the fragrance. Add the stock, salt, pepper, white sugar and vinegar, and braise until the chicken is just cooked. Add the potatoes and green peppers, and continue to braise until all the ingredients are fully cooked. Add MSG, blend well and transfer to a serving dish.

吃法三：凉拌鸡

Couse III: Chicken Salad

食材配方

公鸡半只（约1 000克）　大葱50克　辣椒油100毫升　花椒粉1克　酱油20毫升　白糖2克　醋5毫升　味精2克　芝麻油5毫升

成菜特点

色泽红亮、肉质软糯、麻辣鲜香。

制作工艺

1. 将治净后的公鸡入锅中煮熟，捞出晾凉后斩成约3厘米见方的块；大葱切成长约1. 5厘米的节。
2. 用辣椒油、花椒粉、酱油、白糖、醋、味精、芝麻油调匀成麻辣味汁，再放入鸡块、大葱节拌匀装盘即成。

Ingredients

half a rooster (about 1,000g); 50g scallion; 100ml chili oil; 1g ground Sichuan pepper; 20ml soy sauce; 2g white sugar; 5ml vinegar; 2g MSG; 5ml sesame oil

Features

appealing reddish in color; spicy and savory in taste; tender chicken

Preparation

1.Boil the chicken till cooked through, remove and leave to cool. Cut the chicken into 3cm cubes and scallion into 1.5cm-long sections.
2. Combine in a basin the chili oil, ground Sichuan pepper, soy sauce, white sugar; vinegar; MSG and sesame oi, blend well and add the chicken and scallion. Mix well and transfer to a serving dish.

大蒜青椒鸡

Braised Chicken with Garlic and Green Peppers

食材配方

公鸡肉600克　独头蒜200克　青二荆条200克　郫县豆瓣30克
秘制调料30克　姜片10克　葱段15克　食盐2克　白糖2克　味精3克
酱油5毫升　鲜汤1 000毫升　食用油200毫升

成菜特点

色泽酱红、肉质熟软、咸鲜香辣。

制作工艺

1. 公鸡肉斩成约4厘米见方的块；青二荆条入炭火中烧出“虎皮”纹，再切成长约4厘米的段；独头蒜入150℃的食用油中炸至皱皮后捞出。
2. 锅置火上，入食用油烧至180℃时，放入鸡块煸炒至皮干、吐油，再下郫县豆瓣、秘制调料、姜片、葱段炒香，然后加入鲜汤、食盐、白糖、酱油烧至鸡肉软熟，入独蒜、青二荆条烧至软熟，最后加入味精和匀，起锅装盘后即成。

评鉴

郫县德源大蒜是国家地理标志保护产品，蒜素含量远高于普通大蒜，具有抗菌、抗病毒等保健养生之功。大蒜青椒鸡是德源镇的一道经典农家特色风味菜品，系选用德源大蒜与当地散养的土公鸡烧至而成，在当地的农家乐中多有销售，尤以盛兴源乡村酒店的出品最具特色。

Ingredients

600g chicken(rooster); 200g single-clove garlic; 200g erjingtiao green peppers; 30g Pixian chili bean paste; 30g secret-recipe sauce; 10g ginger, sliced; 15g scallion, segmented; 2g salt; 2g white sugar; 3g MSG; 5ml soy sauce; 1,000ml stock; 200ml cooking oil

Features

golden brown in color; tender chicken; spicy an savory in taste

Preparation

1. Cut the chicken into 4cm cubes. Grill the erjingtiao green pepper in charcoal fire to get the so-called "tiger skin", and cut into 4cm-long section. Deep fry the garlic in 150℃ oil until the skin wrinkles.
2. Heat oil in a wok to 180℃, and stir fry the chicken for extra water contents to evaporate and fat to melt. Add Pixian chili bean paste, secret-recipe sauce, ginger and scallion, stirring to bring out the aroma. Add the stock, salt, white sugar and soy sauce, and braise till the chicken is soft and cooked through. Blend in the garlic and green peppers, and braise till cooked through. Blend in MSG, remove from the heat and transfer to a serving dish.

Reviews

Deyuan Garlic, a national geographic index produce with higher allicin than average garlic produced elsewhere, has better anti-virus and anti-bacteria values. Braised Chicken with Garlic and Green Peppers, a specialty of Deyuan County, uses locally produced garlic and free-range chicken as the main ingredients. This dish can be found in most local farm restaurants, and the most renowned is Shengxingyuan Rural Restaurant.

郫县豆瓣鱼

Chili Bean Paste Flavored Fish

食材配方

草鱼1尾（约600克） 姜片10克 葱段15克
姜米15克 蒜米30克 葱花25克 郫县豆瓣50克
食盐2克 白糖5克 醋6毫升 酱油5毫升 料酒25毫升
胡椒粉0.5克 水淀粉25克 鲜汤500毫升
食用油2 000毫升（约耗100毫升）

成菜特点

色泽红亮、肉质细嫩、咸鲜微辣、香味浓郁。

制作工艺

1. 草鱼治净，在鱼身两侧各剞数刀，用姜片、葱段、食盐、胡椒粉、料酒（10毫升）拌匀，码味15分钟。
2. 锅置火上，入食用油烧至200℃时，放入草鱼炸至表皮起硬膜后捞出。
3. 锅置火上，入食用油烧至100℃时，下郫县豆瓣、姜米、蒜米、葱花（10克）炒香至油呈红色，然后加入鲜汤、草鱼、料酒、白糖、酱油，烧至鱼肉变软、成熟后捞出装入盘中。
4. 将锅中汁水用水淀粉收浓成二流芡，然后放入醋、葱花（15克）搅匀，出锅浇淋在鱼身上即成。

评鉴

郫县豆瓣鱼是郫都区的传统名菜，是鱼与郫县豆瓣碰撞出来的美味，深受大众喜爱，曾荣获“郫县大众喜爱菜品”“成都市郫都区十大川菜菜品”“成都名菜”等称号。其代表性餐厅有印象泰和园、西御园、容和苑、尚河渔庄等。

Ingredients

a grass carp (about 600g); 10g ginger, sliced; 15g scallion, segmented; 15g ginger, finely chopped; 30g garlic, finely chopped; 25g scallion, finely chopped; 50g Pixian chili bean paste; 2g salt; 5g white sugar; 6ml vinegar; 5ml soy sauce; 25ml Shaoxing cooking wine; 0.5g pepper; 25g cornstarch batter; 500ml stock; 2,000ml cooking oil (about 100ml to be consumed)

Features

reddish brown in color, savory, spicy and slightly sweet and sour in taste; tender in texture

Preparation

1. Make several cuts into the body of the carp, mix well with the salt, sliced ginger, pepper, segmented scallion and 10ml cooking wine, and leave to marinate for fifteen minutes.
2. Place a wok over the flame, add oil and heat to 200℃. Deep fry the carp till the skin becomes light brown and slightly hardens.
3. Place a wok over a flame, add oi and heat to 100℃. Stir fry the Pixian chili bean paste, chopped ginger, chopped garlic and 10g chopped scallion until aromatic. Add the stock, fish, cooking wine, white sugar and soy sauce, braise till the fish becomes soft and cooked through. Remove the fish and transfer to a serving plate.
4. Add the batter to the wok to thicken the sauce, blend in the vinegar and 15g chopped scallion, and pour over the fish.

Reviews

As a traditional dish of Sichuan, Chili Bean Paste Flavored Fish enjoys enormous popularity among diners. It ranks among Most Popular Dishes of Pidu District, Top10 Dishes of Pidu District and Renowned Dishes of Chengdu, etc. Restaurants famous for this dish include Yixiang Taihe Garden, Xiyu Garden, Ronghe Garden and Shanghe Fish Restaurant.

沱砣鱼
Tuo Tuo Fish

食材配方

花鲢1 500克　老豆腐500克　鸡蛋2个　泡萝卜75克　泡豇豆75克　泡仔姜75克　大蒜30克　三年酿造的郫县豆瓣50克　一年酿造的郫县豆瓣50克　特制沱砣鱼料250克　干辣椒50克　红花椒10克　青花椒10克　食盐5克　料酒15毫升　胡椒粉3克　醪糟汁10毫升　白糖5克　醋5毫升　藤椒油20毫升　花椒粉5克　芝麻油2毫升　红薯淀粉60克　葱花20克　芹菜碎20克　香菜碎10克　鲜汤2 000毫升　化猪油200毫升　菜籽油2 000毫升（约耗400毫升）

成菜特点

色泽红亮、鱼肉细嫩、豆腐滑爽、麻辣鲜香。

制作工艺

1. 花鲢剁成长约6厘米、宽约2厘米的大块，用食盐、料酒、胡椒粉拌匀，码味15分钟后洗净，加鸡蛋液、红薯淀粉拌匀，入180℃的食用油中炸至紧皮后捞出。
2. 泡萝卜、泡豇豆、泡仔姜分别切成长约4厘米的条；豆腐切成长约4厘米、粗约2厘米见方的条，入加有食盐的水中焯水后备用。
3. 锅置火上，入菜籽油（200毫升）、化猪油（200毫升）烧至100℃时，下郫县豆瓣炒出香味，然后下泡萝卜、泡豇豆、泡仔姜、大蒜炒香，接着加入鲜汤、特制沱砣鱼料、鱼块、豆腐烧沸，再下醪糟汁、白糖、醋，改用小火焖烧至熟，最后下花椒粉、藤椒油、葱花、芹菜碎、香菜碎炒匀，起锅倒入盛器中。
4. 锅置火上，入菜籽油（200毫升）烧至130℃时，下干辣椒、红花椒、青花椒炸香，出锅后淋在盛器中的鱼身上即成。

评鉴

沱砣鱼是郫都红星饭店结合老成都的火锅麻辣汤料，以及郫都人喜欢吃的豆瓣和泡菜创制而成，“砣”是四川方言，是泛指比较大的块状物；“沱”是指沱江，表达了创制者深厚的乡土情结。此菜麻辣鲜香、诱人食欲，被评为“中国国际旅游美食之旅最受欢迎菜品”“中国川味金轿奖—川味中国.百菜百味榜金榜菜品”和“成都名菜”等。

Ingredients

1,500g spotted silver carp; 500g firm tofu; 2 eggs; 75g pickled radish; 75g pickled yard long beans; 75g pickled tender ginger; 30g garlic;50g Pixian chili bean paste (three-year fermentation); 250g specially-made tuo tuo fish sauce; 50g dried chilies; 10g Sichuan peppercorns; 10g green Sichuan peppercorns; 5g salt; 15ml Shaoxing cooking wine; 3g pepper; 10ml fermented rice juice; 5g sugar; 5ml vinegar; 20ml green Sichuan pepper oil; 5g ground Sichuan pepper; 2ml sesame oil; 60g swwet potato starch; 20g scallion, finely chopped; 20g celery, finely chopped; 20g coriander, finely chopped; 2,000ml stock; 200ml lard; 2,000ml rapeseed oil (about 400ml to be consumed)

Features

appealing reddish in color; tender fish and smooth tofu; spicy and savory in taste

Preparation

1. Cut the fish into 6cm-long, 2cm-wide chunks, mix with salt, cooking wine and pepper, blend well and leave to marinate for fifteen minutes. Rinse the fish chunks, and mix well with beaten egg and sweet potato starch. Deep fry the fish chunks in 180℃ oil until the skin tightens.
2. Cut the pickled radish, yard long beans and tender ginger into 4cm strips. Cut the tofu into 4cm-long, 2cm-thick strips, and blanch in salted water.
3. Heat 200ml rapeseed oil and 200ml lard in a wok to 100℃, and stir fry Pixian chili bean paste until fragrant. Blend in the pickled radish, yard long beans, tender ginger and garlic, and continue to stir to bring out the aroma. Add the stock, fish sauce, fish chunks and tofu, bring to a boil, and add the fermented rice juice, white sugar and vinegar. Turn down the flame, and simmer over low heat till the ingredients are cooked through. Blend in the ground Sichuan pepper, green Sichuan pepper oil, chopped scallion, celery and coriander, remove from the heat and transfer to a serving dish.
4. Heat 200ml rapeseed oil in a wok to 130℃, fry dried chilies, Sichuan peppercorns and green Sichuan peppercorns until fragrant, and pour over the fish.

Reviews

This dish is an invention of Red Star Restaurant in Pidu District, combining the spicy hot pot broth of Chengdu with tofu and pickles, two ingredients most favored by people in Pidu District. The first character of its name, Tuo, means big chunks, and the second character, synonym of the first one, refers to the Tuo River which flows through the hometown of the dish inventor. This delicate, savory and spicy fish has won a number of titles including Most Popular Dish of International Culinary Tour of China, Jinjiao Medal of Sichuan Cuisine, Top Dish of Sichuan Cuisine and Renowned Dish of Chengdu .

回锅鱼

Twice-Cooked Fish

食材配方

净鱼肉300克　青蒜苗60克　甜椒50克　青椒50克　独头蒜10克　鸡蛋液60毫升　郫县豆瓣40克　豆豉6克　酱油5毫升　白糖2克　玉米淀粉60克　料酒5毫升　胡椒粉0.2克　食盐1克　姜葱水10毫升　葱油5毫升　菜籽油1 500毫升（约耗80毫升）

成菜特点

色泽红亮、形整不烂、外酥内嫩、咸鲜微辣。

制作工艺

1. 净鱼肉片成厚约0.2～0.3厘米的片，用食盐、胡椒粉、料酒、姜葱水拌匀，码味10分钟。
2. 青蒜苗切成马耳朵形；甜椒、青椒切成菱形块；独头蒜切片；豆豉剁细；将鸡蛋液与玉米淀粉调匀，加入葱油搅匀成全蛋淀粉糊。
3. 锅置火上，入菜籽油烧至150℃时，将鱼片裹上全蛋淀粉糊入锅炸至定型后捞出，待油温回升至180℃时，再放入鱼片复炸至外酥内嫩后捞出。
4. 锅置火上，入菜籽油烧至150℃时，下甜椒、青椒滑熟后捞出。
5. 锅置火上，入菜籽油烧至120℃时，先下郫县豆瓣、蒜片炒香，再下豆豉、青蒜苗炒香，然后放入鱼片、甜椒、青椒、酱油、白糖炒匀，起锅装盘即成。

评鉴

回锅鱼是从经典川菜名菜“回锅肉”演变而来，其区别是用鱼肉代替猪肉，不但营养价值得以提升，而且色彩、口感也更加丰富，是一道老少皆宜的佳肴，曾获得“成都名菜”“郫都区十大川菜菜品”等称号。

Ingredients

300g fish; 60g baby leeks; 50g bell peppers; 50g green peppers; 10g garlic; 60ml beaten eggs; 40g Pixian chili bean paste; 6g fermented soy beans; 5ml soy sauce; 2g white sugar; 60g cornstarch; 5ml Shaoxing cooking wine; 0.2g pepper; 1g salt; 10ml ginger-scallion juice; 5ml scallion oil; 1,500ml rapeseed oil (about 80ml to be consumed)

Features

appealing lustrous color; savory and slightly spicy fish tender on the inside and crispy on the outside

Preparation

1. Cut the fish into 0.2-0.3cm thick slices. Mix well with salt, pepper, cooking wine and ginger-scallion juice. Leave to marinate for ten minutes.
2. Cut the baby leeks into horse-ear slices, and the bell peppers and green peppers into diamond slices. Slice the garlic. Finely chop the fermented soy beans. Mix beaten eggs with cornstarch, add scallion oil, and blend well to make a paste.
3. Put the wok on fire, add rapeseed oil and boil it to 150℃, then wrap the fish fillet with egg starch paste and fry it until it is set. When the oil temperature rises to 180℃, fry the fish fillet again until it is crisp outside and tender inside.
4. Heat oil in a wok to 150℃, and stir fry bell peppers and green peppers till just cooked.
5. Heat oil in a wok to 120℃, and stir fry Pixian chili bean paste and garlic till fragrant. Add fermented soy beans, and continue to stir to bring out the aroma. Add the fish slices, bell peppers, green peppers, soy sauce and white sugar, blend well and transfer to a serving dish.

Reviews

This dish derives from Twice-Cooked Pork, where fish is used as a substitute for pork. This derivative increases nutritional values and presents a colorful dish with layered taste. Popular among old and young, Twice Cooked Fish is honored as Famed Chengdu Dish and ranked among Top10 Dishes of Pidu District.

水煮鱼片

Boiled Fish Slices in Chili Sauce

食材配方

草鱼一尾（约1 000克） 青笋尖100克 芹菜80克 蒜苗80克 香菜10克 郫县豆瓣60克 蒜米20克 食盐5克 酱油6毫升 胡椒粉1克 味精3克 白糖1克 料酒20毫升 芝麻油5毫升 鸡蛋清30毫升 干淀粉30克 水淀粉20克 葱花10克 刀口辣椒80克 鲜汤600毫升 菜籽油150毫升

成菜特点

色泽红亮 鱼片嫩滑 麻辣鲜香。

制作工艺

1. 草鱼宰杀后洗净，片成厚约0.4厘米的片，用食盐（2克）、料酒（10毫升）、胡椒粉（0.5克）拌匀，码味10分钟后再加入鸡蛋清、干淀粉拌匀。
2. 青笋尖切成长约10厘米的条；芹菜、蒜苗分别切成长约10厘米的段。
3. 锅置火上，入菜籽油烧至120℃时，下郫县豆瓣炒香，再下蒜米（10克）炒香，然后掺入鲜汤，下食盐（3克）、酱油、胡椒粉（0.5克）、味精、白糖、料酒（10毫升）调好味，接着下青笋尖、芹菜、蒜苗煮断生，捞出后装入大碗中垫底，再放入鱼片煮熟，用水淀粉勾成薄芡，下芝麻油和匀后起锅倒在碗中，最后依次放入葱花、蒜米（10克）、刀口辣椒，淋上200℃的热油，撒上香菜即成。

评鉴

此菜是在传统经典川菜“水煮牛肉”的基础上演变而来。郫都区地处都江堰灌区，水系发达，养殖鱼类众多，且品质优良，从而使郫都区的鱼类美食成为一大亮点，也是许多农家乐的主打菜品。

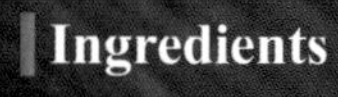

Ingredients

1 grass carp (about 1,000g); 100g asparagus lettuce tips; 80g celery; 80g baby leeks; 10g coriander; 60g Pixian chili bean paste; 20g garlic, finey chopped; 5g salt; 6ml soy sauce; 1g pepper; 3g MSG; 1g white sugar; 20ml Shaoxing cooking wine; 5ml sesame oil; 30ml egg white; 30g cornstarch; 20g cornstarch batter; 10g scallion; 80g blade-minced chilies; 600ml stock; 150ml rapeseed oil

Features

appealing red in color; tender fish slices; hot, numbing, delicate and aromatic in taste

Preparation

1. Slaughter, scale and gut the fish. Rinse and cut into 0.4cm-thick slices. Mix well with 2g salt, 10ml cooking wine and 0.5g pepper. Leave to marinate for ten minutes. before adding egg white and cornstarch. Mix well.
2. Cut asparagus lettuce tips into 10cm strips. Cut celery and baby leeks into 10cm sections.
3. Heat oil in a wok to 120℃, and stir fry Pixian chili bean paste till fragrant. Blend in 10g garlic, and continue to stir till aromatic. Add stock, 3g salt, soy sauce, 0.5g pepper, MSG, white sugar and 10ml cooking wine. Blend in the asparagus lettuce tips, celery and leeks, boil till just cooked, remove and transfer to a big serving bowl. Add the fish slices to the wok, boil till fully cooked, and pour in the batter to thicken the sauce. Add sesame oil, stir well, and remove from the heat. Pour the contents of the wok into the serving bowl, and sprinkle with chopped scallion, 10g chopped garlic and blade-minced chilies. Pour 200℃ hot oil over the fish and garnish with coriander.

Reviews

This dish derives from Boiled Beef in Chili Sauce. Benefiting from the Dujiangyan Irrigation System, Pidu District abounds with rivers and different fish species, making fish dishes the specialties of numerous local farm restaurants.

家常糖醋脆皮鱼

Home-Style Crispy Sweet and Sour Fish

食材配方

草鱼1尾（约750克） 泡辣椒丝10克 葱丝20克
姜片5克 葱段15克 食盐5克 料酒10毫升
泡辣椒末50克 姜米8克 蒜米15克 葱花20克
白糖40克 酱油3毫升 醋18毫升 味精2克
胡椒粉0.5克 芝麻油3毫升 水淀粉60克
鲜汤250毫升 食用油2 000毫升（约耗120毫升）

成菜特点

色泽红亮、外酥内嫩、甜酸微辣。

制作工艺

1. 草鱼治净，在鱼身两侧各剞5～6刀，用食盐（2克）、料酒（5毫升）、姜片、葱段拌匀，码味15分钟。
2. 用食盐（3克）、白糖、酱油、醋、味精、料酒（5毫升）、胡椒粉、芝麻油、水淀粉、鲜汤调成芡汁。
3. 锅置火上，入食用油烧至180℃时，将鱼身挂上水淀粉放入锅中炸至定型后捞出，待油温回升至200℃时，将鱼入锅复炸至色泽金黄、外酥内熟后捞出装盘。
4. 锅置火上，入食用油烧至120℃时，先下泡辣椒末、姜米、蒜米、葱花炒至出香，再倒入芡汁，待收汁浓稠呈糊芡状且亮油后，出锅淋在鱼身上，撒上葱丝和泡辣椒丝即成。

评鉴

此菜是在川菜经典菜品“糖醋脆皮鱼”的基础上发展而来，因加有泡辣椒末而使菜肴的味感更加丰富，还可去腥增色，成为郫都当地鱼肴烹制的一种独特风味。

Ingredients

1 grass carp (about 750g); 10g pickled chilies, cut into slivers; 10g scallion, cut into slivers; 5g ginger, sliced; 15g scallion, segmented; 5g salt; 10ml Shaoxing cooking oil; 50g pickled chilies, finely chopped; 8g ginger, finely chopped; 15g garlic, finely chopped; 20g scallion, finely chopped; 40g white sugar; 3ml soy sauce; 18ml vinegar; 2g MSG; 0.5g pepper; 3ml sesame oil; 60g cornstarch batter; 250ml stock; 2,000ml cooking oil (about 120ml to be consumed)

Features

golden brown in color; tender on the inside but crispy on the outside; sweet, sour and slightly spicy in taste

Preparation

1. Rinse the fish. Make five or six cuts in each side of the fish, and mix well with 2g salt, 5ml cooking wine, sliced ginger and segmented scallion. Leave to marinate for fifteen minutes;
2. Combine in a bowl 3g salt, white sugar, soy sauce, vinegar, MSG, 5ml cooking wine, pepper, sesame oil, batter and stock. Stir well to make the seasoning sauce.
3. Heat oil in a wok to 180℃, coat the fish with batter, deep fry to fix the shape and remove. Reheat the oil to 200℃ and deep fry the fish for a second time until it becomes golden, crispy on the outside, fully cooked through on the inside. Remove and transfer to a serving dish.
4. Heat oil in a wok to 120℃, and blend in chopped chili pickles, ginger, garlic and scallion, stirring until fragrant. Pour in the seasoning sauce, and wait till the sauce thickens and the oil becomes clear. Remove from the flame and pour over the fish. Garnish with pickled chili and scallion slivers.

Reviews

This dish is an upgraded version of the traditional Crispy Sweet and Sour Fish. Pickled chilies are added to enrich and enhance the tastes and remove the fishy smell, presenting a peculiar home-style flavor of fish dishes in Pidu District.

鸿运鱼头

Good Luck Fish Head

食材配方

花鲢鱼头1个（约1 750克）　小米辣椒200克　美人椒400克　泡辣椒100克　泡仔姜30克　食盐20克　味精3克　胡椒粉1克　醋10毫升　化猪油20毫升　化鸡油20毫升

成菜特点

色泽红亮、质地软糯、入口爽滑、酸辣鲜香。

制作工艺

1. 花鲢鱼头洗净，沥干水分后装入大凹盘中。
2. 小米辣椒、美人椒、泡辣椒、泡仔姜分别剁碎后入盆，加入食盐、味精、胡椒粉、醋、猪油、鸡油拌匀成剁椒浇在鱼头上，入笼用旺火蒸15分钟至鱼头成熟后取出。

评鉴

鸿运鱼头色泽红艳，故以此寓意命名。此菜选用黑龙滩生长期长达3～5年的深水大花鲢鱼头制作而成，烹制手段借鉴了剁椒鱼头的方法，并加入了泡辣椒和仔姜这两种四川地区的特色调料，既可去腥解腻，又能使口味更为鲜香醇厚、辣而不燥。

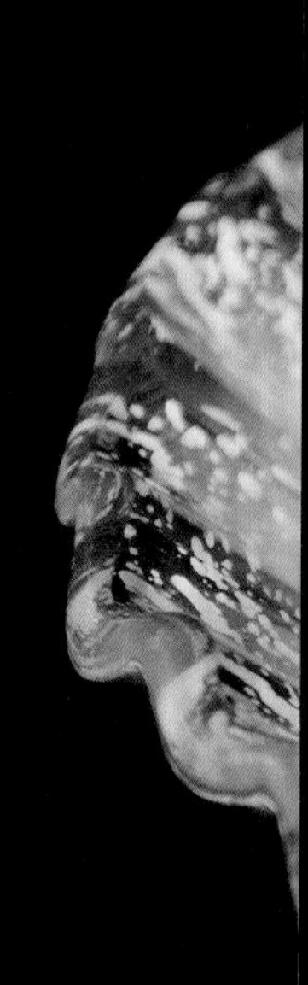

Ingredients

1 head of spotted silver carp (about 1,750g); 200g birds' eyes chilies; 400g red long peppers (also called beauty chilies); 100g pickled chilies; 30g pickled tend ginger; 20g salt; 3g MSG; 1g pepper; 10ml vinegar; 20ml lard; 20ml chicken oil

Features

lustrous appealing color; soft and glutinous fish; sour and spicy, fresh and fragrance taste

Preparation

1. Clean the carp fish head, drain and move to a big deep plate.
2. Chop the birds' eye chilies, red long peppers, pickled chilies and pickled tender ginger respectively and remove into a big bowl. Add salt, MSG, pepper, vinegar, lard and chicken oil and blend well. Pour the mixture over the fish head. Steam the fish head over high heart for fifteen minutes until fully cooked.

Reviews

The dish is red in color, an auspicious symbol in Chines culture, hence its name Good Luck Fish Head. Drawing on the cooking method of Fish Head with Chopped Chilies and using the head of spotted sliver carp that has lived in deep waters of Heilongtan (Black Dragon Lake) for three to five years, the dish added pickled chilies and ginger peculiar to Sichuan in order to enhance the flavors and remove the fishy smell.

食材配方

净鲜鱼肉500克　郫县豆瓣30克　姜片10克　葱段15克　姜米7克　蒜米15克　葱花25克　食盐2克　白糖3克　味精2克　料酒10毫升　生粉100克　水淀粉20克　鲜汤100毫升　菜籽油1 500毫升（约耗70毫升）

成菜特点

色泽红亮、形状美观、外酥内嫩、咸鲜爽口、酸甜略辣、豆瓣香浓。

制作工艺

1. 先在净鲜鱼肉上剞“十”字花刀，再切成约5厘米见方的鱼花块，用食盐、姜片、葱段、料酒拌匀，码味15分钟。
2. 锅置火上，入菜籽油烧至180℃时，将鱼花块均匀地裹上生粉，入锅炸至定型后捞出，待油温回升至200℃时，将鱼花块再次入锅复炸至外酥内嫩时捞出，装入盘中待用。
3. 锅置火上，入菜籽油烧至120℃时，先下郫县豆瓣炒香，再下姜米、蒜米炒香，然后加入鲜汤、白糖、味精、水淀粉收汁成二流芡，最后入葱花和匀，出锅淋在鱼花上即成。

评鉴

豆瓣鱼花是郫县豆瓣与鱼肉的又一次碰撞，成形美观、酱香浓郁、营养丰富，有增强机体免疫力等作用。此菜曾荣获“郫都区名菜”“郫都区大众喜爱菜品”等称号。

豆瓣鱼花

Fried Fillet in Chili Bean Sauce

Ingredients

500g Chinese sturgeon fillet; 30 Pixian chili bean paste; 10g ginger, sliced; 15g scallion, segmented; 7g garlic, finely chopped; 25g scallion, finely chopped; 2g salt; 3g white sugar; 2g MSG; 10ml Shaoxing cooking wine; 100g cornstarch; 20g cornstarch batter; 100ml stock; 1,500ml rapeseed oil (about 70ml to be consumed)

Features

bright dark brown in color; appealing presentation; tender fish crispy on the outside; savory and slight spicy, sweet and sour taste; rich chili bean paste aroma

Preparation

1. Make cross cuts into the sturgeon fillet, and slice into 5cm chunks. Combine fish chunks with salt, sliced ginger, scallion segments and cooking wine, mix well and leave to marinate for fifteen minutes.
2. Heat rapeseed oil in a wok to 180℃, Coat the fish with corn starch, deep-fry to fix the shape and then ladle out. Reheat the oil to 200℃, and deep fry the fish for a second time till crispy outside and tender insider. Remove the fish to a serving plate.
3. Heat rapeseed oil in a wok to 120℃, add Pixian chili bean paste, and stir fry till aromatic. Add chopped ginger and garlic, and continue to stir fry to bring out the aroma. Add the stock, white sugar, MSG and cornstarch, stirring till the sauce is thick. Blend in chopped scallion, stir well and pour over the fish.

Reviews

Fried Fillet in Chili Bean Sauce, a combination between Pixian chili bean paste and fish, is beautiful in presentation and rich in nutrients. Chili bean paste not only enhances the flavors of the dish but boosts the immune system as well. The dish has won honorable titles such as Famous Dish in Pidu District and Favorite Dish of Pidu people.

Scorching Stir-Fried Tench

食材配方

丁桂鱼1条（约750克）　自制辣椒碎250克　花椒20克　花生碎20克　熟芝麻15克　酥黄豆20克　大头菜粒20克　葱花30克　一品鲜酱油15毫升　醋15毫升　郫县豆瓣20克　食盐3克　料酒10毫升　葱段15克　姜米5克　蒜米10克　白糖15克　鸡精2克　味精2克　菜籽油1 500毫升（约耗50毫升）

成菜特点

色泽棕红、鱼肉酥香、香辣不燥。

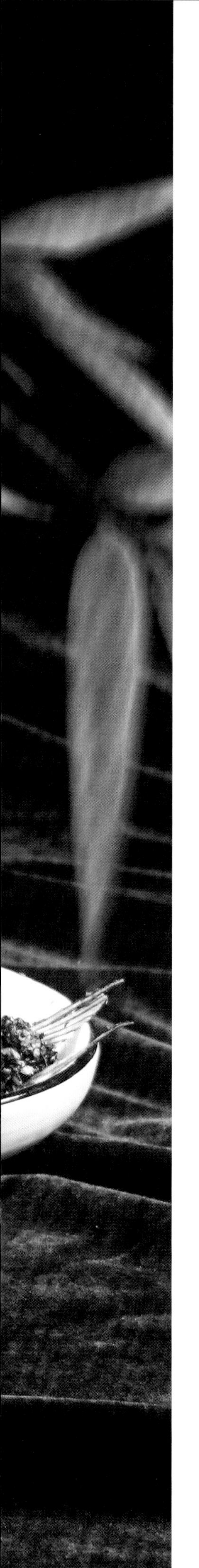

制作工艺

1. 丁桂鱼宰杀后洗净，加入食盐、料酒、葱段码味15分钟。
2. 锅置火上，入菜籽油烧至180℃时，放入丁桂鱼炸至外酥内熟后捞出备用。
3. 锅内留油，依次放入姜米、蒜米、花椒、郫县豆瓣、自制辣椒碎炒出香味，再下丁桂鱼、白糖、鸡精、味精、一品鲜酱油、醋，将鱼炝炒入味后捞出装盘，淋上锅内味汁，撒上大头菜粒、酥黄豆、花生碎、熟芝麻、葱花即成。

评鉴

此菜是根据传统“炝锅鱼”略加改动而成，鱼肉入口更为酥香，辣而不燥，深受年轻人喜爱，是一道适宜搭配啤酒消夏而食的美味。

Ingredients

1 tench (about 750g); 250g homemade chopped chilies; 20g Sichuan peppercorns; 20g crispy peanuts, finely chopped; 15g roasted sesame seeds; 20g crispy soybeans; 20g preserved kohlrabi, finely chopped; 30g scallion, finely chopped; 15ml Yipinxian soy sauce; 15ml vinegar; 20g Pixian chili bean paste; 3g salt; 10ml Shaoxing cooking wine; 15g scallion, segmented; 5g ginger, finely chopped; 10g garlic, finely chopped; 15g white sugar; 2g chicken essence; 2g MSG; 1,500ml rapeseed oil (about 50ml to be consumed)

Features

red and lustrous in color; crispy fish; fragrant and spicy tastes

Preparation

1. Slaughter and rinse the tench, and marinate with salt, cooking wine and scallion segments for fifteen minutes.
2. Heat rapeseed oil in a wok to 180℃, deep fry the tench till cooked through and crispy outside.
3. Pour out most of the oil, add chopped ginger, garlic, Sichuan peppercorns, Pixian chili bean paste, homemade chopped chilies in order and stir fry till fragrant. Add the tench, white sugar, chicken essence, MSG, soy sauce and vinegar, scorching for the fish to absorb the flavors of seasonings. Remove the fish to a serving plate, drizzle with the sauce in the pot and sprinkle with chopped kohlrabi, crispy soybeans and chopped scallion.

Reviews

This dish is an upgraded version of the traditional Scorching Fish. The savory, crispy and spicy fish is a summer delicacy that goes well with beer and enjoys immense popularity among the young.

食材配方

仔鲶800克　独头蒜100克　泡红辣椒100克　姜米10克　葱段30克　食盐2克　郫县豆瓣50克
白糖12克　酱油5毫升　醪糟汁15毫升　料酒20毫升　味精2克　芝麻油3毫升　醋10毫升
水淀粉30克　鲜汤600毫升　胡椒粉0.5克　化猪油300毫升　菜籽油1 500毫升（约耗100毫升）

成菜特点

色泽红亮、鱼肉细嫩、咸鲜微辣、略带酸甜。

制作工艺

1. 仔鲶治净，用刀尖在脊背上横斩一刀使其骨断肉连，用食盐、料酒（10毫升）、胡椒粉拌匀，码味15分钟。
2. 泡红辣椒去籽，取一半切成长约8厘米的段，另一半剁成细末。
3. 锅置火上，入化猪油烧至150℃时，放入独头蒜炸至皱皮后捞出装入碗中，然后注入鲜汤，上笼蒸至断生。
4. 锅置火上，入菜籽油烧至200℃时，将仔鲶入锅炸至紧皮后捞出。
5. 锅内留油，先下郫县豆瓣、泡椒末炒出香味，再下姜米、泡红辣椒段、葱段炒香，然后加入鲜汤、酱油、白糖、醪糟汁、料酒（10毫升）、仔鲶、独蒜烧沸，改用小火将仔鲶烧至入味后捞出装入盘中，用独头蒜围边。
6. 锅中汁水用水淀粉勾成二流芡，入味精、芝麻油、醋搅匀，出锅淋在仔鲶上即成。

评鉴

此菜是源自清末犀浦当地的一道名菜，至今已逾百年历史，在抗战时期，曾受到寓居成都的张大千先生等文化名人的盛赞，被载入《中国菜谱》和《川菜烹饪事典》。此菜采用仔鲶红烧而成，火候考究，既要将鱼肉烧至软熟入味，又要保持鱼形完整，鱼肉细嫩。历来都是游客来犀浦旅游时的必点菜品，曾荣获"成都名菜"称号。

Ingredients

800g baby catfish; 100g single-clove garlic; 100g pickled red chilies; 10g ginger, finely chopped; 30g scallion, segmented; 2g salt; 50g Pixian chili bean paste; 12g white sugar; 5ml soy sauce; 15ml fermented glutinous rice juice; 20ml Shaoxing cooking wine; 2g MSG; 3ml sesame oil; 10ml vinegar; 30g cornstarch batter; 600ml stock; 0.5g pepper; 300ml lard; 1,500ml rapeseed oil (about 100ml to be consumed)

Features

bright brown color; tender fish; savory, slightly spicy, sweet and sour tastes

Preparation

1. Rinse the baby catfish, make one cut into its back bone with a leaver point to break the spine, but the meat still connects. Transfer the fish to a big bowl, and blend well with salt, 10ml cooking wine and pepper. Leave to marinate for fifteen minutes.

犀浦鲶鱼

Xipu Catfish

2. Deseed the pickled red chilies. Cut half into 8cm-long segments, and finely chop the other half.
3. Heat lard in a wok to 150℃, fry the single-clove garlic until the skin wrinkles, and remove to a bowl. Add stock to the bowl, and steam till the garlic is just cooked.
4. Heat rapeseed oil in a wok to 200℃,and deep fry the catfish till the skin tightens up.
5. Pour out most the oil, and stir fry Pixian chili bean paste and minced chili pickles till fragrant. Add chopped ginger, pickled chili segments and scallion sections, stirring till aromatic. Pour in stock, soy sauce, white sugar, fermented rice juice, 10ml cooking wine, baby catfish and single-clove garlic, bring to a boil and turn down the heat. Simmer over low heat for the fish to absorb the flavors of seasonings. Remove the fish to a serving plate, and surround with single-clove garlic.
6. Add cornstarch batter to the wok, stir well and blend in MSG, sesame oil and vinegar. Pour the sauce over the catfish.

Reviews

This dish originated from a famous dish in Xipu District, which has a history of more than 100 years. It has been highly praised by celebrities such as Zhang Daqian who lived in Chengdu during the Anti-Japanese War. Therefore, it was recorded in *Chinese Recipe* and *Sichuan Cuisine Records*. It uses brown braising to cook the baby catfish, and the key is the control of heat and time so that the fish is tender and savory but still retains its original shape. It has been honored as Renowned Dish of Chengdu, a must-have for visitors to Xipu.

苗苗鱼火锅

Miaomiao Fish Hot Pot

食材配方

①食材：三角峰1000克　自制火锅底料300克　酸菜150克　姜片30克　葱节50克　干辣椒节300克　花椒50克　鸡精10克　味精10克 香菜30克　食盐10克　料酒80毫升　菜籽油1500毫升　鸡油250克　清水1500毫升

②碟料：蒜米10克　香菜碎5克　大头菜粒5克　葱花5克　蚝油3毫升　味精1克　芝麻油5毫升

成菜特点

色泽红亮、鱼肉鲜嫩、咸鲜麻辣、香味浓郁。

制作工艺

1. 三角峰宰杀后治净。
2. 锅置火上，入菜籽油、鸡油烧至120℃时，下姜片、酸菜、葱节、干辣椒节、花椒炒香，再下自制火锅底料炒香，接着加入清水、料酒、味精、鸡精、食盐烧沸，入三角峰煮沸3分钟后再焖3分钟至鱼成熟，最后放入香菜。
3. 将蒜米、香菜碎、大头菜粒、葱花、蚝油、味精、芝麻油入碗，再舀一勺锅中的原汁兑成味碟蘸食。食完三角峰后，还可点火烫涮其他食材。

评鉴

苗苗鱼为四川方言，是指长度为10～12厘米的小鱼。苗苗鱼火锅是郫都区三道堰最著名的美食，食材包括三角峰、丁桂、仔鲶、裸斑等，客人可根据自己喜好任意选择。烹制此菜所用的火锅底料，是采用独家秘制的豆瓣调料，再辅以菜籽油和多种辛香料熬制而成，味道咸中带鲜、鲜中带辣、辣中带麻、麻中带香，让人印象深刻、回味无穷。

Ingredients

Food Materials 1,000g sanjiaofeng catfish; 300g specially-made hot pot soup base; 150g pickles; 30g ginger, sliced; 50g scallion, segmented; 300g dried chilies; 50g Sichuan peppercorns; 10g chicken essence; 10g MSG; 30g coriander; 10g salt; 80ml Shaoxing cooking wine; 1,500ml rapeseed oil; 250ml chicken oil; 1,500ml water

Dipping Sauce 10g garlic, finely chopped; 5g coriander, finely chopped; 5g preserved kohlrabi, finely chopped; 5g scallion, finely chopped; 3g oyster sauce; 1g MSG; 5ml sesame oil

Features

lustrous red in color; tender fish; spicy, savory and aromatic in taste

Preparation

1. Slaughter sanjiaofeng catfish and rinse.
2. Heat rapeseed oil and chicken oil in a wok to 120℃, stir fry ginger slices, pickles, segmented scallion, dried chilies and Sichuan peppercorns till fragrant. Add hot pot soup base, water, cooking wine, MSG, chicken essence and salt, bring to a boil and add the fish. Bring to a second boil, and continue to boil for three minutes. Turn off the heat, cover the pot and leave the fish in the pot for three more minutes until the fish is fully cooked. Add coriander to the pot.
3. Combine chopped garlic, coriander, kohlrabi, scallion, oyster sauce, MSG and sesame oil in a bowl, add a ladle of the fish soup and stir well to make the dipping sauce. Other ingredients can be added and boiled after the fish is consumed.

Reviews

Miaomiao fish, a Sichuanese dialect, refer to those small fish about four to five inches in length. Miaomiao Fish Hot Pot is the most famous dish in Sandaoyan of Pidu District, and a variety of small fishes are available for diners to choose from. The specially-made hot pot soup base uses secret-recipe chili bean paste which is stir fried and then simmered with various spices, presenting a peculiar salty, savory, spicy taste.

豆瓣藿香鳝片

Chili Bean Flavored Eels with Huoxiang Herbs

食材配方

鳝鱼片500克　鲜笋100克　青笋100克　藿香20克　独头蒜50克　姜片10克　姜米10克　香葱段15克　郫县豆瓣70克　白糖2克　食盐2克　陈醋5毫升　胡椒粉2克　料酒20毫升　高汤500毫升　化鸡油100毫升　菜籽油1 000毫升（约耗230毫升）

成菜特点

色泽红亮、鳝鱼软糯、咸鲜麻辣、藿香浓郁。

制作工艺

1. 鳝鱼片切成长约12厘米的段，用姜片、香葱段、料酒（10毫升）腌制片刻，再入180℃的菜籽油中炸至皱皮后捞出。
2. 藿香切为粗丝；鲜笋、青笋分别切成长约8厘米的薄片，焯水后装入凹盘中垫底。
3. 锅置火上，入菜籽油（200毫升）、鸡油烧至120℃时，放入独头蒜、郫县豆瓣、姜米炒香，再加入高汤、鳝鱼片、白糖、食盐、陈醋、胡椒粉、料酒（10毫升），将鳝鱼片烧至软糯、入味后，连汤带汁倒入凹盘中，撒上藿香丝即成。

评鉴

鳝鱼又名黄鳝，富含不饱和脂肪酸，有增强记忆力等作用。此菜以稻田中的鳝鱼为主要食材，用郫县豆瓣和藿香共同烹制而成，营养丰富，口感Q弹，味道醇厚、浓香。

Ingredients

500g eels; 100g fresh bamboo shoots; 100 spring bamboo shoots; 20g huoxiang herb (agastache rugosus); 50g single-clove garlic; 10g ginger, sliced; 10g ginger, finely chopped; 15g scallion, segmented; 70g Pixian chili bean paste; 2g white sugar; 2g salt; 5ml vinegar; 2g pepper; 20ml Shaoxing cooking wine; 500ml stock; 100ml chicken oil; 1,000ml rapeseed oil (about 230ml to be consumed)

Features

lustrous red in color; soft and glutinous eel; savory and spicy in taste with a strong huoxiang aroma

Preparation

1. Cut the eels into 12cm sections, add the sliced ginger, scallion and 10ml cooking wine, and leave to marinate. Deep fry the eel sections in 180℃ rapeseed oil until the skin wrinkles, and remove from the oil.
2. Sliced the huoxiang herb into thick slivers. Cut the fresh bamboo shorts and spring bamboo shoots into 8cm-long slices, blanch and transfer to a deep platter.
3. Heat a wok over the flame, add 200ml rapeseed oil and the chicken oil, and heat to 120℃. Add the garlic, Pixian chili bean paste and chopped ginger, and stir fry till fragrant. Add the stock, eels, white sugar, salt, vinegar, pepper and 10ml cooking wine, and braise till the ingredients are soft are soft. Pour the contents in the wok into the platter, and sprinkle with huoxiang slivers.

Reviews

Eels, also called yellow eels, are rich in unsaturated fatty acid, which helps to enhance memory and relieve muscle strength decrease. This savory and delicate dish, using local rice eels as the main ingredients and Pixian chili bean paste as the main seasoning, is both nutritious and springy.

食材配方

耗儿鱼750克　自制底料250克　姜片150克　芝麻15克　大蒜瓣100克　葱节150克　啤酒500毫升　清水200毫升　干辣椒节75克　花椒50克　鸡精3克　味精2克　泡酸菜100克　土豆片100克　魔芋100克　水发木耳100克　藕片100克　蚝油300毫升　芝麻油30毫升　食盐15克　花椒油30毫升　香菜30克　熟芝麻10克　食用油2 000毫升（约耗100毫升）

成菜特点

色泽红亮、鱼肉细嫩、麻辣味浓。

制作工艺

1. 耗儿鱼治净，用姜片（20克）、葱节（30克）、啤酒（50毫升）、食盐（5克）拌匀，码味15分钟。
2. 锅置火上，入食用油烧至200℃时，将耗儿鱼入锅炸至表面皱皮时捞出。
3. 锅内留油烧至120℃时，依次放入大蒜瓣、姜片（130克）、干辣椒节、花椒、泡酸菜、蚝油炒香，再下自制底料炒出颜色、香味，然后倒入剩下的啤酒和清水烧沸，在放入耗儿鱼煮熟后，下鸡精、味精、花椒油、芝麻油，接着放入煮熟的土豆片、藕片、魔芋、木耳烧沸，出锅后倒入盛器内，撒上香菜、熟芝麻即成。

评鉴

此菜是从火锅演变而来的一道冷锅菜品，麻辣、刺激，深受年轻人喜爱，吃完锅中的耗儿鱼和配菜后，还可点火再涮食其他一些素菜。

极品耗儿鱼

Deluxe Filefish

Ingredients

750g filefish; 250g self-made sauce; 150g ginger, sliced; 15g sesame seeds; 100g garlic cloves; 150g scallion, cut into sections; 500ml beer; 200ml water; 75g dried chilies, segmented; 50g Sichuan peppercorns; 3g chicken essence; 2g MSG; 100g pickles; 100g potatoes, sliced; 100g conjak jelly; 100ml water-soaked wood ear fungus; 100g lotus roots; 300ml oyster sauce; 30ml sesame oil; 15g salt; 30ml Sichuan pepper oil; 30g coriander; 10g roasted sesame seeds; 2,000ml cooking oil (about 100ml to be consumed)

Features

lustrous red in color; tender fish; hot and numbing in taste

Preparation

1. Rinse the filefish, combine with 20g ginger, 30g scallion, 50ml beer and 5g salt, blend well and leave to marinate for fifteen minutes.
2. Heat oil in a wok to 200℃, and deep fry filefish till the skin wrinkles.
3. Pour out most of the oil, heat the rest of the oil to 120℃, and add garlic cloves, 130g ginger, dried chilies, Sichuan peppercorns, pickles and oyster sauce, stirring until fragrant. Blend in the self-made sauce, and continue to stir till fragrant. Pour in the rest the beer and water, bring to a boil and add the fish. Boil till the fish is cooked through, and blend in chicken essence, MSG, Sichuan pepper oil, sesame oil, precooked potato slices, lotus root slices, conjak jelly and wood ear fungus. Bring to a boil, pour into a serving pot and sprinkle with coriander and sesame seeds.

Reviews

This is a cold pot dish derived from hot pot. When the ingredients in the pot have been consumed, vegetables can be added and boiled. It is especially popular among the young for its spicy taste.

百姓烩菜

Multi-Ingredient Folk Stew

食材配方

唐元韭黄200克　五花肉30克　猪肝20克
猪腰25克　鸡胗20克　青椒10克　甜椒10克
木耳10克　竹笋10克　熟青豌豆10粒　洋葱10克
芹菜10克　鲜香菇3克　郫县豆瓣5克　泡辣椒末5克
泡辣椒段3克　姜片2克　蒜片2克　酱油6毫升
醋4毫升　白糖5克　胡椒粉1克　芝麻油3毫升
花椒油3毫升　水淀粉10克　食用油60毫升

成菜特点

色泽红亮、咸鲜略辣、口感丰富。

制作工艺

1. 唐元韭黄切成长约3厘米的段；青椒、甜椒、木耳、竹笋、洋葱、芹菜、鲜香菇分别切成长约3厘米的条；五花肉、猪肝分别切片；鸡胗先剞花刀，再切成鸡冠花形；猪腰先剞花刀，再切成腰花。
2. 锅置火上，入食用油烧至180℃时，下五花肉略炒片刻，再入郫县豆瓣、泡辣椒末炒出香味，接着下鸡胗花、肝片、腰花炒至断生，然后下唐元韭黄、青椒、甜椒、木耳、竹笋、洋葱、芹菜、鲜香菇、熟青豌豆炒至断生，之后入酱油、白糖、胡椒粉、醋调味，出锅前加入水淀粉收汁浓稠，最后下芝麻油、花椒油炒匀，起锅装盘即成。

评鉴

此菜是选用郫都区的特色蔬菜“唐元韭黄”等原料烹制而成，用料多样，色彩、味感丰富，价廉物美，是深受当地民众喜爱的特色菜品，曾荣获“郫都区餐饮名菜”称号。

Ingredients

200g Tangyuan Chinese chives; 30g pork belly; 20g pork livers; 25g pork kidneys; 20g chicken gizzards; 10g green chili peppers; 10g bell peppers; 10g wood ear fungus; 10g bamboo shoots; 10 cooked green peas; 10g onions; 10g celery; 3g shiitake mushrooms; 5g Pixian chili bean paste; 5g pickle chilies, finely chopped; 5g pickled chilies, cut into sections; 2g ginger, sliced; 2g garlic, sliced; 6ml soy sauce; 4ml vinegar; 5g white sugar; 1g pepper; 3ml sesame oil; 3ml Sichuan pepper oil; 10g cornstarch batter; 60ml cooking oil

Features

lustrous red in color; salty, savory and slightly spicy in taste; a multitude of textures

Preparation

1. Cut the Tangyuan Chinese chives into 3cm segments. Cut the green chili peppers, bell peppers, wood ear fungus, bamboo shoots, onions, celery and shiitake into 3cm strips. Slice the pork belly and pork livers. Make cross cuts into chicken gizzards and pork kidneys so that they look like flowers.
2. Heat oil in a wok to 180℃, stir fry pork belly briefly and blend in Pixian chili bean paste and chopped chili pickles, stirring until fragrant. Add chicken gizzards, pork livers and pork kidneys, stir fry till just cooked, and add Chinese chives, green peppers, bell peppers, wood ear fungus, bamboo shoots, onions, celery, shiitake and green peas, stirring till just cooked. Blend in soy sauce, sugar, pepper, vinegar and cornstarch batter. Stir till the sauce is thick, add sesame seeds and Sichuan pepper oil, stir well and transfer to a serving dish.

Reviews

This dish uses a local specialty, Chinese Chives of Tangyuan, and various other ingredients to present a multitude of colors and textures. It is popular among the average folks for its reasonable price, and has won the honor of Renowned Dish of Pidu District.

食材配方

①食材：牛肉片80克　鸡胗片80克　午餐肉片80克　鸡肉片80克　鸭肉片80克　藕片80克　土豆片80克　莴笋条80克　冬瓜片50克　香菇片50克　金针菇50克　白菜50克　生菜80克　木耳80克　笋子50克　番茄80克

②调料：冒菜底料400克　鲜汤5 000毫升　姜片50克　干辣椒100克　花椒80克　芝麻油10毫升　辣椒油30毫升　香菜30克　蒜泥30克　葱花20克　芹菜花20克

成菜特点

色泽红亮、麻辣味浓、食材丰富。

制作工艺

1. 在汤料锅中加入鲜汤、冒菜底料、姜片、干辣椒、花椒熬煮至出香。
2. 将各种经刀工处理好的荤、素食材分别装入小篓筐，放入汤料锅中冒煮两分钟至熟后倒入大汤碗中，舀一勺原汤入碗，再放入蒜泥、芝麻油、辣椒油、芹菜花、葱花、香菜即成。

唐元冒菜

Tangyuan Maocai

评鉴

“冒”系四川方言，作为一种烹饪方式，其意是指将食材放入锅中的汤料中短时间煮制成熟，与“氽”的意思比较接近，但用时略微偏长。“冒菜”是把多种不同食材放入特制的汤料里煮熟后倒入碗中，先舀一勺锅中的原汁，再加入多种调料后食用。其类型可大致分为火锅冒菜、卤水冒菜和白水冒菜三大类，尤以火锅冒菜最具特色，有“冒菜是一个人的火锅，火锅是一群人的冒菜”之说。“冒菜”是四川地区的特色菜肴，起源于成都，唐元冒菜是其分支之一，以麻辣口味为主，特色突出。

Ingredients

Food Materials 80g beef, sliced; 80g chicken gizzards, sliced; 80g luncheon meat; 80g chicken, sliced; 80g duck, sliced; 80g lotus roots, sliced; 80g potatoes, sliced; 80g stem lettuce, cut into strips; 50g winter melon, sliced; 50g shiitake mushrooms, sliced; 50g golden needle mushrooms; 50g Chinese cabbage; 80g lettuce; 80g wood ear fungus; 50g bamboo shoots; 80g tomatoes

Seasonings 400g maocai soup base; 5,000ml stock; 50g ginger, sliced; 100g dried chilies; 80g Sichuan pepper; 10ml sesame oil; 30ml chili oil; 30g coriander; 30g garlic, crushed; 20g scallion, finely chopped; 20g celery, finely chopped

Features

red and lustrous in color; a variety of ingredients; spicy and savory in taste

Preparation

1. Combine stock, maocai soup base, ginger, chilies and Sichuan peppercorns in a pot, bring to a boil, and continue to boil until fragrant.
2. Process the ingredients, transfer to a small bamboo basket, and boil in the pot for two minutes until all the ingredients are cooked through. Remove the contents of the basket to a big bowl, ladle up some soup from the pot and pour into the bowl. Add garlic, sesame oil, chili oil, celery, scallion and coriander, and serve.

Reviews

Mao, a Sichuanese dialect, means boiling ingredients very briefly. Maocai means boiling mixed ingredients in specially-made soup till cooked through, pouring into a bowl, adding a ladle of soup and topping with seasonings. Maocai is divided into three categories according to the soups: hot pot soup, seasoned broth and plain water, among which the most popular is the hot pot soup. The saying goes that maocai is hot pot for one person while hot pot is the maocai for a group of people. Maocai is a Sichuan specialty with its origin in Chengdu. Tangyuan Maocai is a branch deriving from Chengdu with especially hot and numbing tastes.

宫保杏鲍菇

Gongbao King Oyster Mushrooms

食材配方

杏鲍菇600克　干辣椒节20克　花椒3克　姜片10克　蒜片15克　葱丁25克　腰果35克　食盐3克　白糖30克　酱油3毫升　醋10毫升　干淀粉75克　水淀粉30克　食用油1 500毫升（约耗75毫升）

成菜特点

色泽棕红、质地微糯、口感软香、麻辣酸甜。

制作工艺

1. 杏鲍菇洗净，切成约1. 5厘米见方的丁，用干淀粉拌匀，入150℃的食用油中炸至定型（约七成熟）后捞出。
2. 用食盐、白糖、酱油、醋、水淀粉调成芡汁。
3. 锅置火上，入食用油烧至150℃时，下干辣椒节、花椒、姜片、蒜片、葱丁炒香，再入杏鲍菇炒熟，然后烹入芡汁至收汁亮油，最后加入腰果和匀，起锅装盘即成。

评鉴

杏鲍菇是一种大型肉质伞菌，菌肉肥厚、结实，质地脆嫩，色泽乳白，有杏仁的香味、鲍鱼的口感，适合多种烹制方式。此菜选用郫都区战旗村种植的优质杏鲍菇，采用“宫保鸡丁”的传统做法，素菜荤做，特别适合老年人及素食消费者食用。

Ingredients

600g king oyster mushrooms; 20g dried chilies; 3g Sichuan pepper; 10g ginger, sliced; 15g garlic, sliced; 25g scallion, chopped; 35g cashew nuts; 3g salt; 30g white sugar; 3ml soy sauce; 10ml vinegar; 75g cornstarch; 30g cornstarch batter; 1,500ml cooking oil (about 75ml to be consumed)

Features

reddish brown in color; soft and slightly glutinous in texture; spicy, sour and sweet in taste

Preparation

1. Rinse the mushrooms, and cut into 1.5cm cubes. Combine with cornstarch, stir well and deep fry in 150℃ oil until the mushrooms are medium done and their shape is fixed.
2. Combine salt, sugar, soy sauce, vinegar and batter in a bowl, stir well to make the seasoning sauce.
3. Heat oil in a wok to 150℃, and blend in dried chilies, Sichuan peppercorns, ginger, garlic and scallion, stirring until fragrant. Add the mushrooms, stir fry until fully cooked, and pour in the seasoning sauce. Stir until the sauce is thick and oil becomes clear, add cashew nuts, stir well and transfer to a serving dish.

Reviews

King oyster mushroom is a large quality fungus with a firm and crispy texture, milky color and a peculiar almond aroma, suitable for a variety of cooking methods. This dish takes the premium quality king oyster mushrooms grown in Zhanqi Village of Pidu District as the main ingredients and employs the cooking method used in the famous dish Gongbao Chicken. This dish is especially popular among the old and the vegetarian.

食材配方

豆腐150克　血旺150克　黄凉粉150克　牛肉末20克　食盐5克　味精3克　白糖5克　老抽6毫升　郫县豆瓣30克　干辣椒节5克　花椒3克　水淀粉20克　花椒油5毫升　蒜苗花10克　蒜末10克　姜末10克　鲜汤250毫升　食用油70毫升

成菜特点

色泽红亮、形整不烂，咸、鲜、麻、辣、香、烫兼而有之。

制作工艺

1. 豆腐、血旺、黄凉粉分别切成约2厘米见方的大丁。
2. 锅置火上，注入清水和食盐（2克）烧沸，放入豆腐焯水后捞出备用。
3. 锅置火上，入食用油（20毫升）烧至120℃时，放入牛肉末炒香，再下老抽（1毫升）炒成棕红色后捞出。
4. 锅置火上，入食用油（50毫升）烧至150℃时，下蒜末、姜末、郫县豆瓣、干辣椒节、花椒炒香，再加入鲜汤、食盐（3克）、味精、老抽（5毫升）、豆腐、血旺、凉粉烧沸，煮3分钟至熟后，再下牛肉末、花椒油和匀，然后烹入水淀粉勾芡，起锅装盘后撒上蒜苗花即成。

评鉴

从字义上讲，“烘”的本意是“用火使东西变熟、变热或干燥”，在菜肴烹调中，是将锅中烹制的食材采取“烘”的方式将汁水烘烧至干。郫县农家张三烘最经典的菜式有三样，分别是烘血旺、烘豆腐和烘凉粉，故称“三烘”，其凉粉软嫩滑爽、豆腐细嫩麻辣、血旺鲜嫩入味，各具特色。

Ingredients

150g tofu; 150 pork blood curd; 150g yellow pea jelly; 20g beef, minced; 5g salt; 3g MSG; 5g white sugar; 6ml dark soy sauce; 30g Pixian chili bean paste; 5g dried chilies; 3g Sichuan pepper; 20g cornstarch batter; 5ml Sichuan pepper oil; 10g baby leeks, finely chopped; 10g garlic, finely chopped; 10g ginger, finely chopped; 250ml stock; 70ml cooking oil

Features

appealing reddish in color; savory, fresh, numbing, spicy, aromatic and hot pea jelly

Preparation

1. Cut tofu, blood curd and pea jelly into 2cm cubes.
2. Heat a wok over the flame, add water and 2g salt, bring to a boil and blanch the tofu cubes.
3. Heat 20ml oil in a wok to 120℃, and stir fry beef mince till fragrant. Blend in 1 ml dark soy sauce, continue to stir until the beef becomes dark brown, and remove from the wok.

4. Heat 50ml oil in a wok to 150℃, and blend in garlic, ginger, Pixian chili bean paste, dried chilies and Sichuan peppercorns, stirring until fragrant. Pour in the stock, add 3g salt, MSG, 5ml soy sauce, tofu, blood curd and pea jelly, bring to a boil, and continue to boil for three minutes. Add beef mince and Sichuan pepper oil, blend well and stir in the cornstarch batter to thicken the sauce. Transfer to a serving dish, and sprinkle with chopped scallion.

Reviews

Zhang San Hong (Zhang's Three Dry Braisings if literally translated) Farm Restaurant in Pidu District serves three classic dishes: dry braised pork blood curd, dry braised tofu and dry braised pea jelly, hen the name "three dry braisings". Pea jelly is soft, tender and smooth; tofu tender and spicy; pork blood delicate and savory.

张三烘 Zhang's Dry Braising Trio

魔方豆花

Magic Cube Tofu Pudding

食材配方

黄豆500克　石膏粉35克　清水3 000毫升　食盐100克　味精100克
鸡精100克　白糖100克　酱油100毫升　醋100毫升　蚝油100毫升
香辣酱100克　芝麻酱100克　海鲜酱100克　沙拉酱100克
花椒粉20克　藤椒油100毫升　芝麻油100毫升　辣椒油300毫升
馓子100克　花生碎100克　酥黄豆100克　熟芝麻100克
酥豆豉碎100克　大头菜粒100克　榨菜碎100克　折耳根碎100克
蒜米100克　青椒碎100克　小米辣碎100克　泡菜碎100克
葱花100克　香菜碎100克　芹菜碎100克

成菜特点

豆花洁白、质地细嫩、味道丰富。

制作工艺

1. 黄豆用清水浸泡至透后去皮、洗净，用石磨加清水磨成豆浆后用纱布滤浆取液；石膏用少量清水化开。
2. 把各种调料分别装入小碗、小碟中。
3. 将过滤后的豆浆倒入锅中烧沸，加入石膏水搅匀后加盖、关火，使豆浆凝固成豆花，然后配上各种调料味碟与豆花一同上桌即成。食客可根据自己的口味喜好，搭配不同的调料制成味碟蘸食。

评鉴

魔方豆花是原三源农庄钟大作先生创制的一道郫都特色名菜。制作此菜有两大要点：一是豆花制作，原料选用的是品质优良、出浆率高的优质白毛黄豆，先用石磨磨制成浆，再加入石膏水做成绵软细嫩、韧性良好的豆花。二是"蘸水魔力"，可供蘸食豆花的作料多达四十余种，分放在数十个小碟中，场面蔚为壮观，食用时，客人可根据个人喜好自己动手调制蘸碟味道，正是因为蘸碟之味能像玩魔方一样自由变换、组合，此菜因此而得名。

Ingredients

500g soybean; 35g gypsum powder; 3,000ml water; 100g salt; 100g MSG; 100g chicken essence; 100g white sugar; 100ml soy sauce; 100ml vinegar; 100ml oyster sauce; 100g spicy sauce; 100g sesame sauce; 100g hoisin sauce; 100g salad sauce; 20g ground Sichuan pepper; 100ml green Sichuan pepper oil; 100ml sesame oil; 300ml chilli oil; 100g sanzi (deep-fried twisted noodles); 200g crispy peanuts, finely chopped; 100g crisp soybeans; 100g toasted sesames; 100g crispy fermented soybeans, finely chopped; 100g preserved kohlrabi, finely chopped; 100g zhacai (pickled mustard tuber) finely chopped; 100g fish mint, finely chopped; 100g garlic, finely chopped; 100g green chilies, finely chopped; 100g bird-eye chilies, finely chopped; 100g pickles, finely chopped; 100g scallion, finely chopped ; 100g coriander, finely chopped; 100g celery, finely chopped

Features

pure white in color; tender in texture with diversified flavors; Diners can mix different seasonings the way they like

Preparation

1. Soak the soybean in water, then remove the hull and rinse before grinding in a stone mill with clean water to produce soy milk, then strain in cheese cloth; mix gypsum powder in a spoonful of clean water.
2. Contain the seasonings in small bowls or saucers.
3. Boil the soybean milk in a pot and blend well with gypsum solution. Put the lid on and turn off the heat; let the soybean milk set and coagulate into Tofu pudding. Serve it with the seasoning bowls or saucers, and the diners can mix their preferred seasonings as the dipping sauce for the Tofu pudding.

Reviews

Magic Cube Tofu Pudding is a specialty dish in Pidu District created by Zhong Dazuo from Sanyuan Farm Restaurant. There are two key points to make this dish: Firstly, tofu pudding should be made from premium quality soybeans; grind the soybean with stone mill to produce an elastic, velvety and tender texture. The second is the magic power of the dipping sauce. More than 40 seasonings exquisitely placed in saucers form a magic cube pattern, and the dinners can DIY the dipping sauce so the taste can vary in an innumerable combinations like magic cube, and that is why the dish uses magic cube as its name.

第三节　面点小吃

Section 3 PASTRY AND SNACKS

食材配方

肉末100克　抄手皮100克（约15张）　鸡蛋清1个　干淀粉30克　姜葱水50毫升　姜米5克　葱花10克　花椒粉0.5克　鸡精3克　味精2克　食盐8克　特制豆瓣酱50克

成菜特点

色泽棕红、皮薄馅嫩、酱香微辣。

制作工艺

1. 肉末入盆，分两次加入姜葱水将肉末搅散，再依次加入鸡蛋清、姜米、食盐、干淀粉搅拌均匀成馅心。
2. 在抄手皮上放入馅心，对叠成三角形，再将左右两个角尖向中折叠粘合成菱角形，制成抄手生坯。
3. 将特制豆瓣酱、花椒粉、鸡精、味精放入碗中调成味料。
4. 锅置火上，入清水烧沸，下抄手生坯煮至面皮发亮、起皱时，捞出沥干余水，放入装有味料的碗中，撒上葱花即成。

评鉴

豆瓣抄手成功的秘诀是特制豆瓣酱，它是由两种酿造年份不同的郫县豆瓣与豆豉、秘制香料粉、白糖、芝麻等一起炒制而成。此小吃用它调味，色泽红亮、酱香微辣，具有浓郁的郫都特色，是一道非常美味的面食小吃，曾荣获“成都市知名创新小吃”称号，还得到四川电视台等多家媒体的重点推介。

Ingredients

100g pork mince; 100g wonton wrapper (about 15 pieces); 1 egg white; 30g cornstarch; 50ml scallion-ginger juice; 5g ginger, finely chopped; 10g scallion, finely chopped; 0.5g ground Sichuan pepper; 3g chicken essence; 2g MSG; 8g salt; 50g specially-made chili bean sauce

Features

red brown in color; thin wrapper with tender filling; slightly spicy with agreeable chili bean sauce aroma

Preparation

1. Put the pork mince in a big bowl, and add the scallion-ginger juice twice, whisking thoroughly before adding egg white, ginger, salt and corn starch to make the filling.
2. Put the filling into the center of the wonton wrapper and fold into a triangle shape, then press the right and left edges together to form a water caltrop shaped raw wonton.

豆瓣抄手

Wontons in Chili Bean Sauce

3. Combine in a serving bowl the specially-made chili bean sauce, ground Sichuan pepper, chicken essence and MSG to make the seasoning sauce.
4. Heat water in a pot, bring to a boil, add the wontons and boil till the wrapper gets translucent and wrinkled. Transfer the wontons to the serving bowl.

Reviews

The key to success in wonton cooking is the special bean-based sauce made of two kinds of broad beans harvested in different years in Pidu District, and fried with fermented beans, special spice powder, sugar and sesame. The special bean paste gives the dish a bright red color and slightly spicy paste aroma, which makes it a unique and delicious local specialty in Pidu District. The dish has been recognized as a famous and innovative cuisine in Chengdu and widely covered by multiple medias including Sichuan TV.

玉笼轩芙蓉包

Yulongxuan Hibiscus Buns

食材配方

五花肉末500克　面粉500克　老面50克　白糖20克　碎米芽菜180克　葱花70克　食盐3克　老抽2毫升　姜末5克　鸡精5克　胡椒粉2克　花椒粉2克　秘制酱料30克　料酒10毫升　小苏打5克　清水325毫升　芝麻油10毫升　化猪油120毫升

成菜特点

皮白光滑、皮薄馅多、味道咸鲜、酱香味浓。

制作工艺

1. 将面粉、老面、清水、白糖调制成团，待其充分发酵后，加入小苏打扎成正碱面团。
2. 锅置火上，入化猪油烧至120℃时，先下五花肉末炒散籽，再下食盐、料酒、老抽、白糖、姜末、胡椒粉、碎米芽菜、鸡精、秘制酱料炒香，待其自然冷却后，加入芝麻油、花椒粉、葱花拌匀制成馅料。
3. 将面团搓成条，下成每个重约13克的剂子，再擀成直径约8厘米的圆皮，包入约22克的馅料，然后捏成16～18个细褶花纹的包子生坯，放入刷有食用油的蒸笼内，用旺火、沸水蒸熟即成。

评鉴

此点是郫都区的特色风味小吃，始创于1927年。玉笼轩芙蓉包采用传统老面发酵工艺，皮薄光滑、馅多鲜香，曾荣获“成都名菜”“舌尖上的天府‘金辣椒奖’”称号。

Ingredients

500g minced pork belly; 500g wheat flour; 50g sourdough; 20g white sugar; 180g yacai (pickled mustard greens, minced); 70g scallion, finely chopped; 3g salt; 2ml dark soy sauce; 5g ginger, finely chopped; 5g chicken essence; 2g pepper; 2g ground Sichuan pepper; 30g secret-recipe sauce; 10ml Shaoxing cooking wine; 5g baking soda; 325ml water; 10ml sesame oil; 120ml lard

Features

white and smooth bun skin; abundant fillings in a thin wrapper; salty and rich in flavor

Preparation

1. Mix wheat flour, sourdough, water and white sugar, blend well and knead into a dough. Set aside to prove thoroughly before adding baking soda.
2. Heat the lard in a wok to 120℃ and add the minced pork, stirring so that the pork pieces do not cling together; add salt, cooking wine, soy sauce, sugar, ginger, pepper, yacai, chicken essence and secret-recipe sauce, stir fry till fragrant, and remove from the heat. Leave to cool, and add sesame oil, ground Sichuan pepper and chopped scallion, stirring well to make the filling
3. Roll and knead the dough into long strips and cut into portions of around 13g each; roll with a rolling pin each dough piece into a round wrapper about 8cm in diameter. Wrap 22g fillings for each bun and pinch 16-18 pleats on top. Place the buns into steamer basket slightly brushed with oil at the bottom and steam on high heat until fully cooked.

Reviews

As a specialty of Pidu District, Yulongxuan Hibiscus Buns were created in 1927, using sourdough preferments. The wrapper is thin and smooth, filled with aromatic and tasty fillings. The dish has won the honor of Chengdu Specialty Dish and "Golden Chilli Award".

食材配方

面粉1 000克　酵母10克　泡打粉10克　白糖20克　猪肉末500克　韭菜碎500克　食盐15克　花椒粉15克　酱油5克　鸡蛋2个　淀粉50克　芝麻油10毫升　姜米30克　鸡精15克　味精10克　温水560毫升　清水50毫升

成菜特点

形状美观、面薄韭香、咸鲜滋润。

制作工艺

1. 酵母用少量温水泡开，加入面粉、泡打粉和温水调匀揉成面团，静置醒发1小时。
2. 将猪肉末、花椒粉、食盐、酱油、鸡精、味精、鸡蛋、淀粉、姜米、清水调匀，再加入韭菜碎、芝麻油搅拌均匀，制成韭菜肉馅。
3. 面团揉好后下成剂子，再压成扁饼状，包入韭菜肉馅，制成提花包子生坯，静置醒发15分钟。
4. 将包子生坯放入笼中，入蒸锅内用大火蒸10分钟至熟即成。

评鉴

韭菜又名长生韭，寓意长久、长寿，在郫都区的种植历史很悠久，其中，唐元韭黄是国家地理标志产品，也是远近闻名的郫都区特产蔬菜。由于韭菜具有特殊的强烈气味，用于面食中，大多制作成韭菜馅的饺子，用韭菜做包子则较为少见，这一打破传统的做法，却没让人失望，由于其皮薄汁多、味道独特而受到人们的喜爱。

Ingredients

1,000g wheat flour; 10g yeast; 10g baking soda; 20g white sugar; 500g pork mince; 500g chives, finely chopped; 15g salt; 15g ground Sichuan pepper; 5ml soy sauce; 2 eggs; 50g cornstarch; 10ml sesame oil; 30g ginger, finely chopped; 15g chicken essence; 10g MSG; 560ml warm water; 50ml water

Features

appealing look; thin wrapper with aromatic and tasty chive fillings

Preparation

1. Mix yeast with some water water, blend well and whisk in wheat flour, baking powder and warm water, stirring well; knead the dough for a while and let stand for one hour.
2. Mix well pork mince, ground Sichuan pepper, salt, soy sauce, chicken essence, MSG, eggs, cornstarch, chopped ginger and water before adding Chinese chives and sesame oil; blend well to make the fillings.

韭菜包子 Chive Buns

3 Knead and divide the dough into small portions; press and roll each portion flat, wrap up the fillings, and let rise for fifteen minutes.

4. Put the raw buns into a steamer and steam over high heat for ten minutes.

Reviews

Chives have been grown and harvested in Pidu District since long time ago and the Chinese chives of Tangyuan are on the list of geological indications and a widely renowned specialty vegetable of Pidu District. Chinese chives are also known as long-life vegetable, for its pronunciation in Chinese is a homonym to eternity and longevity. Since the chive has strong aromas, it's usually used in dumplings rather than in buns. This dish is ingeniously innovative and very popular for its unique and juicy flavor.

蒜泥干拌抄手

Wontons in Garlic Sauce

食材配方

猪前夹肉100克　抄手皮120克　姜葱水10毫升　胡椒粉0.2克　食盐3克　芝麻油4毫升　味精2克　白糖20克　酱油10毫升　料酒5毫升　辣椒油60毫升　蒜泥30克　花生碎10克　鲜汤50毫升

成菜特点

色泽红亮、皮薄馅嫩、咸鲜甜辣、蒜香浓郁。

制作工艺

1. 将猪前夹肉剁细成肉糜，加入姜葱水、食盐（1. 5克）、胡椒粉（0.1克）、料酒搅打至肉糜黏稠起胶，再分次加入鲜汤搅打至肉糜松散，然后加入芝麻油（2毫升）拌匀，制成水打馅心。
2. 在抄手皮上放入馅心，对叠成三角形，再将左右两个角尖向中折叠粘合成菱角形，制成抄手生坯。
3. 将食盐（1克）、蒜泥、白糖、味精、酱油、辣椒油、芝麻油（2毫升）调成蒜泥味汁。
4. 锅置火上，入清水烧沸，下抄手生坯煮至面皮发亮、起皱时捞出，沥干余水后放入装有调味汁的碗中，撒上花生碎即成。

评鉴

蒜泥干拌抄手是唐昌镇“杨抄手”最具代表性的品种。抄手即馄饨，四川各地均有制作，但大多为清汤、红汤两大品类，汤汁较多。本款抄手采用蒜泥干拌的形式，虽不太常见，却风味独具，因而深受当地民众和外地食客的喜爱。

Ingredients

100g pork (front upper leg cut); 120g wonton wrapper; 10ml scallion-ginger juice; 0.2g pepper; 3g salt; 4ml sesame oil; 2g MSG; 20g white sugar; 10ml soy sauce; 5ml Shaoxing cooking wine; 60ml chilli oil; 30g garlic, crushed; 10g crispy peanuts, finely chopped; 50ml stock.

Features

bright red in color; thin wrapper with tender filling; sweet and spicy in taste with garlic aromas

Preparation

1. Finely chop the pork, place in a big bowl, and add the scallion-ginger juice, 1.5g salt, 0.1g pepper and cooking wine, whisking until the mixture becomes gelatinous. Add the stock in several portions and continue to whisk until the mixture is loose. Add 2ml sesame oil and blend well to make the filling.
2. Put the filling into the center of wonton wrapper and fold into a triangle shape, then press the right and left edges together to form a water caltrop shaped raw wonton..
3. Combine in a serving bowl 1g salt, garlic, sugar, soy sauce, chilli oil and 2ml sesame oil, stirring well to make the garlic sauce.
4. Heat water in a pot, bring to a boil, add the wontons and boil till the wrapper gets translucent and wrinkled. Transfer the wontons to the serving bowl, and sprinkle with peanuts.

Reviews

Wontons in Garlic Sauceis a signature dish in Yang's wonton restaurant in Tangchang Town. Wonton, known as chaoshou in Sichuan, is widely available in the whole Sichuan region, but mostly served with consomme or spicy red soups. This dish is rare as it's served without soup, but its unique flavor is very appealing to both locals and tourists.

大椿巷葱葱卷

Assorted Vegetable Rolls in Dachun Alley

食材配方

春卷皮500克　青笋300克　胡萝卜300克　白萝卜100克
食盐5克　酱油50毫升　醋20毫升　白糖30克　味精2克
辣椒油100毫升　花椒粉1克　芥末油10毫升　芝麻油10毫升

成菜特点

皮薄色白、软中带脆、麻辣甜酸。

制作工艺

1. 青笋、胡萝卜、白萝卜分别切长约10厘米、粗约0.2厘米见方的细丝，用冷开水淘洗一下，晾干水分备用。
2. 用食盐、酱油、醋、白糖、味精、辣椒油、花椒粉、芥末油、芝麻油调匀成味汁装入味碟中。
3.将青笋丝、胡萝卜丝、白萝卜丝拌匀，先用春卷皮包裹成直径约1. 5厘米粗的长卷形，再切成长约8厘米的段，装盘后配上味碟即成。

评鉴

葱葱卷是用薄面皮包卷上三种素菜丝，再蘸取味汁食用的一种风味小吃，类似于“淋味春卷”。唐昌镇大椿巷的葱葱卷，因特色鲜明而远近闻名。

Ingredients

500g spring roll wrappers; 300g asparagus lettuce; 300g carrot; 100g white radish; 5g salt; 50ml soy sauce; 20ml vinegar; 30g white sugar; 2g MSG; 100ml chilli oil; 1g ground Sichuan pepper; 10ml mustard oil; 10ml sesame oil.

Features

white and thin skin; soft, tender and crispy texture; spicy, sweet and sour in taste

Preparation

1. Cut the lettuce, carrot and radish into 10cm long, 0.2cm thick slivers, rinse in cold boiled water, and drain.
2. Combine salt, soy sauce, vinegar, sugar, MSG, chilli oil, ground Sichuan pepper, mustard oil and sesame oil in a bowl, stir well to make the seasoning sauce, and transfer to a saucer.
3. Blend well the three kinds of slivers, place in a wrapper, and roll up into a cylinder of 1.5cm in diameter.Cut into into 8cm long rolls, put on a serving plate and serve with the saucer.

Reviews

Assorted Vegetable Rolls is a dish made of vegetable slivers wrapped up in thin wrappers and served with the dipping sauce. The rolls made in Dachun Alley of Tangchang Town is famous for its thin and soft wrapper, tender and crispy texture, spicy, sweet and sour taste.

食材配方

米粉1 500克　嫩玉米粒3 500克　白糖750克　泡打粉100克　玉米叶足量

成菜特点

色泽金黄、质地绵软、入口化渣、味道香甜。

制作工艺

1. 嫩玉米粒洗净后加入清水打制成玉米浆；米粉用清水浸泡待用。
2. 将浸泡好的米粉和玉米浆加入白糖、泡打粉搅拌均匀，静置醒发30分钟待用。
3. 将醒发好的玉米粉用玉米叶包制成圆筒形，入笼蒸制40分钟后取出即成。

玉米馍馍
Steamed Corn Buns

评鉴

玉米馍馍是流行于郫都民间的一道粗粮小吃，当地很多农家都能制作，因其口感绵软、风味别具而受到郫都当地民众和外地游客的青睐。

Ingredients

1,500g rice flour; 3,500g tender corn kernels; 750g white sugar; 100g baking powder; sufficient corn husks

Features

golden in color; soft in texture; sweet in taste

Preparation

1. Wash corn kernels clean and put them in a blender. Add water and blend into pulp. Soak rice flour in water.
2. In a basin, mix well-soaked rice flour, corn pulp, white sugar and baking powder. Set for thirty minutes.
3. Scoop an appropriate amount of mixture onto a corn husk and wrap into a cylindrical shape. Repeat the process with the remaining mixture and corn husks. Place into a steamer and steam for forty minutes.

Reviews

Steamed Corn Buns are a healthy coarse grain snack and a specialty in the countryside. The bun looks light golden, tastes spongily soft and palatably sweet. It is favored by men and women of all ages living either in or out of Pidu District.

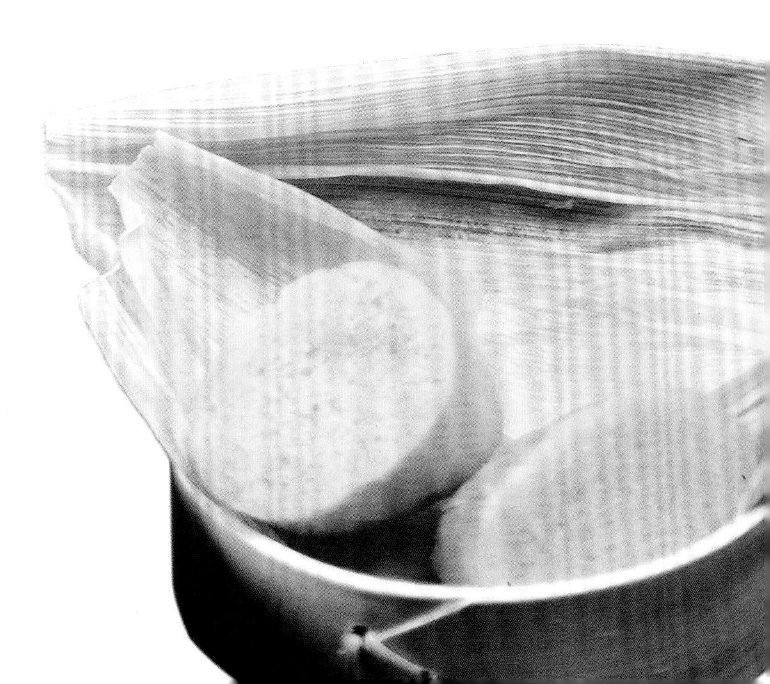

豆浆馍馍

Steamed Green Soybean Buns

食材配方

米粉2 000克　青豆3 000克　白糖750克　泡打粉100克　桑叶足量

成菜特点

色泽青绿、质地绵软、入口化渣、豆香清甜。

制作工艺

1. 米粉用清水泡发制成米浆。
2. 将青豆打磨成浆，加入米浆、泡打粉、白糖和匀，静置醒发30分钟，用桑叶包制成三角形，入笼蒸制40分钟即成。

评鉴

豆浆馍馍是流行于郫都区的传统民间小吃，每逢青豆成熟季节，农村中的很多人家，都会用刚刚上市的青豆做成豆浆馍馍来尝新，既是居家常食之物，也是当季宴请亲朋好友的佳品。

Ingredients

2,000g rice flour; 3,000g green soybeans; 750g white sugar; 100g baking powder; sufficient mulberry leaves

Features

green in color; soft in texture; sweet in flavor

Preparation

1. Add water to rice flour, and leave to ferment to make rice pulp.
2. Grind green soybeans into soy milk. Mix soy milk with rice pulp, baking powder and white sugar thoroughly. Set aside for thirty minutes. Wrap an appropriate amount of mixture with mulberry leaves into a triangle shape. Repeat to finish all the mixture. Place into a steamer, and steam for forty minutes.

Reviews

Steamed Green Soybean Buns are a traditional folk snack in Pidu District. Almost every peasant family will make this bun in the harvest season of green soybeans every year. The appealing green color, spongily soft taste and fragrantly sweet flavor have made it a local delicacy to share with families and guests.

食材配方

荞麦面粉100克　食盐2克　酱油10毫升　醋10毫升　味精1克　芥末油1毫升　辣椒油20毫升　芝麻油5毫升　鲜汤20毫升　葱花5克　清水40毫升

成菜特点

色泽红亮、面质滑爽、酸辣可口。

制作工艺

1. 将食盐、酱油、醋、味精、芥末油、辣椒油、芝麻油、鲜汤入碗调匀成味汁。
2. 荞麦面粉用清水调成面团，放入底部带有小孔的专用木制模具中压制成荞麦面条，直接落入锅中煮熟后捞出，放入装有味汁的碗中，撒上葱花即成。

评鉴

荞麦面条为现场手工制作，因压制而成的荞麦面条直接落入沸水中，犹如面花入水，故而得名荞花面。荞麦富含膳食纤维和铁、锰、锌等多种人体所需的微量元素，具有降糖、降脂、降胆固醇和抗衰等食疗作用。

Ingredients

100g buckwheat flour; 2g salt; 10ml soy sauce; 10ml vinegar; 1g MSG; 1ml mustard oil; 20ml chili oil; 5ml sesame oil; 20ml stock; 5g scallion,finely chopped; 40ml water

Features

bright red in color, smooth in texture, sour and spicy in flavor

Preparation

1. In a serving bowl, mix well salt, soy sauce, vinegar, MSG, mustard oil, chili oil, sesame oil and stock.
2. Mix buckwheat flour with water. Combine and knead into a dough. Put the dough in a special wooden mold. Press the dough through the small holes on the bottom of the mold and the noodles will fall directly into the pot with boiling water. Get cooked noodles out. Transfer to the serving bowl. Sprinkle chopped scallion on top and serve.

Reviews

The buckwheat noodles are handmade on site. When the noodles drop into the pot, they will scatter and swirl in boiling water, which looks like flowers in bloom. This is why people name the dish “Blooming Buckwheat Noodles”.

酸辣荞花面
Sour and Spicy Blooming
Buckwheat Noodles

酥肉豆花

Tofu Pudding with Crispy Pork Topping

食材配方

黄豆500克　石膏粉35克　酥肉200克　馓子200克　酥花生200克　酥黄豆200克　海带200克　大头菜粒200克　葱花200克　食盐30克　辣椒油300毫升　味精20克　花椒粉3克　芝麻油20毫升　红苕水淀粉200克　清水3 500毫升

成菜特点

色泽洁白、质地细嫩、麻辣咸鲜、酥糯兼备。

制作工艺

1. 黄豆用清水浸泡至透，去皮后洗净，用石磨加清水（3 000毫升）磨成豆浆，用纱布滤浆取汁。
2. 石膏粉用水溶化；酥肉切成长约4厘米、粗约0.5厘米的条；馓子加工成长约4厘米的段；海带煮熟，切成长约4厘米、宽约0.4厘米的丝。
3. 锅置火上，入清水500毫升烧沸，慢慢倒入红苕水淀粉搅匀至熟，制成稀糊状的卤汁。
4. 锅置火上，先入豆浆烧沸，再入石膏水搅匀，加盖后关火，使其凝固成豆花。
5. 食用时，先将卤汁舀入碗中，再舀入豆花，然后放入食盐、辣椒油、味精、花椒粉、芝麻油、酥肉、馓子、酥花生、酥黄豆、海带、大头菜粒、葱花即成。

评鉴

酥肉豆花是流行于郫都区的一道著名小吃，色白、质嫩的豆花与酥肉相配，外酥内糯，再辅之以黄豆、花生、馓子的脆生酥香，味道更浓厚，口感更丰富。

Ingredients

500g soybeans; 35g gypsum powder; 200g crispy fried pork; 200g sanzi (deep-fried twisted noodles); 200g crispy peanuts; 200g crispy soybeans; 200g kelp; 200g preserved kohlrabi, finely chopped; 200g scallion,finely chopped; 30g salt; 300ml chili oil; 20g MSG; 3g ground Sichuan pepper; 20ml sesame oil; 200g sweet potato starch batter; 3,500ml water

Features

snow-white in color; tender, crispy and glutinous in texture; spicy and savory in taste

Preparation

1. Soak soybeans thoroughly in water. Rub to peel and rinse. Place peeled soybeans into a stone mill, add 3,000ml clean water, and grind into soy milk. Filter soy milk with a gauze.
2. Dissolve gypsum powder in water. Cut the crispy fried pork into strips with a length of 4cm and a thickness of 0.5cm; Break sanzi into segments of 4cm long; Cut kelp into slivers of 4cm long and 0.4cm thick.
3. Heat a pot over stove fire. Pour 500ml water into the pot and heat to a boil. Slowly pour in the sweet potato starch batter and stir well to make a thin paste.
4. Bring a pot of soy milk to a boil. Add gypsum mixture and stir well. Cover the pot and turn off the fire. Wait for the soy milk to solidify into tofu pudding
5. Ladle seasoned broth into a serving bowl, then the bean curd. Add condiments and other ingredients: salt, chili oil, MSG, ground Sichuan pepper, sesame oil, crispy fried pork, sanzi, crispy peanuts, crispy soybeans, kelp, kohlrabi, scallion. Ready to serve.

Reviews

Tofu Pudding with Crispy Pork Topping is a famous street food created in Pidu District. The tenderness of bean curd, the complex texture of crispiness and glutinousness from crispy fried pork, together with the crunchiness and savoriness of soybeans, peanuts and sanzi, endow this dish with a strong flavor and a rich taste.

天鹅蛋

食材配方

糯米粉150克　澄粉25克　白糖60克　红糖60克　白芝麻20克　清水60毫升　菜籽油2 000毫升（约耗100毫升）

成菜特点

色泽金黄、皮酥内糯、味道香甜。

制作工艺

1. 将澄粉用95℃的水温烫揉均匀，加入糯米粉、清水、白糖揉匀成团，再搓成长条，下成每个重60克的剂子，搓圆后裹上白芝麻制成生坯。
2. 锅置火上，入菜籽油烧至150℃时，先下生坯炸至浮面，再下红糖，用炒勺不断拨动，直到炸至体积极度膨胀、色呈金黄后捞出即成。

评鉴

这里所说的“天鹅蛋”，在成都等地又叫“糖油果子”，因形似天鹅蛋而得名，寓意吉祥、幸福和安康。

Ingredients

150g glutinous rice flour; 25g wheat starch; 60g white sugar; 60g brown sugar; 20g white sesames; 60ml water; 2,000ml rapeseed oil (about 100ml to be consumed).

Features

golden brown in color; crispy outside and hollow inside; sweet in taste

Preparation

1. Add hot water at 95℃ to wheat starch and stir thoroughly. Add glutinous rice flour, water and white sugar to wheat starch mixture. Knead into a dough. Shape the dough into a long bar and cut into portions. Each potion weighs approximately 60g. Shape each portion into a round ball. Sprinkle dough balls with white sesame seeds.
2. Heat a wok over flame. Add rapeseed oil and heat to 150℃. Deep fry the dough balls until they float on the surface. Add brown sugar and stir constantly with a ladle. Keep frying until the dough balls increase in volume and turn golden brown. Transfer to a serving dish.

Reviews

Swan Eggs , also called "tang you guo zi" (meaning Fried Sweet Rice Balls) in Chengdu, are so named for their resemblance to swan eggs in shape. Besides, since swan eggs are a symbol of auspiciousness, happiness and well-being, people name this snack after it for the good connotations. Swan Eggs are golden brown in color, crispy on the outside and chewy on the inside, sweet and delicious in taste, and are suitable for all ages.

食材配方

糯米1 000克　红豆300克　花椒3克　食盐20克　生菜籽油50毫升　粽叶15张　细绳15根

成菜特点

形似宝塔、椒盐味浓、软糯清香。

制作工艺

1. 将糯米、红豆分别用清水浸泡12小时；粽叶洗净。
2. 将浸泡好的糯米、红豆洗净后沥干多余水分，加入食盐、花椒、生菜籽油拌匀成粽子生料。
3. 将粽子生料放入粽叶中，逐个包成四角宝塔形，用细绳缠扎紧实，放入锅中用清水煮一个半小时，关火后再焖20分钟出锅即成。

评鉴

粽子是中国端午节的传统食品，现已扩展为日常美食。郫都区唐昌镇金沙村号称“粽子西施”的故乡，据传，很早以前，男子们外出劳作，女子们就在家集体包粽子卖给远近的客商，由此成为闻名的粽子村。金沙村的粽子品种丰富、味道香甜、质地软糯，大量销往郫都城区及成都等地。

Ingredients

1,000g glutinous rice; 300g red beans; 3g Sichuan pepper; 20g salt; 50ml raw rapeseed oil; 15 pieces of reed leaves; 15 strings

Features

pagoda-like in shape; numbing and salty in taste; soft and glutinous in texture; aromatic in smell

Preparation

1. Presoak glutinous rice and red beans separately in clean water for a whole night. Rinse the reed leaves.
2. Rinse and drain the well-soaked glutinous rice and red beans. Add salt, Sichuan pepper and raw rapeseed oil. Mix well.
3. Scoop up an appropriate amount of mixed ingredients onto reed leaves, wrap into a quadrangle-based pagoda shape, and tie up with strings to make zongzi. Repeat the above process to finish all the ingredients. Put zongzi into a pot and boil for one and a half hours. Turn off the fire and leave on the stove for another twenty minutes. Transfer to a serving plate.

金沙粽子 Jinsha Zongzi

Reviews

Zongzi, the traditional iconic Dragon Boat Festival food, is now a common delicacy for daily consumption. Jinsha Village, located in Tangchang Town, Pidu District, is known as the hometown of "Zongzi Xishi" ("Zongzi Belle"). It is said that in early years local men left their homes in search of employment, while women stayed in the village and gathered together to make zongzi and sell them to guests and vendors near and far. This has made Jinsha Village renowned for Zongzi. Jinsha Zongzi is rich in variety, sweet in taste, and soft and glutinous in texture. It has been sold in large quantities on markets both in Pidu District and other areas.

第五章 CHAPTER FIVE

美食景观 沉浸体验

The Immersion Experience of Gourmet Attractions

郫都区饮食文化底蕴深厚，美食品种丰富，产业集中，文旅融合十分紧密。这里是中国农家乐旅游的发源地，农家乐、乡村酒店遍地开花，乡村美食休闲游如火如荼；这里特色美食琳琅满目，美食街区竞相涌现，现代时尚与古朴典雅融合共生，街镇美食品鉴游方兴未艾；这里拥有中国唯一的川菜产业城，中国川菜博览馆、中国・川菜文化体验馆、川菜博物馆交相辉映，川菜产业发展与文化创新传承相得益彰，川菜文化体验游蓬勃发展。郫都区将美食与旅游完美融合，能在这里亲身体验沉浸式的深度休闲，更是令人身心陶醉、流连忘返。

The food culture in Pidu has profound foundation. There are various types of gourmet food and concentrated industries. The culture and tourism are closely integrated. Pidu is the cradle for Chinese happy farmhouse tourism. Happy farmhouses and rural hotels flourish everywhere as recreational travels for garden gourmet food grow vigorously. The specialties are dazzling here with emerging food blocks combining modern and classical features. Food tasting tour in the streets and towns is in the ascendant. Owning the unique Sichuan Cuisine Industrial City, Pidu has witnessed Sichuan Cuisine Exhibition Hall of China, China • Sichuan Cuisine Cultural Experience Museum and Sichuan Cuisine Museum adding radiance to each other, Sichuan cuisine industry development and cultural innovation complementing each other and cultural experience tourism of Sichuan cuisine prospering. The deep integration of gourmet food and tourism in Pidu marks a terrific immersion experience of gourmet attractions.

第一节 乡村美食休闲游

Section One Recreational Travels for Rural Gourmet Food

郫都区是中国农家乐的发源地。这里诞生了中国第一家农家乐——徐家大院，开创了当今中国乡村美食休闲游的先河，造就了全国农家乐发展的“郫都模式”。所谓农家乐，是指在承袭古蜀先民喜宴乐游的风俗基础上，为满足乡村游乐市场的需求，而将乡村美食与游乐相互融合的一种新型消费形态。“吃农家饭、住农家屋、干农家活、享农家乐”成为民众体验农家生活的主要目的。如今，郫都区农家乐的规模化、品牌化、标准化水平正日益提高，其转型、升级步伐也持续加快，高品质、综合性乡村休闲度假区——“乡村酒店”

不断兴起。总之，郫都区乡村美食休闲游的持续繁荣，为保护、传承乡村文明，促进郫都美食的产业发展，推动乡村振兴都发挥着越来越重要的作用。

郫都农家乐的乡村美食（陈　燕/摄）

Pidu District is the birthplace of Chinese happy farmhouse. The first Chinese happy farmhouse, the Courtyard of the Xu's, was born here, creating the garden gourmet food and rural tourism and forming "Pidu Model" for the development of happy farmhouses nationwide. The so-called happy farmhouse is a new consumption product integrating rural gourmet food with tourism to satisfy demands for village recreation market based on the previous custom that forefathers in Ba and Shu enjoyed banquets and recreation. It is the main purpose for people who experience the happy farmhouse to "eat homemade cook, live in the farmer's hut, live their lives and enjoy natural fun". So far, the happy farmhouse in Pidu has been larger in scale, brand-oriented, standardized and modernized. At the same time, its transformation and upgrading pick up momentum. The high quality and comprehensive rural leisure resorts, namely rural hotels, continue to spring up. The recreational travels for rural gourmet food in Pidu District are growing, which play an important role to protect and inherit the rural civilization as well as promote development of the Pidu food industry and rural revitalization.

一、郫都农家乐的发展历程及影响

（一）郫都区农家乐的发展历程

（1）第一阶段（1985年～1995年）：农家乐起步期

这一阶段的特点是郫都当地农民自发创办农家乐，由此开启了一种乡村餐饮的全新模式。在20世纪80年代，当全国大力发展粮食生产、破解农民温饱难题的时候，成都郫都区农科村积极发展花卉产业，形成了“一户一品，一品一景”的鲜明特色，成为远近闻名的花卉种植专业村。当时的农科村经常接待前来挑选苗木、花卉的外地客商，不少成都市民和游客也喜欢到此游玩、赏花，当地村民从中发现了难得的商机，于是纷纷在自家院落为远到的客人提供餐食和游乐场所。徐纪元就是其中之一，他从花木生意起家，并在平常热情的留客吃饭中得到启发，于1986年在位于郫都区友爱镇、汉代大儒扬雄故里的农科村创办了“徐家大院”，率先推出农家饮食与乡村旅游相结合的新型餐饮形式，由此诞生了“中国农家乐第一家”，其他村民纷纷效仿，从此开始，农家乐逐渐在郫都区农科村遍地开花。

I. The Development Course and Influence of Pidu's Happy Farmhouses

1. The Development Course of Pidu's Happy Farmhouses

The First Stage (1985-1995): Initial Period of the Happy Farmhouses

At this stage, peasants of Pidu spontaneously started the happy farmhouse, creating a new model of catering. In the 1980s, when the whole nation strenuously developed crop production in order to address hunger problems of peasants, Nongke Village of Pidu District energetically expanded the flower industry, featured by "one household with one distinctive product which displays one beautiful scene". It has become a specialized village for planting flowers known far and wide. At that time, traveling merchants always came to Nongke Village to select nursery stocks. Quite a few Chengdu citizens and tourists also enjoyed themselves and flowers there at weekends. Peasants in Pidu discovered the commercial opportunities and provided catering and recreation place in their own courtyards. Xu Jiyuan was one of them. He started to do vegetation business and then found opportunities when inviting customers to stay for dishes. In 1986, he created the Courtyard of the Xu's in Nongke Village, Youai Town of Pidu District, which is also the hometown of Yang Xiong. The new catering model combining homemade cook with rural tourism was firstly promoted. Thus "the first happy farmhouse in China" came into being. Later, happy farmhouses gradually increased as other villagers imitated Xu's practice.

这一阶段，郫都村民通过利用自家的宅基地和承包地种植花卉、苗木营造环境，同时改造自家房屋、增加简单的接待服务设施，从事农家乐经营活动。“吃农家饭、住农家屋、观农家景、购农家物、干农家活、享农家乐”成为农家乐的主要形式，也成为许多游客选择郊野休闲的主要方式，并逐渐形成一种饮食消费热潮。此时，郫都区比较具有代表性的农家乐大都集中在友爱镇农科村和花园镇筒春村等地。

At this stage, peasants in Pidu operated the business by making use of homesteads, creating good environment by planting flowers and nursery stocks in contracted lands, renovating their own houses and adding simple reception facilities. It is the main form of happy farmhouse to "eat homemade cook, live in the farmer's hut, view the beautiful rural scenery, buy agricultural products, experience peasants' lives and enjoy natural fun". It is also a main method for numerous people's suburban recreation and gradually becomes a fashion of diet consumption. At this time, the typical happy farmhouses in Pidu District concentrated at Nongke Village of Youai Town and Tongchun Village of Huayuan Town.

2. 第二阶段（1996年～2005年）：农家乐蓬勃发展期

这一阶段的特点是郫都农家乐作为当地的特色产业得到良好培育，并且延伸出现了乡村酒店，促进了乡村旅游和农民增收。此时，在政府的大力扶持、引导和规范之下，郫都区农家乐的规模和数量不断扩大，软硬条件不断改善，农家的自然环境更加美化，配套设施日趋完善，经营能力和服务水平也得到很大提高。一些农家乐还把瓦房改建为楼房，将经营范围从农家饮食向住宿延伸，由此产生了为数不少的乡村酒店，并与农家乐形成优势互补的格局。这一时期，全区共有近10个农家旅游重点村，农家乐发展到500家，仅农科村开办的农家乐就近百家，对促进乡村旅游、社会经济发展和农民增收起到了极大的推动作用。

The Second Stage (1996-2005): Booming Period of the Happy Farmhouses

The happy farmhouse in Pidu was well cultivated as local featured industry at this stage and rural hotels emerged which helped promote rural tourism and increase peasants' incomes. At that moment, since the government regulated and guided the industry tirelessly, happy farmhouses in Pidu continually expanded in terms of scale and number, increasingly upgraded their infrastructures, beautified the natural environment, improved their supporting facilities and greatly raised their operation capability and service. Meanwhile, some happy farmhouses replaced tile-roofed houses by storied houses and expanded their operation scales from farmhouse catering and tourism to accommodation. And rural hotels were therefore born and complemented each other's advantages with happy farmhouses. During the period, nearly 10 key villages of rural tourism and 500 happy farmhouses were established. There

郫都区红光街道白云村千人坝坝宴 （杨 健/摄）

were nearly 100 happy farmhouses in Nongke Village. All of these enormously propelled rural tourism, economic and social development as well as increase of peasants' income.

3. 第三阶段（2006 年～2010年）：农家乐快速提升期

这一阶段的特点是外来企业积极进入，与郫都农家乐、乡村酒店共同努力，推动郫都农家乐产业不断提升和乡村旅游发展。此时，以“鹿野苑”“梦桐泉”等为代表的外来企业先后进入郫都，建立了主题乡村度假酒店，而传统的农家乐代表，如徐家大院、刘家大院、泰和园、竹里湾等则不断扩大规模，完善内部功能，加快产品升级换代，向特色化、精品化、标准化方向发展，并且启动现代农业观光旅游项目，促进农家乐和乡村旅游进入到快速发展期。

The Third Stage (2006-2010): Rapid Progress Period of the Happy Farmhouses

In this time, external enterprises entered and they joined hands with happy farmhouses of Pidu and rural hotels to accelerate the continual improvement of the happy farmhouse industry and the development of rural tourism. The external enterprises represented with “Luyeyuan” and “Mengtongquan” entered Pidu one after another and established the theme rural resort hotels. The traditional happy farmhouse like the Courtyard of the Xu's, the Courtyard of the Liu's, Tiaheyuan Restaurant and Zhuliwan Hotel expanded the scale, improved internal functions, quickly upgraded products and developed towards feature, good quality and standard. They launched modern agriculture sightseeing tourism project to drive happy farmhouses and rural tourism into a rapid progress stage.

4. 第四阶段（2011年至今）：农家乐转型升级期

这一阶段的特点是涌现出大量的高档次精品农家乐、星级乡村酒店，成功创建了AAAA级旅游景区，促进农家乐与乡村旅游走上了转型升级和集团化、组团式发展之路。此时，郫都区针对市场发生的较大变化和人们消费需求的升级而主动应变，在原有农家乐优势资源的基础上，进一步改良、完善和转型升级，政府通过多种举措予以大力引导，从而催生出一批外部自然化、内部品质化的精品主题农家乐，发展了一批生态文化新体验的“野奢乐活”农家乐，推出了一批具有家族文化传承风格的庄园式农家乐，同时不断加大基础设施投入，持续提升景观质量及管理、服务水平，精心打造高品质国际乡村休闲度假旅游区。其中，农科村占地面积2.6平方千米，花卉、苗木种植面积达到154公顷，并于2012年成功创建国家AAAA级旅游景区，到2018年，已拥有乡村酒店40余家，其中星级乡村酒店9家，全村人员

农科村——农家乐发源地（郫都区融媒体中心/供）

中，直接从事乡村旅游的就达到3 000余人，全年共接待游客122.5万人次，旅游收入超过1.2亿元。此外，农科村还先后获得“中国农家乐旅游发源地”“全国农业旅游示范点”“全国文明村”“中国十大最有魅力休闲乡村”等称号。目前，郫都区遍布农家乐、星级酒店，拥有许多著名景区，除农科村外，战旗村、青杠树村等也各有特色和高知名度，全区四星级及以上农家乐和乡村酒店达18家，大大满足了人们对美好生活的追求。

The Fourth Stage (2011-): Transformation and Upgrading Period of the Happy Farmhouses

Numerous high-end happy farmhouses and star-rated rural hotels mushroomed and AAAA tourist attractions were created, which drove the happy farmhouse and rural tourism to transform and upgrade, established groups and formed clusters. Pidu District took proactive actions directed at huge changes in the market and upgrading consumption demands of people. It improved, transformed and upgraded the industry based on resource superiority of the happy farmhouse. By multiple measures, the government led to form and encourage a batch of boutique theme happy farmhouses with magnificent natural environment and internal quality facilities, luxuriously wild lifestyle happy farmhouses with the new and fun experience of ecological culture and manor-styled happy farmhouses inheriting family culture. At the same time, the government inputted more in infrastructure and improved the quality, management and services of attractions to build the international rural resort tourist area of high quality. Nongke Village occupies an area of 2.6 square kilometers with flowers and nursery stocks of 154 hectares, becoming a national AAAA tourist attraction in 2012. In 2018, the village has already owned 40 rural hotels, among which 9 were star-rated. The rural tourism industry directly employed over 3,000 people, received 1.225 million tourists and gained 120 million yuan. Besides, Nongke Village was titled as “Cradle of Happy Farmhouse Tourism in China”, “National Agricultural Tourism Demonstration Site”, “National Civilized Village” and “Top Ten Charming Leisure Village”. At present, happy farmhouses and star-rated hotels spread all over Pidu District. There are many well-known scenic spots besides Nongke Village, like Zhanqi Village and Qinggangshu Village with distinctive features and high reputation. The happy farmhouses and rural hotels above 4-star level amount to 18, which meet the people’s growing demands for a better life.

（二）郫都区农家乐发展的影响与作用

郫都区农家乐的诞生与发展，对推动川菜创新及中国餐饮产业的布局与文化构成，乃至对中国各地的“三农”发展、乡村振兴和旅游振兴等许多方面，都产生了极大的影响和促进作用，这里仅阐述三个方面。

2. The Influence and Function of the Development of Pidu's Happy Farmhouses

The birth and development of Pidu happy farmhouses play an important and active role in Sichuan cuisine and Chinese catering industry and culture, development of agriculture, rural areas and peasants as well as rural revitalization and tourism. Three aspects are to be elaborated here.

1. 郫都区农家乐开创了当今中国新型餐饮与乡村旅游形式的先河

郫都区农家乐的产生与快速发展，造就了中国新型餐饮和乡村旅游的郫都模式，可以说，这一形式的确立，具有里程碑式的意义。由于郫都区农家乐在创新餐饮与旅游形式、促进“三农”发展和农民增收、乡村振兴等多个方面均取得了显著成就，使得成都市、四川省，乃至全国各地的政府、协会、企业等，纷纷组团前来参观、考察、学习、借鉴。在以农科村农家乐为代表的示范带动下，农家乐这种新型模式迅速在成都市、四川省及全国众多地区广泛展开，并且进行了本土化发展，衍生出新的系列，各地不仅出现了众多的农家乐，还发展出彝家乐、藏家乐、羌家乐，以及渔家乐、牧家乐等，乡村酒店（度假村）、体验农庄、观光农业区等也遍地开花，一时之间，全国乡村旅游精彩纷呈。此外，美国、德国、日本、荷兰等四十多个国家的友人也前来参观考察，并进行对比研究、学习和借鉴，在一定程度上丰富了国际乡村旅游的内涵。

1. A New Model: Integrating Catering with Rural Tourism

The birth and rapid development of Pidu happy farmhouses forms “Pidu Model”, a new model integrating catering with rural tourism. Therefore, it is something of a milestone. Pidu happy farmhouses have witnessed the significant progress in innovative catering and tourism model, development of agriculture, rural areas and peasants, increased income of peasants and rural revitalization. Governments, associations and enterprises of Chengdu City, Sichuan Province and even other places of China cluster there for visit and study. Motivated by happy farmhouses of Nongke Village as a success, this new catering and tourism model rapidly expands to other places of Chengdu, Sichuan and even the whole China. The happy farmhouses derive to the new types after localization. Not only happy farmhouses mushroomed, but other types including happy Yi-people's houses, happy Tibetan houses, happy Qiang-people's houses, fish fun houses and happy ranch houses, rural hotels (resorts), farm villages for experience and sightseeing agricultural areas bloomed everywhere. The rural tourism all over the nation is magnificent. What's more, visitors from 40 countries like America, Germany, Japan and Dutch also come here for investigation, comparative study and learning, which enriches the connotation of international rural tourism.

2. 郫都区农家乐有力地促进了“三农”发展和乡村振兴

郫都农家乐的诞生和发展，是川菜与乡村休闲旅游的完美结合。它加快了农产品有效流入市场，延伸了农业产业链，促进了第一、第三产业的有机融合，农民的自有资源得到了最有效的利用，农民自身内在的发展动力也得以极大的发挥，实现了农民就近就业和增收致富。1991年农家乐刚起步不久，郫都区农民人均收入仅800余元。随着农家乐的不断发展和转型升级，消费人群不断增加，到2015年，郫都区农民的人均收入已超过2万元，大大缩小了城乡居民的收入差距。同时，农家乐的转型升级，还推动了农村公共服务设施的改善和环境质量的提升，不仅满足了城市人对生态环境和饮食生活日益增长的需求，也有效促进了“三农”发展和乡村振兴。随着农家乐在全国各地的兴起、推广和不断升级，它已成为中国餐饮与乡村旅游的一道特色风景线，持续推动着农村第一、第三产业的互动发展，引领着中国农业产业结构的调整和乡村振兴发展的时代方向。

2. A Powerful Contributor to the Development of Agriculture, Rural Areas and Farmers as Well as Rural Revitalization

The birth and development of Pidu happy farmhouses is the good combination of Sichuan cuisine and rural recreational tourism. It makes agricultural products to enter in the market more quickly, extends the industrial chain of agriculture and promotes the organic integration of the primary industry and the tertiary industry. The peasants' resources are utilized in a most effective way, the internal development driving force among peasants is given full play and they are able to find jobs nearby and get paid more. In 1991, when the happy farmhouses just started, per capita income of Pidu's peasants amounted to 800 yuan only. As the happy farmhouses developed, transformed and upgraded, the consumption group expanded. By 2015, the per capita income of peasants in Pidu has crossed 20,000 yuan, tremendously narrowing the income gap between urban and rural residents. At the same time, the transformation and upgrading of the happy farmhouses helped improve the public service facilities and environment quality in the rural areas. It not only satisfied the growing demands of citizens for eco-environment and catering, but also accelerated the development and revitalization of rural areas. As the happy farmhouse expands to places all around China and keeps being upgraded, it has become characteristic scenery of China's catering and rural tourism, driving interactive development of the primary and tertiary industry in rural areas. It guides the course in an era for agricultural industry restructuring and rural revitalization.

3. 郫都区农家乐有效地促进了乡土文化的传承和传播

农家乐及其衍生形态，就其本质而言，是农民以自有院落空间为载体，主要以家庭为单位向游客提供餐饮、游乐和住宿等服务。“吃农家饭、住农家屋、观农家景、购农家物、干农家活、享农家乐”不仅是农家乐的主要形式，也是城市民众郊野休闲的主要方式，更是他们了解、体验乡土文化最具吸引力的窗口。随着郫都区农家乐的诞生和不断发展，以及在全国各地出现的农家乐和渔家乐、牧家乐等衍生系列，都吸引了数以万计的城市居民前来消费和体验，在满足他们口福、眼福和身心愉悦同时，也用特别的方式传承、传播了当地的饮食文化和其他乡土文化。可以说，农家乐已成为留住乡愁、传承和发展乡土文化的最好载体。

3. An Effective Boost to the Inheritance and Spreading of Local Culture

In nature, the happy farmhouse and its derived types provide services including catering, recreation and accommodation to tourists with peasants' own courtyards as carriers. It is the main form of happy farmhouse to "eat homemade cook, live in the farmer's hut, view the beautiful rural scenery, buy agricultural products, experience peasants' lives and enjoy natural fun". It is also a main method for numerous citizens' suburban recreation and even the most attractive way for people to experience local culture including food. The birth and development of Pidu happy farmhouses as well as emerging happy farmhouses, fish fun houses and happy ranch houses around the nation attract thousands of citizens for consumption and experience. When satisfying people's taste and view as well as physical and mental pleasure, the local culture including food is inherited and spread in a special way. We can say that the happy farmhouse has become the best carrier to keep one's homesickness and inherit and develop local culture.

（三）郫都区农家乐的产生及发展原因

农家乐发源于郫都区，开创了当今中国新型餐饮与乡村旅游形式的先河，为川菜及中餐发展、乡村振兴都做出了重要贡献。农家乐在郫都区的诞生并非偶然，而是有着深厚的历史渊源，可以说，它是富足优渥的自然环境、千年传承的饮食游乐习俗与川菜包容创新、大众平民的文化基因等多重因素共同作用的结晶。

3. Secrets of the Occurrence and Development of Pidu's Happy Farmhouses

Originating in Pidu, the happy farmhouse has created a novel form of catering and rural tourism, making great contribution to the development of Chinese cuisine, Sichuan cuisine as well as the rural revitalization. It is not an accident. Instead, the abundance of natural environment, the thousand-year-long food culture and enjoyment-valued customs, the inclusiveness and innovation of Sichuan cuisine, as well as the popular gene rooted in the culture together contributed to the development of happy farmhouses.

郫都区乐林艺术村 （李长清/摄）

1. 优越的自然条件

郫都区地处天府之国腹心地带，位于成都市西北部。这里气候宜人，冬无严寒、夏无酷暑、温暖湿润，常年风调雨顺；这里是川派盆景之乡，绿树成林、青草茵茵、花木繁多、争奇斗艳；这里八河并流、土地肥沃，禽、畜、鱼、蔬众多，素有“膏腴”之誉。郫都区四季常青，其优越的自然环境，优美的田园风光，丰富的食材资源，是都市人远离喧嚣，安享休闲时光和乡村美食的理想去处，从而为郫都区农家乐的产生，提供了优越的自然条件和坚实的物质基础。

Superior Natural Conditions

Located at the heart of “the Land of Abundance” and northwest of Chengdu, Pidu District has a mild and humid weather without severe heat or cold. There are so many trees and flowers in Pidu, the home of Sichuan bonsai. Besides, with eight rivers flowing through, Pidu is called “a place of fertility” with countless livestock, poultry, fish, fruits and vegetables. Such advantageous natural environment, beautiful idyllic scenery and abundant food materials have made Pidu an ideal destination for the citizens to get away from the hustle and bustle and enjoy the comfort and leisure as well as the gourmet food in the countryside, providing natural conditions and laying material foundation for the happy farmhouses in Pidu District to take shape.

2. 必要的人文条件

四川人自古以来就崇尚美食和喜好游乐。早在汉晋时期，蜀中宴饮、休闲之风便初见端倪，据汉代成都人扬雄《蜀都赋》载，蜀中宴饮则“置酒乎荥川之闲宅，设座乎华都之高堂，延帷扬幕，接帐连冈”。晋代左思《蜀都赋》也言：“乐饮今夕，一醉累月。”唐宋时期，巴蜀宴饮、游乐之风已闻名全国，游宴、船宴尤为兴盛，苏东坡写有“蜀人游乐不知还，闲适尚以蚕为市”的诗句。直到清代末年，蜀中游宴依然盛行，据傅崇矩《成都通览》载，“成都之筵宴之所”就有城内筵宴、城外筵宴两类。新中国成立后，特别是改革开放以来，随着社会经济的发展，社会和谐安定，人民的生活水平不断提高，在这样的社会环境下，四川人喜好美食和游乐的习俗，得以有效传承和发扬光大，一时之间，享田园之乐、品农家美食，顿时成了当代四川人休闲、游乐与美食结合的一种时尚追求，从而为郫都区农家乐的产生提供了必要的人文条件。

Necessary Cultural Conditions

Since ancient times, the people in Sichuan have learned how to enjoy delicacy and entertaining activities. As early as Han and Jin periods, people in Sichuan began holding feasts. As recorded in the *Story of Chengdu* by Yang Xiong, when people in Sichuan gather for a feast, "they would prepare good wine in a leisure house near the river or a proper hall in the city. Sometimes, they held feasts outdoors with so many guests that the host has to keep adding tables, some of which even reach to the hills nearby. The article with the same title written by Zuo Si in the Jin Dynasty also illustrated that "people who have fun on this one night will feel hangover in the next days or even months." During the Tang and Song periods, the fame of Sichuan for holding feasts has already been known all over the country, among which travelling feasts and boat feasts were the most popular ones. Su Shi, a famous poet in Song Dynasty, once wrote "the people in Sichuan never cease to play. They would sell silkworms in trade fairs held in spring when they are spare". Even when it entered late Qing Dynasty, people in Sichuan still enjoyed their feasts. Fu Chongju wrote in *An Overview of Chengdu*, "the places for people in Chengdu to hold feasts" could be either inside the city or outside the city. Since the foundation of New China, especially after reform and opening up, the economy of China continued to grow. Meanwhile, the society has become more and more harmonious and stable and the living standards of the people kept climbing. Under such circumstance, the tradition of people in Sichuan to enjoy delicacy and amusement has been inherited and carried forward. Realizing such tradition in the countryside, specifically at the farmhouses, has become a fashionable pursuit today, which provides the cultural conditions for the happy farmhouses in Pidu District to come into being.

3. 丰沛的动力源泉

川菜在其漫长的历史发展过程中，逐步形成了大众化、包容性与创新求变的文化属性，从而造就了川菜不断发展、变化的文化基因。自20世纪80年后，随着社会经济的不断发展和人们生活水平的不断提高，越来越多的四川城市民众乐意到乡村和郊野去休闲、游玩，于是产生了吃农家菜的饮食需求。受川菜传统基因的影响，勤劳、智慧的郫都区农民为了满足市民的新需求，充分利用当地的特色食材，采用多种烹饪方式，为远道而来的游客提供了一道道绿色生态、地道朴实、乡土特色突出的农家菜。于是，农家乐在郫都区应运而生，既满足了大众对乡村美食休闲游的多重需求，又极大地丰富了川菜的文化内涵、进一步壮大了川菜产业。可以说，川菜特有的文化属性，为郫都区农家乐的产生提供了丰沛的动力源泉，而郫都区农家乐的产生与发展，也很好地诠释和传承了川菜的三大文化属性。

Abundant Driving Force

The popularization, inclusiveness and innovativeness of Sichuan cuisine were the result of its long-time development, which have been deeply rooted in the cultural gene of Sichuan cuisine. Since the 1980s, as the economy and society continued to develop and the living standards kept being raised, more and more citizens in Sichuan chose the countryside for a place of leisure, creating new needs for food at the farmhouses. Based on the original features of Sichuan cuisine, diligent and talent people of Pidu have made full use of characteristic local materials and applied various cooking methods to provide green, natural, authentic, homely farmhouse food with outstanding rural features for the citizens. As a result, the happy farmhouses were born in Pidu, well satisfying the people's needs for recreational travels for garden gourmet food, enriching the cultural connotation of the Sichuan cuisine and further expanding the industry of Sichuan cuisine. So to say, the Sichuan cuisine's cultural features of popularization, inclusiveness and innovativeness serve as the driving force for the development of Pidu's happy farmhouses, which in turn demonstrates the three major cultural features of Sichuan cuisine.

二、郫都区特色农家乐及乡村酒店纵览

郫都区农家乐及乡村酒店发展至今，数量众多、分布广泛、特色突出，它们共同培植了郫都区餐饮乐园和旅游百花园的绚烂锦绣。畅游其中，小桥流水、鸟语花香、青草如茵、白墙黛瓦，一幕幕优美如画的田园风光让人流连忘返；绿色生态、口味独到、特色鲜明、营养丰富，一道道香气四溢的农家美食令人味蕾绽放；别具一格、服务周到、品质优良、环境温馨，一家家与众不同的乡村酒店让游客宾至如归。到郫都区观农家景、吃农家饭、住农家屋、享农家乐，就是充分享受一次身心愉悦的田园美食休闲之旅。由于篇幅所限，这里仅选取郫都区四星级及以上农家乐、乡村酒店进行简要介绍。

II. An Overview of the Characteristic Happy Farmhouses and Rural Hotels in Pidu District

After so many years of development, there have been numerous happy farmhouses and rural hotels of various kinds with different characteristics all over Pidu District. Such diverse happy farmhouses and rural hotels demonstrate the catering and tourist characteristics of Pidu District, vigorously competing with each other. Walking here, people will find themselves in such a beautiful and tranquil countryside picture with green grass, singing birds, fragrant flowers, small bridges and flowing water, making them even forget to leave. With the motto of quality first, these fully functional farmhouses and hotels have distinctive themes and considerate services, making tourists feel at home, who are lured by the various green products and countryside delicacies with rich nutrition. A tour in Pidu with the finest scenery, delicacies and dwelling experience in the countryside will definitely leave the guests unforgettable memories of an all-round, immersive and extraordinary recreational travels for garden gourmet food. Below are brief introductions to the happy farmhouses and rural hotels rated as 4-star or above.

（一）徐家大院

徐家大院位于郫都区友爱镇国家AAAA级景区——农科村，拥有“中国农家乐第一家”的称号，现已发展成为集餐饮、住宿、商务会议、观光体验旅游为一体的五星级乡村酒店。

1. Courtyard of the Xu’s

Located at the national AAAA scenic area of Nongke Village in Youai Town, Pidu District, the Courtyard of the Xu’s is honored as “The First Happy Farmhouse of China”, which has developed into a 5-star rural hotel integrating functions of catering, accommodation, business meetings, sightseeing and traveling, etc.

徐家大院占地约5.4公顷，拥有130多间格调温馨的精品客房，可容纳1 200人的商务会议楼，以及可同时接待2 000人的农家风味餐厅与阳光餐饮区。酒店建筑以中式庭院、田园风格为主调，环境优美，其建筑风貌与附属景观，共同构成了徐家大院川西园林式的建筑群体。院中浓缩了中国五大盆景流派之特色，种植了近百个花木品种，充分彰显了中国盆景之乡的绚丽多彩。迂曲折回的茶廊与自然环境融为一体，巧妙地烘托出徐家大院“乡村与酒店”互为表里的优美意境。

Covering an area of about 5.4 hectares, the Courtyard of the Xu’s has more than 130 cozy and characteristic hotel rooms, a meeting building that can host 1,200 people and two

徐家大院

dining areas (respectively the countryside-characteristic dining hall and the sunshine dining area) that can receive 2,000 people. Its buildings are mainly Chinese-style yards with countryside characteristics, and the whole hotel together with surrounding decorations form a western-Sichuan garden architecture complex. Within the hotel, there are nearly 100 varieties of flowers and trees with features of the 5 major bonsai genres, demonstrating the gorgeous and varied scenery of the home to Chinese bonsai. The magnificent tea gallery merges with the nature, constituting a beautiful picture of the distinctive combination of "countryside and hotel" of the Courtyard of the Xu's.

徐家大院经历并见证了中国农家乐发展的全部历程，从一栋红砖青瓦房到现今的五星级乡村酒店，先后经过了30余年的奋斗、五代升级发展。第一代农家乐始建于1986年，为徐家老宅——溯源居；第二代农家乐建于1995年，以两层蓝色仿古别墅小洋楼为主，由盆景、松竹、廊院簇拥；第三代农家乐建于2004年，建成的别墅群巧布于川西庭院精美的山水亭台之中；第四代农家乐建于2011年，提档升级为集商务、酒店、会议、餐饮、拓展、休闲、度假、体验为一体的五星级园林式乡村生态酒店；第五代农家乐建于2016年，在川派盆景博览园中建有汇景小院，院中，盆景、园艺错落有致，奇花、草树布局精巧，环境宁静、优雅，院中共建有八个独立主题院落，是川内极为少见的一院一主题、一景一美食的宴饮休闲胜景。

The Courtyard of the Xu's witnessed the development of Chinese happy farmhouse. After more than 30 years' efforts and upgrading for five times, it develops from a single red-brick and gray-green-tile house to a 5-star rural hotel. In 1986, the first-generation of Courtyard of the Xu's was built on the basis of the original old house of the Xu's called Suyuanju. In 1995, the second-generation farmhouse began to be built, which was mainly a small blue 2-story western-style building surrounded by a bonsai garden of pine and bamboo. In 2004, the third-generation farmhouse began to be built, forming a complex of villas scattering in exquisite pavilions of a western-Sichuan style courtyard. In 2011, the farmhouse was upgraded into a 5-star garden-style rural ecological hotel, combining functions of business meeting, accommodation, catering, leisure, vacation, rural experience, etc. In 2016, the newest hotel began to be built. A scenery yard was built in the Sichuan-style bonsai gallery, where the bonsai and rare plants form a beautiful picture. Besides, there are 8 independent themed yards with distinctive sceneries and delicacies, which are pretty rare places for banquets and leisure in Sichuan.

（二）梦桐泉生态酒店

梦桐泉生态酒店位于郫都区红光镇宋家林301号，是一家集会议、住宿、餐饮、运动健身、娱乐、园艺观赏、品茗、度假于一体的五星级乡村酒店。酒店占地约16公顷，园林绿化面积达70%，湖面景观占10%。

2. Mengtongquan Ecological Hotel

Located at No.301, Songjialin Village, Hongguangzhen Town, Pidu District, Mengtongquan Ecological Hotel is a 5-star rural hotel which provides services including meeting, accommodation, catering, sports, gardening sightseeing, tea appreciation, vacation, etc.

酒店餐饮规模庞大，档次齐全，特色鲜明。拥有可容纳800人同时就餐的中餐厅、自助西餐厅和豪华宴会包间。酒店以养生文化为特色，努力开发中华二十四节气菜。酒店菜肴采取川粤结合，不仅有取自当地新鲜食材制作的农家土菜，还有采用雪域高原生态食材烹制的菜肴，如高原牦牛肉、雪域松茸、藏雪莲等，其代表性菜肴有梦桐松鼠鱼、梦桐如意竹荪、青菜焖饭等。

With the greening rate of 70%, the hotel covers an area of about 16 hectares, 10% of which is the lake. The catering area of the hotel is large and characteristic with categories at different consumption level, where there is a Chinese dining hall that is capable of accommodating 800 people at the same time, a western-style buffet dining hall as well as a luxury banquet room. With a health-oriented culture, Mengtongquan Ecological Hotel is working on developing new dishes in accordance with the 24 solar terms. The cuisine of the hotel combines the characteristics of Sichuan and Guangdong cuisine. The guests can not only enjoy fresh and simple countryside dishes made of local materials, but also taste the delicacies made with the original ecological ingredients from the snow-covered plateau, such as Plateau Yak Meat, Plateau Matsutake, Tibetan Saussurea Involcucrata and representative dishes Mengtong Squirrel Fish, Mengtong Bamboo Fungus, Green Vegetables on Rice, etc.

（三）印象泰和园

印象泰和园是郫都区餐饮名店、五星级农家乐，也是成都国际美食之都“餐饮品牌100强”，现有望从路店和太清路店两家门店。

3. Taiheyuan Restaurant

Famous in Pidu District, Taiheyuan Restaurant is a 5-star happy farmhouse and one of the “Top-100 Catering Brand of Chengdu, the Capital of International Gourmet Delicacies”. It has two branches respectively located at Wangcong Road and Taiqing Road.

望从路店始建于1998年，是一家蜀文化主题酒店。该店建筑风格以川西民居为基调，亭台楼阁、雕花木窗、小桥流水、天井回廊、古色古香，充分展示了古蜀文化的地域特性。该店拥有32个不同风格的蜀文化主题包间和宴会厅，可容纳1 000人同时就餐。太清路店始建于2014年，占地约两公顷，交通便利，距离地铁6号线终点站2.5千米，是集餐饮、休闲、婚礼堂于一体的一站式婚礼文化庄园，也是郫都区唯一一家五星级婚礼文化宴会庄园。该店装修个性突出，以东南亚及简欧风、海岛风为主，婚宴设备、设施众多，有高清LED显示屏、梦幻旋转升降舞台、3D动画和一流的灯光音响，可开展新式、古式、中式、西式等不同风格的婚宴接待、模特走秀和各种大型庆典活动。每个宴会厅均配有独立仪式小院和厨房，除传统用餐外，还提供户外自助餐，分餐制，以及泰式、东南亚菜、养生锅、西餐等多元化用餐体验。太清路店是新一代生态乡村旅游餐饮企业的典型代表，菜品特色突出，代表性菜品有八宝鸭、印象豆瓣鱼、家常鳜鱼、芙蓉鲍鱼仔、燕麦煮海参、菌王炒鹅肝等。

Established in 1998, Taiheyuan Restaurant at Wangcong Road is themed with Sichuan culture. The buildings are featured with the characteristics of residential houses in western Sichuan, with the elements like pavilions, terraces, open halls, carved decorations, window paper-cut, small bridges and flowing water, patios, winding corridors, etc., which fully demonstrate the culture of ancient Sichuan. With 32 ancient-Sichuan-culture-themed private rooms and banquet rooms, Taiheyuan can accommodate 1,000 people for dining at the same time. Established in 2014, Taiheyuan Restaurant at Taiqing Road covers an area of about 2 hectares, and is only 2.5 kilometers away from the terminal of Metro Line 6, providing a convenient commuting experience for the guests. As the only 5-star wedding hotel, Taiheyuan Restaurant is designed for one-stop wedding ceremony, with complete functions of catering, leisure, wedding, etc. The decorations of the hotel mainly

印象泰和园

include Southeast Asian style, simple European style and island style, and the wedding halls are all fully equipped with HD LED displays, rotating lifting stages, 3D animation displays and first-class lighting and sound equipment. Therefore, both modern and traditional and both Chinese-style and western-style weddings, catwalk shows and other large-scale celebrations are available here. Each banquet hall is equipped with an independent celebration yard and a kitchen. Besides traditional dining needs, outdoor buffet, separate meals, Thai food, Southeast Asian dishes, healthy diets, western food, etc. are also provided. The Taiqing Road branch is a representative of the new-generation ecological catering businesses combining rural tourism. There are many distinctive dishes at the Taiheyuan Restaurant, among which the representative ones are Duck Stuffed with "Eight Treasures", Chili Bean Paste Flavored Fish, Homemade Mandarin Fish, Baby Abalone with Egg Custard, Sea Cucumber Boiled with Oat, Stir-fried Foie Gras with Mushrooms, etc.

刘氏庄园

（四）刘氏庄园

刘氏庄园位于郫都区农科村迎宾路与东环线交界处，成灌IT快速通道旁，是一家集餐饮美食、游览观赏、休闲娱乐、园艺设计于一体的综合性农家旅游庄园，为五星级农家乐。

4. Liu's Manor

Located at the junction of Yingbin Road and East Loop Line, Nongke Village, Pidu District, near the IT Road of Chengdu-Dujiangyan Expressway, Liu's Manor is a comprehensive rural tourism manor integrating catering, sightseeing, recreation and gardening design into one. It is a five-star happy farmhouse.

园内环境幽静、雅致，清新自然，紫薇盘扎成双龙戏珠篱笆墙，梅花枝条编就玲珑小亭，海棠树丫盘成拱形画廊，金桂飘香，银杏挺立，罗汉松迎宾。园如其名，于幽静中得袭人香气。扬雄文化博览园坐落其中。扬雄塑像矗立于中央，背后为草庐"玄草亭"，园内东边有一长廊为"子云廊"。

The environment is quiet, fresh and natural. The crape myrtle branches are shaped as two dragons frolicking with a pearl to be the wattled wall. The plum flower branches fabricate exquisite pavilions. The cherry-apple tree forks are woven into arched galleries. The laurels emit its fragrance, ginkgoes stand upright and yaccas greet guests. The quiet farm is of sweet odors. Yang Xiong Culture Expo Park settles here. The statue of Yang Xiong stands erect in the center, behind which is a thatched cottage named "Xuancao Pavilion". In the east of the park is a corridor named "Ziyun Corridor".

餐厅古朴典雅，飞檐翘角，金铎叮当，大厅面积近5 000平方米，可同时容纳800人就餐，另有十余个雅间。餐饮特色鲜明，尤以草原土猪系列菜肴最受游客青睐。2005年10月承办了成都世界美食节郫县分会场的千人坝坝宴，备受民众好评。

The catering is quite featured. The decoration of restaurant is elegant with classic simplicity. There are angled cornices and jingling golden bells. The lobby covers nearly 5,000 square meters, which can accommodate 800 people. There are another a dozen of private rooms. The dishes highlight individuality, among which grassland rural pigs series is the most popular. In October, 2005, it held one thousand-people Babayan Feast in Pixian Session of Chengdu International Gourmet Festival.

（五）红星饭店

红星饭店

红星饭店位于郫都区老成灌公路的太清路口，是一家集特色餐饮、休闲娱乐、商务会议于一体的四星级农家乐，先后获得“四川餐饮名店”“成都餐饮百强企业”“郫县餐饮名店”等称号。红星饭店始创于1994年，占地面积约1.3公顷。该店与时俱进，不断创新发展，就餐环境和硬件设施持续改善。餐厅可容纳1 500余人同时就餐，拥有66个雅间和豪华包间，拥有8个可容纳10～20桌的独立宴会厅，若干个3～5桌的包间，以及茶楼、露天溪畔茶坊、露天啤酒广场、露天烤全羊等餐饮经营场所，还拥有两个可容纳百余人的多功能会议室。红星饭店历来重视菜品创新，并为此制订了“每月一创、两季一新”的发展规划，不断推进传统菜品改良和创新研发，重点强化菜品的文化内涵，提高菜品的营养价值，菜点特色非常鲜明。其“明厨亮灶”的经营模式，透明、卫生，顾客可直观点菜，现点现做，代表性菜品有沱砣鱼、老福坛、沱砣肉等。

5. Hongxing Restaurant

Hongxing Restaurant settled at the end of Taiqing Road of the old Chengdu-Dujiangyan Highway. It is a 4-star happy farmhouse integrating specialty catering, recreation and entertainment and business conference. It is titled as “Famous Restaurant in Sichuan”, “Top-100 Enterprise of Chengdu”, “Famous Restaurant in Pixian”, etc. Founded in 1994, Hongxing covered about 1.3 hectares. It advances with the times, commits to innovation and development as well as improves its dining environment and facilities. The restaurant can accommodate 1,500 people at the same time, with 66 private rooms and deluxe rooms, 8 independent banquet halls of 10-20 dining tables and several rooms of 3-5 tables. It also owns tea houses, outdoor tea rooms by the stream, outdoor beer squares and outdoor places to roast whole lamb. There are 2 multifunction conference rooms which can hold more than a hundred people. Hongxing Restaurant has always attached great importance to dish innovation, and formulated the development plan of "creation every month and innovation every two quarters", constantly promoting the improvement of traditional dishes and innovative research and development, focusing on strengthening the cultural connotation of dishes, and improving the nutritional value of dishes. Under the bright kitchen model, customers can directly order the dishes while seeing the fresh food ingredients. The dishes can be cooked immediately in the kitchen, which is bright, transparent, safe and clean. The representatives include Tuo Tuo Fish, Stewed Thick Soup with Earthen Pot, Tuo Tuo Meat, Braised Baby Carp with Chili Sauce, etc.

（六）逸亭

逸亭位于风景秀丽的江安河畔，地处郫都区与温江区的交界处，是一家集休闲娱乐、餐饮于一体的四星级农家乐，先后荣获“成都市十大魅力农家乐”“中国乡村旅游金牌农家乐”等荣誉称号。

6. Yiting Happy Farmhouse

Located at the side of Jiang’anhe River and the crossing of Pidu District and Wenjiang District, Yiting Happy Farmhouse is a 4-star one integrating recreation, entertainment and catering. It has been awarded as one of Chengdu’s Top-ten Glamorous Farmhouse, and China Rural Tourism Gold-medal Farmhouse, etc.

逸亭三面环水，绿道纵横，占地面积约0.67公顷，绿化面积达80%。各季的鲜花、水果、野菜都有特设的采摘区域，可供游客亲身体验，一年四季都能感受到川西田园的乡土风情。逸亭的餐饮接待能力很强，服务功能完善，既可承接各种室内宴会，还可举办草坪婚礼、团队活动、户外冷餐会等。此外，逸亭还在园区内配备了棋牌娱乐室、健身运动区、孩童玩乐园、绿道骑游休息处等。逸亭菜品以鱼鲜为主，配以时令鲜香的农家菜、野菜，其招牌菜品炝锅刺婆，以独特的制作工艺和口味而赢得了食客的广泛赞誉。

Bounded by water on three sides and filled with greenways, it occupied an area of about 0.67 hectares with over 80% green space. In each season, the flowers, fruits and edible wild vegetables can be picked at certain areas, so people can experience by themselves. So to say, people can feel the rural scene in the countryside of western Sichuan all the time in a year. Yiting Happy Farmhouse does well in catering reception with its continually improved services. It can hold many indoor parties, lawn weddings, team activities, and outdoor buffet parties, etc. Meanwhile, it is equipped with chess and card rooms, fitness area, children's play area, and rest area for people after cycling through the greenways. The food in Yiting Happy Farmhouse is mainly the fresh aquatic products, coupled with seasonal farm vegetables and wild vegetables. The specialty dish is Spicy Fried Mandarin Fish, which is praised widely by the customers for its unique cooking method and flavor.

（七）望阳阁

望阳阁位于郫都区三道堰镇的“头堰”与“二堰”地段，是一家集中餐、茶坊、住宿、会议、野外篝火于一体的综合性四星级农家乐。

7. Wangyangge Happy Farmhouse

Located at the first weir and the second weir, Sandaoyan Town, Pidu District, Wangyangge Happy Farmhouse is a 4-star comprehensive happy farmhouse integrating catering, tearoom, accommodation, meeting, and outdoor bonfire.

望阳阁始创于2003年，占地约3.4公顷，整体建筑为仿古式，亭台楼榭、斗拱飞檐、古色古香，加之临水而立，亲水长廊紧连着笼篼杩槎“导水堤”，将古镇、古堰、古桥、古阁，与水车、水牛、水溪、水景融为一体，相映成趣，漫步其中，令人心旷神怡，流连忘返。望阳阁的餐饮规模较大，其多功能宴会厅可同时容纳1 000多人就餐，包间装修精致、风格多样。入夜后有篝火烧烤，夜啤酒饮，可让食客大饱口福。除餐饮外，音乐晚会、棋牌、垂钓、游玩等休闲娱乐项目也十分丰富。望阳阁的特色餐饮为干锅、农家菜和富有本地特色的“水豆花”。

Wangyangge Happy Farmhouse was founded in 2003 and occupies an area of about 3.4 hectares. The whole architecture style is archaic and shows the antique beauty, with elegant pavilions as well as upturned eaves and angled cornices. Moreover, it stands by the water with a corridor linking to the embankment for water diversion built with the triangular wooden frame closely. The ancient town, weir, bridge, pavilion and the waterwheel, buffalo, stream, waterscape are delightful and fused together. When people wander there, they will be free of mind and even linger on without any thought of leaving. Wangyangge is capable of accommodating a large number of guests. Its multi-purpose banquet hall can hold over 1,000 people to have meals and its private rooms are decorated prettily with distinctive styles. At night, it prepares bonfires, barbecues and beer for the customers' pleasure. At the same time, people are offered with many entertainment activities and

望阳阁

facilities, such as music party, mahjong and chess room, fishing, strolling and so on. The food provided by Wangyangge is diversified, with the specialties like griddle cooked dishes, homemade dishes and "bean curd pudding" with local characteristics.

（八）杨鸡肉（陌上人家）

杨鸡肉位于郫都区太清路成灌高速释迦收费站附近，是集餐饮、会议、休闲、娱乐于一体的四星级农家乐。其创始人杨远福被国家旅游局授予"中国乡村旅游致富带头人"的荣誉称号。

8. Yang's Chicken (Moshangrenjia Happy Farmhouse)

Located at the Taiqing Road of old Chengdu-Dujiangyan Highway, and next to the Shijia toll station of present Chengdu-Dujiangyan Expressway, Yang's Chicken (Moshangrenjia Happy Farmhouse) is a 4-star happy farmhouse integrating functions of catering, meeting, recreation and entertainment. Its founder Yang Yuanfu has been awarded with the title of "Chinese Bellwether of Acquiring Wealth through Developing Rural Tourism" by National Tourism Administration.

杨鸡肉（陌上人家）

杨鸡肉占地约两公顷，建筑面积4 000平方米，庭园风格偏向苏州园林，园中小桥流水，曲径通幽，绿树、花草点缀其间，亭台阁榭错落有致；建筑设计体现出川西民居的风格，室内装修民俗风情浓郁，给人以成都平原特有的和谐、温馨之感。杨鸡肉拥有布局合理的大、中餐厅和休闲区，能同时接待800余人就餐，其主餐厅可容纳500余人同时就餐。杨鸡肉主打菜品为鸡肉，代表性菜肴有红味凉拌鸡、白味鸡、风味鸡杂、麻辣凤爪、茶香鸡翅、水晶锅巴鸡、葱香鸡蛋干等。此外，该店制作的"九大碗"系列菜、鱼鲜系列菜及鸡汤菜肴也颇具特色，能充分满足不同人群的消费需求。

Yang's Chicken occupies an area of about 2 hectares and the floor area of 4,000 square meters. If one walks through the whole courtyard, the scene around is always changing, as if you were in Suzhou Classical Gardens. Inside the courtyard, water flows beneath the bridge and the winding paths lead to some secluded quiet place embellished with green trees and wild flowers. The pavilions are well arranged. Such architecture designing reflects the style of folk houses in western Sichuan, with the interior decoration showing a strong feeling of folk customs, thus leaving people the special sense of harmony and warmth of Chengdu Plain. Yang's Chicken is equipped with well-spaced big and middle-sized restaurants and the recreation facilities, which allows more than 800 people to have meals at a time. The main restaurant can accommodate more than 500 people. Its food is diverse with chicken as the main material. Typical dishes include Chicken Salad in Hot Sauce, Boiled Chicken with Chili Sauce, Fried Chicken Giblets with Shredded Pickled Pepper, Chicken Claw Salad in Spicy Sauce, Fried Tea-Flavored Chicken Wing, Fried Chicken with Crust of Cooked Rice, Fried Sliced Dried Egg with Scallion Puddings, etc. Meanwhile, its "nine dishes" and fresh seafood dishes as well as the dishes boiled in pure chicken soup are quite characteristic, and could meet the different consumption demands of the customers at all levels.

（九）铜壶苑泰兴休闲庄

铜壶苑泰兴休闲庄位于郫都区太清路中段，是一家集主题宴席、特色餐饮、休闲娱乐、茶坊会议于一体的四星级农家乐，被评为“四川省乡村旅游精品特色业态经营点”。

铜壶苑泰兴休闲庄

9. Tonghuyuan (Taixing) Leisure Farm

Located at middle of Taiqing Road, Pidu District, Tonghuyuan (Taixing) Leisure Farm is a 4-star happy farmhouse integrating functions and facilities of themed feasts, featured dinning, recreation, tea rooms and meeting and is chosen as the “Sichuan Rural Tourism Operator with the Quality and Distinctive Business form”.

铜壶苑泰兴休闲庄始建于2015年，为铜壶苑餐饮服务公司旗下的第三家餐饮企业。该店大门高大宏伟，两边是渗透式围墙。一进大门的广场，即是一个大型铜壶造型，铜壶微倾，壶水下注，似向客人鞠躬问好。其服务设施着重围绕主题宴席进行设计，拥有整体升降舞台，主楼一楼主要承接大、中型婚宴、生日宴等各种主题宴。铜壶苑泰兴休闲庄的饮食特色以鱼鲜为主，有高、中、低档鱼类20余种供顾客选择，可做成麻辣、糖醋、家常等多种鱼肴美味。此外，该店的代表性菜肴还有大刀耳片、火鞭牛肉、百姓烩菜、粗粮糊糊等。

Founded in 2015, Tonghuyuan (Taixing) Leisure Farm is the third catering business affiliating to Tonghuyuan Catering Service Company. Its gate is tall and magnificent, with completely transparent fences on two sides. Once entering the square through the gate, people can see a large teapot-shaped decoration, which tilts slightly with water pouring down, seeming to make a bow and extend greetings to the customers. Its service facilities are designed for the service system of the themed party, and are equipped with lifting stage. The first floor of the main building is used to hold various theme parties like big and middle wedding party and birthday party. The food served in Tonghuyuan (Taixing) Leisure Farm is mainly the fresh aquatic products. There are over 20 kinds of fish of high, middle and low grade for people to choose from, which can be cooked to spicy, sweet and sour flavors or homemade flavors. Meanwhile, the typical dishes include Stewed Sliced Pig Ears with Chili Oil, Firecracker Beef, Multi-Ingredient Folk Stew, Porridge Cooked with Coarse Food Grain, etc.

（十）西御园乡村酒店

西御园乡村酒店位于老成灌公路郫都城区往都江堰方向1.5千米处，是一家集餐饮、住宿、会议、娱乐于一体的四星级乡村酒店，先后获得“四川省最佳会议度假酒店”“郫都区餐饮名店”等称号。

10. Xiyuyuan Rural Hotel

Located at the previous Chengdu-Dujiangyan Road and 1.5 kilometers in the mid from Pidu to Dujiangyan City, Xiyuyuan Rural Hotel is a 4-star rural hotel integrating catering, accommodation, meeting, recreation and tea time, and is awarded with the titles of “Best Conference Resort of Sichuan” and “Famous Catering Spot of Pidu”.

西御园乡村酒店

该酒店占地6.7公顷，环境幽雅，有一泓池塘清水，环池为临水茶座，树林蓊蓊郁郁，林间小路曲径通幽。该酒店拥有可容纳1 500人的大、中、小型会议厅10个，有标准间、单间、套房和别墅房170多间，多功能、多媒体设备齐全，能满足各种会议需求，餐厅拥有3个豪华大厅和数十个包间，可同时接待2 000人进餐。其代表菜品有苕菜狮子头、太极蚕豆羹、龙井红烧肉等。

The hotel occupies an area of 6.7 hectares with graceful environment. There is a clear pond surrounded by flourishing trees with seats around for people to enjoy tea. The paths among the green trees are tranquil and interesting. The hotel is equipped with all facilities and devices needed, and can accommodate 1,500 people for meetings with 10 small, middle and large meeting halls. Also, there are over 170 twin rooms, single rooms, suites and villas, equipped with necessary multi-function and multimedia devices to guarantee all meetings of different levels. The hotel has 3 luxurious halls and a dozen of private rooms, which can accommodate 2,000 people to have dinner at a time. Its typical dishes include Lion's Head Meatballs with Shaocai, Taichi Broadbean Soup, and Braised Pork in Brown Sauce with Longjing Tea, etc.

（十一）竹里湾乡村酒店

竹里湾乡村酒店位于郫都区友爱镇释迦桥村，是一家集餐饮、住宿、会议、商务宴请、休闲品茗于一体的四星级乡村酒，先后获得“川西林盘保护单位”“四川十大品牌农家乐”等称号。

11. Zhuliwan Rural Hotel

Located at Shijiaqiao Village, Youai Town, Pidu District, Zhuliwan Rural Hotel is a 4-star rural hotel integrating functions of catering, accommodation, meeting, business dinner, recreation and tea time. It is awarded with the title of "Protected Unit of Woods in Western Sichuan", "Sichuan's Top 10 Branded Happy Farmhouse", etc.

竹里湾乡村酒店

竹里湾紧邻岷江下游，地处清水环绕的河心半岛上，占地面积约两公顷。作为川西林盘的典型代表，竹里湾将院落、竹林、河湾、田园有机融合，水车、水榭、水岸、吊桥相映成趣，营造出一片清幽雅静的休闲乐园。园内苗木景观设置精巧，银杏林遮天蔽日、郁郁葱葱，春夏一林碧绿，秋冬一片金黄，是游客观赏、拍摄、绘画的理想之选。该酒店拥有大型宴会厅1个，特色包间、小院20个，茶坊、民宿客房49间。菜品特色突出，招牌菜有菊花鱼、养颜豆汤、虾仁捞锅巴、青豆箭鸭、家常红沙鱼、滋味头菜、麻辣鲍鱼仔、烧牦牛头等。

Zhuliwan encircles the peninsula of Minjiang River, and closely neighbors the river islet surrounding by clear water on the downstream, which occupies an area of 2 hectares. As the typical representative of the rural settlements surrounded by bamboo forest, Zhuliwan integrates the courtyard, bamboo forest, river bend and corp fields together, with the waterwheel, waterside pavilion, river bank, suspension bridge form a delightful picture, creating a quiet and beautiful recreation paradise. The layout of nursery-grown plants inside is full of creativity. The ginkgo forest covers all the sky with the green and dense branches and leaves. In spring and summer, the leaves are dark green. While in fall and winter, the leaves will become golden, then it becomes the ideal place for people to admire, take pictures and draw the fine view. Its capability for catering and accommodation reception is strong. The hotel has a large banquet hall, 20 featured private rooms and small yards, as well as 49 tea rooms and guest rooms. Its dishes are distinctive with specialties like

Chrysanthemum Fish, Bean Soup for Beauty Maintaining, Stir-fried Shrimp with Crust of Cooked Rice, Fried Duck with Green Peas, Homemade Braised Golden Pomfret in Soy Sauce, Fried Sliced Cabbage, Fried Baby Abalone with Chili Sauce, Braised Tibetan Yak Head, etc.

（十二）观景沅

观景沅位于郫都区农科村，是一家集餐饮、书吧、茶室、会议、棋牌于一体的四星级乡村酒店。

12. Guanjingyuan Rural Hotel

Located at Nongke Village, Pidu District, Guanjingyuan Rural Hotel is a 4-star rural hotel integrating catering, study, tea rooms, meeting, chess and card rooms.

观景沅

该酒店占地面积约两公顷，庭院结构紧凑而精致，古色古香的传统四合院内收藏了大量川西风格的家具、用具和民俗古旧物品；院内花草、树木错落有致，婀娜多姿，桩头盆景造型奇巧，玉兰、海棠争奇斗艳，银杏、罗汉松遒劲挺拔，环境淡雅而清爽。登楼远眺，全村胜景尽收眼前。观景沅拥有可接待50～150人的多种会议室，可容纳100人住宿，餐饮服务周到、全面，拥有大型宴会厅和各式包间，能同时容纳800人用餐，其代表性菜肴为极品耗儿鱼。

The hotel occupies an area of bout two hectares, with a compact and delicate structured courtyard. A number of typical furnishings, utensils and old folk items in western Sichuan displayed within the classical traditional quadrangle courtyard. Its environment is quietly elegant and refreshing, with pretty and charming trees and flowers scattering inside the yard. The bonsai is exquisite in appearance. The mangnolia and cherry-apple compete with each other for beauty of looks. The ginkgo and yacca are sturdy, tall and straight. Overlooking from the top of the building, one can see the view of the whole village. The guest rooms could accommodate 100 people, with various meeting rooms holding 50 to 150 people respectively. Its catering service is considerate and comprehensive. 800 people can have meals at the same time with large banquet hall and private rooms equipped. Its representative dish is made of filefish.

（十三）崇宁映像酒店

崇宁映像酒店位于郫都区唐昌镇，距离唐昌公交总站100米，是一家集餐饮、会议、住宿、茶坊于一体的四星级乡村酒店。

13. Chongning Yingxiang Rural Hotel

Located at Tangchang Town and 100 meters away from Tangchang bus terminal, Chongning Yingxiang (which means Image of Chongning) Rural Hotel is a 4-star rural hotel integrating functions and facilities of catering, meeting, accommodation and tea rooms.

该酒店占地面积约8公顷，建筑面积11 000平方米，以挖掘、体现唐昌镇（原崇宁县）的传统文化为特色。酒店设有能同时容纳600人就餐的大型宴会厅，和能容纳300人就餐的中型宴会厅，还有川西民居风格的餐厅包间13个。“崇宁故事”包间弥漫着浓烈的传统文化气息，四周的墙上挂满了历时久远的老照片。一楼大会议室为“崇德尚宁”厅，厅内的装饰主题为崇宁八景，山墙上有唐昌镇自公元677年建县至1958年11月撤县的大事件；一楼小会议室为“遗风”厅，里面大多是唐昌镇及原崇宁县市民生活的老照片，茶坊墙壁上塑有原崇宁老县城的浅浮雕图案。该酒店菜肴特色鲜明，鱼肴尤为出彩，烹饪方法以烧、炸、熘、炒为主，其代表性菜品有熘鱼片、五香熏鱼、鱼松等。

The hotel occupies an area of about 8 hectares and a floor area of 11,000 square meters, featured by exploring and showing the culture of ancient Tangchang, namely Chongning (old way of referring to Tangchang) culture. It is equipped with the large banquet hall which could accommodate 600 people to have dinner and the middle banquet hall which could accommodate 300 people as well as 13 private rooms of folk house style in western Sichuan. The big private room named “Chongning Gushi” (which means Chongning story) shows a strong feeling of Chongning culture with old pictures hanging on the surrounding walls. There is a large meeting room named “Chongde Shangning” (which means worshiping high moral standards and a peaceful life) in first floor, with the themed decoration of eight sceneries in old Chongning indoors and the written introduction of the events happening in Tangchang from its foundation in 677 A. D. to its cancellation in November, 1958 on the external wall. The small meeting room in the first floor is named “Yifeng”, inside which are the old pictures of urban life in Tangchang and the former Chongning County, arousing the tourists to recollect the good old days. There are pictures of relief culture center in the old county seat of the former Chongning County on the external walls of the tea rooms. Its dishes are of distinctive features, especially the fish. The cooking methods are mainly braising, deep-frying, quick-frying and frying. The typical dishes are Quickly Fried Fish Slices, Fried Fish with Spiced Sauce, and Fish Floss, etc.

（十四）临水轩

临水轩位于郫都区农科村，是一家集餐饮、住宿、休闲、会议、品茗于一体的四星级乡村酒店。

14. Linshuixuan Rural Hotel

Located at Nongke Village, Pidu District, Linshuixuan Rural Hotel is a 4-star rural hotel with functions of catering, accommodation, leisure, meeting, tea appreciation, etc.

临水轩

临水轩乡村酒店占地面积约1.4公顷，园内川西民居与仿古建筑融为一体，叠山理水，颇具苏州园林的韵味，茂林修竹、桂树成片、生机勃勃；处处字画满壁、翰墨留香，文化气息浓郁。该酒店拥有景色宜人的休闲区10个、住宿区4处、各具特色的温馨客房28间；还设有汉服体验馆，可供游客体验、拍照。临水轩餐厅拥有大小包间7个，菜品特色突出，其代表性菜品有家常拌土鸡、风车脊骨、白果煨土鸡、豆瓣鱼、家常鱼、豆浆馍馍等。

Covering an area of about 1.4 hectares, Linshuixuan Rural Hotel combines the characteristics of residential houses in westernSichuan and antique buildings. There are rockeries, streams, bamboos and osmanthus trees within the hotel, reminding the guests vibrant and beautiful charm of Suzhou gardens. Calligraphy works and paintings are hung everywhere in the hotel, revealing a strong cultural atmosphere. Moreover, there are 10 leisure areas, 4 accommodation areas and 28 cozy rooms with distinctive characteristics, as well as a Han Chinese clothing experience hall for tourists to take photos. In terms of the catering services, Linshuixuan Rural Hotel has outstanding features. There are 7 private rooms of all sizes in total. The specialty dishes are Homemade Free-Range Chicken in Chili Sauce, Cold Stewed Ribs, Simmered Free-Range Chicken with Ginkgo Fruit, Chili Bean Paste Flavored Fish, Homemade Fish, Steamed Green Soybean Buns, etc.

（十五）绿上源

绿上源位于郫都区德源街道东林村，是一家集餐饮、会议、住宿、休闲娱乐于一体的四星级乡村酒店。

15. Lvshangyuan Rural Hotel

Located at Donglin Village, Deyuan Sub-district, Pidu District, Lvshangyuan Rural Hotel is a 4-star rural hotel with functions of catering, meeting, accommodation, leisure, etc.

绿上源环境优美、安静，小溪潺潺、白鹭翻飞、绿树成荫、百花绽放，是欣赏自然风光、放松心情、远离喧嚣的理想去处。该酒店拥有中餐宴会大厅及数十个特色主题包间，主打自贡盐帮菜，游客还可亲身参与烤兔、烤全羊等美食趣味体验活动，其代表性菜品有干锅鸡、干锅鸭掌、剁椒皮蛋等。

The hotel has a beautiful and quiet environment with gurgling streams, flying egrets, tall trees and blooming flowers, making it an ideal place to appreciate the natural scenery, have a relax and get away from the hustle and bustle. There is a major dining hall for Chinese food and dozens of themed private rooms, providing characteristic catering services for the guests. The hotel mainly provides the dishes from Zigong City, the “Capital of Salt”. Guests can also participate in making roast rabbit, roast lamb and other delicacies to experience the fun of cooking. Representative dishes are Chicken in Dry Pot, Duck Palm in Dry Pot, Preserved Egg with Chopped Pepper, etc.

第二节 街镇美食品鉴游

Section Two A Tour for Street Snacks

郫都区美食街区分布较广、各具特色，大致可归为四大类型：一是历史文化内涵深远、传统特色美食集中的街区，如三道堰惠里特色美食街、唐昌大椿巷特色小吃街；二是集现代生活综合服务于一体、餐饮品牌林立、美食产业较发达的街区，如石犀里特色商业美食街区、郫都区BLOCK街区；三是依托地区饮食资源优势而形成的美食旅游特色街区，如战旗村旅游美食聚集区、太清路美食街；四是烟火气息浓厚、夜市经济突出的美食街区，如犀浦夜市街区、“川菜世纪”市集等。这些美食街区风格独特，优势互补，既是郫都美食的重要聚集区、郫都美食文化的集中展现地，也是人们观光游玩、品鉴美食的理想去处。这些美食街区，不仅推动了郫都区餐饮产业的发展、促进了乡村振兴，而且还极大地满足了人民群众对美好生活的需要。因篇幅所限，这里仅从中遴选出几处最具郫都特色的美食街区进行介绍。

There are various gourmet blocks widely spread in Pidu District with distinctive features, which can be divided into at least 4 types: the first type is the ones with long history and abundant culture, where the visitors can find plenty of traditional gourmet food with characteristics. Such blocks include Huili Gourmet Street at Sandaoyan Town, Dachun

Alley Snack Block at Tangchang Town; the second type is the blocks with sufficient modern life services, catering brands and well-developed gourmet industry, including Shixili Business and Gourmet Block, Pidu BLOCK, etc.; the third type is the ones based on the advantages of catering resources, with traveling and gourmet food as characteristics. The representative ones are Tourism and Gourmet Food Cluster of Zhanqi Village, Taiqing Road Gourmet Street, etc.; the fourth ones are the down-to-earth gourmet blocks with night market economy as the characteristics, such as Xipu Night Market Block, "Chuancai Shiji (namely Century of Sichuan Cuisine)" Market, etc. Each of these gourmet blocks has different styles and complementary advantages, making them places to gather the food of Pidu and demonstrate its related culture. Moreover, they are also places for people to go sightseeing and enjoy gourmet food, promoting the development of Pidu District's catering business, boosting the rural revitalization, as well as continuously meeting the people's needs for a better life. Due to limited space, only the representative gourmet blocks are to be introduced here.

一、太清路美食街

太清路美食街位于郫筒街道太清路，成灌高速石家立交桥出口处，地理位置优越、交通便利，于2018年开街营业，是郫都区著名的农家乐旅游美食街。

I. Taiqing Road Gourmet Street

Located at the cross of Shijia Overpass Exit of Chengdu-Dujiangyan Expressway and Taiqing Road, Pitong Subdistrict, Taiqing Road Gourmet Street commenced business in 2018 and became a famous traveling and gourmet destination for happy farmhouses thanks to its superior location and convenient transportation

这条美食街拥有众多美食品牌，汇集了大小餐饮店30多家，既有印象泰和园、杨鸡肉（陌上人家）、铜壶苑、汇鑫苑、太清锦宴、鄢鸡肉等知名星级农家乐，也有宜宾特色鲫鱼、海棠里耗儿鱼、肥肠鸡等特色美食店。这里环境幽雅、四季花香、流水潺潺、亭台错落，是许多餐饮店家的共同特色；这里菜品丰

太清路（郫都区融媒体中心/供）

富、特色鲜明，如郫县豆瓣鱼、苕菜狮子头、火鞭牛肉、大刀耳片、杨鸡肉等，均远近闻名；这里美食业态丰富，特色宴饮活动多样，如杨鸡肉（陌上人家）推出的“夜食”“夜游”消费新模式，印象泰和园着力打造的婚礼文化宴会庄园等。

There are and more than 30 eateries of numerous gourmet brands here, including famous star-rated happy farmhouses like Taiheyuan Restaurant, Yang's Chicken (Moshangrenjia Happy Farmhouse), Tonghuyuan Leisure Farm, Huixinyuan Happy Farmhouse, Taiqingjinyan Happy Farmhouse, Yan's Chicken, etc. as well as characteristic eateries like Crucian Carp with Yibin Characteristics, Haitangli Filefish, Spicy Chicken and Pork Intestines, etc. Many of them have a tranquil environment with floral fragrance, singing birds, flowing water under the bridges as well as picturesque pavilions scattered in the gardens; the specialty dishes are of various kinds, Chili Bean Paste Flavored Fish, Lion's Head Meatballs with Shaocai, Firecracker Beef, Stewed Sliced Pig Ears with Chili Oil, Yang's Chicken; these catering businesses also have various forms of operation and distinctive feasts. For example, Yang's Chicken (Moshangrenjia Happy Farmhouse) works hard to build itself into a new consumption destination for "night dining" and "night traveling"; Taiheyuan Restaurant has become a 5-star themed wedding hotel, etc.

太清路美食街所处的郫筒街道，历史底蕴深厚、文化氛围浓郁，其所在区域，有传为纪念古蜀望帝、从帝而修建的望丛祠，是中国西南地区唯一一座“一祠祭二主”的帝王陵冢，也是国家AAAA级旅游景区。望丛祠民俗文化活动较多，每年清明前后，这里都要举办赛歌会，男女老少均可参加，规模宏大、人数众多、影响广泛，已成为郫都区的一大民间风情盛会。此外，“望丛古蜀文化节”业已成为郫都区的一大民间节日品牌，节日期间，来自四面八方的群众纷纷涌入望丛祠祭拜望、从二帝，气氛祥和、热闹非凡，既为追根寻祖，又为传承古蜀文化。

Pitong Street, where Taiqing Road Gourmet Street is located, has profound historical deposits and rich cultural atmosphere. The Wang Cong Shrine built to memorize King Wang and King Cong of the ancient Shu Kingdom is the only royal mausoleum in the Southwest as well as an AAAA tourism area. There are many folk activities in Wang Cong Shrine. Every year around the Qingming Festival, there is a large-scale singing contest, when all people are allowed to participate. It has become a major folk festival in Pidu District with huge influence. Besides, as Wang Cong Ancient Shu Cultural Festival becomes another festival brand, people from all places flock in Wang Cong Shrine to worship the two kings, in order to find their roots and inherit the culture of ancient Shu Kingdom, constituting an extremely lively scene.

二、郫都区BLOCK街区

郫都区BLOCK街区位于郫都区郫筒街道海骏达商圈，由海骏达美食城、金融中心、智趣汇·袋鼠集夜市三部分组成，是郫都区重点打造的具有国际潮流的特色街区。

II. Pidu BLOCK

Located at Haijunda business area, Pitong Subdistrict, Pidu District, Pidu BLOCK is composed of Haijunda Gourmet City, Financial Center and Zhiquhui-Kangroo Night Market, which is a characteristic block with visual effects in Pidu District.

该街区定位于为电子信息产业功能区人才提供配套服务，聚焦服务青春客户群体，围绕“年轻”“地标”“国际”三个关键词，从设计到服务都秉承高颜值要求，以创智北环路为中心，按照“3+1+N”社区商业模式，引进人脸识别技术、24小时无人店等新型服务业态，打造年轻客户群体社交圈，以及集合休闲、娱乐、餐饮、购物等多种主题业态于一体的国际化快节奏消费体验中心。其中，海骏达美食城又名玩味潮流街，长1 100米，建筑面积5万平方米，共3层，涵盖了旅游、餐饮、休闲、娱乐等业态，来自省内外的餐饮名店、城市首店、音乐酒吧，如巴国布衣、何师烧烤、胡桃里、大千精酿酒馆等，均汇集于此，每周还会举办特色主题文化旅游活动。每当夜幕降临，音乐、美食、美酒、表演精彩纷呈，是郫都夜间潮流的打卡地。智趣汇·袋鼠集夜市是郫都首个集装箱风情美食夜市，于2021年2月开业。该夜市共设置集装箱固定摊位54个、创意移动小型摊位18个，利用“魔方”“七巧板”“YES”与现代钢结构相结合，突出“智趣汇”主题，经营业态以特色小吃、休闲美食为主，国际潮流范儿十足，其特色店有禾风寿司、首尔姐姐的厨房、炒酸奶、玉笼轩等。智趣汇·袋鼠集夜市作为特色网红街区，不但给郫都国际社区注入了新鲜活力，也为郫都夜生活的经济发展按下了“加速键”。

The block is aimed at providing supporting services for talents in the Electronic Information Industrial Area, especially the young people. Guided by the three keywords, namely “young”, “landmark” and “international”, the block has upheld high standards for outlook and high demands for service. With Chuangzhi North Ring Road as the center and according to the “3+1+N” community business model, the block is trying to build itself into a group social circle for young customers and an international fast-tempo consumption experience center integrating leisure, entertainment, catering and shopping among various themed commercial activities. Haijunda Gourmet City (Fun and Fashion Street), 1,100 meters long and three floors high with a construction area of 50,000 square meters, is a must-visit destination in Pidu. It accommodates so many commercial activities including traveling, catering, leisure, entertainment as well as numerous famous catering brands, first branches of chain stores and music bars, such as He’s BBQ, Hutaoli, Daqian Craft Beer, etc. Every week there will be cultural traveling events with characteristic themes, and every night there will be music, gourmet food, good wines, excellent shows, etc. Opened in February, 2021, Zhiquhui-Kangroo Night Market is the first container-style gourmet night market. With 54 fixed container-style booths and 18 small movable creative booths, the night market combines cubic-shaped, tangram-shaped decorations and English letters “YES” with modern steel construction, which demonstrates the theme of “Zhiquhui (a combination of wisdom and delight)”. These shops mainly serve characteristic snacks including Hefeng Sushi, Kitchen of Seoul’s Miss, Yogurt Ice Cream, Yulongxuan, etc. As an Internet-celebrity block, Zhiquhui-Kangroo Night Market provides fresh energy for the Pidu international community, and speeds up the development of nightlife economy in Pidu.

三、“川菜世纪”市集

“川菜世纪”市集位于有“郫县豆瓣之乡”美誉的安德街道，成灌高速安德立交桥出口处，于2021年5月开市，是郫都区以“川菜体验”为主题的一条特色商业街区。

III. “Chuancaishiji” Market

Located at Ande Overpass Exit of Chengdu-Dujiangyan Expressway, Ande Subdistrict, “Birthplace of Pixian Chili Bean Paste” — “Chuancaishiji” Market was opened in May 2021 as a characteristic business block with the theme of “Experience Sichuan Cuisine”.

“川菜世纪”市集通过功能叠加和价值开发，因地制宜地将特色美食文化与民俗文化、地理标志保护产品、非物质文化遗产相结合，着力植入社区商业文化新业态、新载体、新体验的消费场景，共吸引了270余户商家入驻，营造出浓厚的市井烟火氛围。这里美食纷呈，汇集了天南海北众多特色美食，如关东煮、螺蛳粉、寿司、烤肉、苕皮豆干、包浆豆腐、海鲜、小龙虾等，令人大饱口福。此外，“川菜世纪”市集还努力打造多元化的乡镇消费场景，不仅有美食，还有精品服装、特色商品、水幕电影、文化演出等活动，极大地满足了游客的多元化需求，有力地带动了安德的夜市经济，也是成都乐游、乐食、乐购、乐享的夜间消费新去处。另外，该市集周边还有中国·川菜文化体验馆、中国川菜体验园景区及中国川菜产业城，川菜文化氛围浓厚，川菜产业发达，不仅能提供川菜美食品鉴，还可亲身感受川菜“研、学、游”

"川菜世纪"市集（刘刚钰/摄）

这一全方位、沉浸式的川菜体验之旅。

With functions added and value developed, it combines characteristic cuisines with folk culture, geographical indication protection products and intangible cultural heritages. Besides, with the consumption scenes of new community business activities, new carriers and new experiences, it has attracted more than 270 businesses, creating a lively environment for nightlife consumption. Here are a variety of characteristic gourmet food, including Japanese oden, rice noodles with river snails, sushi, barbecue, roasted sweet potato slice and roasted dried bean curd, crispy tofu, seafood, spicy crayfish, etc., serving a real treat for everyone. Meanwhile, "Chuancaishiji" Market is working hard to build diversified countryside consumption scenes with delicate clothes, specialty goods, water curtain movies and cultural shows besides gourmet food, which can satisfy the diversified demands of guests and effectively boost the nightlife economy, making it a new destination for delighted traveling, catering, shopping, entertainment as well as nightlife consumption in Chengdu. In addition, with the China • Sichuan Cuisine Cultural Experience Museum, China's Sichuan Cuisine Experience Park, China's Sichuan Cuisine Industrial City in the surrounding area, there is a strong atmosphere of Sichuan cuisine culture and developed Sichuan cuisine industry in the region. Therefore, the market can not only provide opportunities for tasting Sichuan cuisine, but also satisfy the demands for researching Sichuan cuisine, contributing an all-around and immersive experience for Sichuan cuisine.

四、菁蓉镇特色商业街区

菁蓉镇特色商业街区位于郫都区德源街道，是一条集创客服务、夜间消费、商务研学、旅游、餐饮于一体的创客服务型特色商业街区，曾获得2020年上半年成都"最美街道"第二名。

IV. Jingrong Town Characteristic Business Block

Located at Deyuan Subdistrict, Pidu District, Jingrong Town Characteristic Business Block integrates entrepreneur services, nightlife consumption, business research, traveling and catering, which has won the second place of "The Most Beautiful Block" in the first half of 2020.

菁蓉镇是全国著名的创客小镇，全国首批28个"双创"示范基地及成都市3个众创空间引领区之一。菁蓉镇特色商业街区总面积约1万平方米，总长度约800米，厚植电子信息产业特色，围绕全球创客、创新人才等消费群体，布局创客业态区、轻餐饮区、特色餐饮区、生活服务区，拥有川粤汇小厨、食庐餐厅、筷子米线、刘氏小吃、御恋烧仙草、红嘴巴奶茶等30余家餐饮企业和店铺。此外，街区还建有创意设计IP

吉祥物宇航员“菁菁”“蓉蓉”，以及具有网红特质的泡泡树、彩虹部落等科技互动装置，另外还设有用艺术叠加重组集装箱打造的集展示、创意、漫时休闲等功能于一体的创意屋，并有外摆、彩绘、夜景灯光系统交错的体验化场景。目前，该商业街区正努力打造成为服务电子信息产业功能区的生活高地。

Jingrong Town is a nationally famous entrepreneur town, one of the first 28 Innovation & Entrepreneurship Demonstration Bases and one of the three pilot zones for public innovation. Covering an area of about 10,000 square meters, the block is 800 meters long. With electronic information as the main feature and international entrepreneurs and innovation talents as main customers, the block strives to build an entrepreneur area, a light meal area, a specialty food area and a life service area. More than 30 businesses have entered the block: Chuanyuehui Canteen, Shilu Restaurant, Kuaizi Rice Noodle, Liu's Snack, Yulian Herbal Jelly, Hongzuiba Milk Tea, etc. Moreover, there are innovative IP mascots “Jingjing” and “Rongrong”, high-tech interaction devices like “bubble trees” and “rainbow tribes”, multifunction innovation houses (integrating showcases, innovation, leisure, etc.) made of artistic movable containers and experience scenes containing decorations, colored drawing and night view light system. At present, the block is working to build itself into a leading life zone serving the Electronic Information Industrial Area.

五、石犀里特色商业街区

石犀里特色商业街区位于郫都区犀浦街道华都路120号，成都地铁2号线、6号线，有轨电车蓉2号线在此交汇，交通便利，于2018年8月开业，是集特色餐饮、休闲娱乐、运动健身、文化创意、人文主题为一体的大型现代综合性商业街区。其周边汇集了西南交通大学（犀浦校区）、四川大学锦城学院、成都纺织高等专科学校等高校，人气颇高。

V. Shixili Business and Gourmet Block

Shixili Business and Gourmet Block is located at No.120 Huadu Road, Xipu Subdistrict, Pidu District with convenient transport for Metro Line 2, Metro Line 6 and Tram Line 2 intersect here. Starting since August, 2018, it is a large modern comprehensive business block integrating the functions of specialty catering, recreation and entertainment, sports, cultural creativity as well as humanistic elements. It is very popular and is surrounded by colleges and universities such as Southwest Jiaotong University (Xipu Branch), Jincheng College of Sichuan University and Chengdu Textile College.

润扬双铁广场——石犀里（郫都区融媒体中心/供）

石犀里商业街区主题特色突出，注重现代时尚和历史底蕴相结合，打造了千年石犀文化主题美陈、樱花文创市集系列美陈、石犀艺术展示空间等专题项目。石犀里美食业态丰富，品种琳琅满目，既有麦当劳、星巴克、乡村基、豪客来、茶百道、味源餐厅等知名餐饮品牌，也有泰式火锅、韩式烤肉、重庆火锅、海鲜烧烤等特色餐饮，以及开元米粉、宜宾燃面、剁椒拌饭、串串香、烤大串、担担面、烤五花等面点小吃。开街当天达到6万人次，此后，日均达到2万人次前来游玩、品鉴，是郫都区新晋的网红打卡地，很好地诠释了年轻人的生活美学观念，正引领和创造郫都区崭新的生活价值空间。

With prominent features in its theme, it focuses on a combination of modern fashion and historical foundation, creates the major projects like Themed Display of One-thousand-year Shixi Culture, Display of Sakura Cultural Creation Collections Fair and Display Space of Shixi Art. There are various delicacies in Shixili block, where famous catering brands like McDonald's, Starbucks Coffee, CSC, Houcaller, Chabaidao, Weiyuan Restaurant and featured catering stores like Tai-style hotpot, Korean barbecue, Chongqing hotpot, seafood barbecue gather together. There are also snacks like Kaiyuan Rice Noodle, Yibin Ranmian Noodles, Mixed Rice with Chopped Chili Sauce, skewers in hotpot, roasted skewers, Dan Dan Noodles and roasted streaky pork. On the day when it began to operate, 60,000 people came. Then, an average 20,000 people came to the block for strolling each day. It becomes an emerging popular place in Pidu District, interpreting the life aesthetics of young people very well and guiding and creating a brand-new life value in Pidu District.

六、犀浦夜市街区

犀浦夜市街区位于郫都区犀浦街道，主要由犀浦夜市、围城南路美食街组成。该街区美食地域范围广，美食聚集度高，人气超旺，为郫都区夜市经济的典型代表和最接地气的美食街区之一。

VI. Xipu Night Market Block

Located at the Xipu Subdistrict, Pidu District, the Xipu Night Market Block consists of Xipu Night Market and Weicheng South Road Gourmet Street. The block gathers a wide range of food from different areas. With super popularity, it is a typical example of night market economy of Pidu and one of the most down-to-earth food blocks.

犀浦夜市街区（郫都区商务局/供）

犀浦夜市从开市至今已近20年，是该美食街区的核心地带，夜市经营方式灵活，主要以流动摊贩为主，商家类型多样，既包括各种小吃店，又有服饰、化妆品及各种生活用品商铺，可满足民众多样化的生活需求。其中，美食是犀浦夜市的最大亮点，既有钵钵鸡、冷锅串串、狼牙土豆、凉粉、冰粉、砂锅米线、苕皮豆干等川内美食，也有寿司、巴西烤肉、韩国拌饭、东北烧烤、广东烧腊等天南海北的各种美味，价格亲民，现做现吃，口味地道，可给消费者带来奇特、难忘的味蕾体验。

Xipu Night Market is the core area of the food block and has been open for nearly 20 years. Its operation modes are diversified with mobile vendors as the main part. The merchant types are various, including snacks stores, clothes stores, cosmetics stores and household goods stores, satisfying people's different life needs. In Xipu Night Market, the food is the highlight. There are Sichuan delicacies including earthen bowl chicken, cold pot skewers, spike potatoes, bean noodles in chili sauce, ice jelly, casserole rice noodle, roasted sweet potato slice and roasted dried bean curd as well as all kinds of delicious food from different areas like sushi, churrasco, Bibimbap, barbecue with Northeast China characteristics and roasted meat of Guangdong. The dishes are cooked at the scene, delicious and authentic at a lower price, which brings unforgettable taste experience for customers.

围城南路美食街紧邻犀浦夜市，是一条餐饮特色突出的美食街，长约800米，美食店铺沿街而立，尤以各地著名美食为主，如新疆大盘鸡、重庆烧鸡公、云南天麻火腿鸡、丽江风味腊排骨、绵阳鱼汤米粉、石棉火烤烧烤、西昌火盆烧烤等，菜品物美价廉、装修整洁、服务用心，因而广受欢迎。

Next to Xipu Night Market, Weicheng South Road Gourmet Street is with outstanding catering features. Food stores stand along the 800-meter street, which offer well-known delicacies from different areas, such as Xinjiang Big Plate Chicken, Chongqing Stewed Rooster, Yunnan Stewed Chicken Soup with Gastrodia Elata and Ham, Lijiang Flavored Air-dried Rib, Mianyang Rice Noodles in Fish Soup, Shimian Barbecue with Fire and Xichang Braizer Barbecue. The specialty restaurants have cheap and fine dishes, universally popular among the people. Their decorations are clean and tidy and the services are good.

七、三道堰惠里特色美食街

惠里特色美食街位于郫都区三道堰镇，是集特色美食、娱乐休闲、旅游观光为一体的综合性特色美食街区。

VII. Huili Gourmet Street at Sandaoyan Town

Huili Gourmet Street at Sandaoyan Town of Pidu District is a comprehensive characteristic gourmet block combining specialty, entertainment and sightseeing tourism.

三道堰是国家AAAA级旅游景区，原名“三导堰”，是因用竹篓装卵石截水，做成三道距离相近的堰头导水灌田而得名，已有一千余年历史，曾是著名的水陆码头和商贸之地，被誉为“天府水乡”，先后荣获“全国环境优美乡镇”“省级城乡环境综合整治环境优美示范镇”“省级园林城镇”“成都市十大魅力城镇”和“成都市十佳休闲旅游城镇”等荣誉称号。惠里特色美食街是三道堰的核心景区之一，全长2 000米，占地5万平方米，依托优质生态资源，以清末民初水乡码头文化为主题，运用传统川西古建筑与徽派建筑风格相结合的建筑手法，以老街老巷、字号商铺、码头水景和庭院记忆为核心着力点，着力重塑珍贵的历史遗迹，丰富历久弥新的文化内涵，写意式还原三道堰作为川西重镇的繁荣景象。美食街共分为三期：一期以三道堰传统美食餐饮为主，有石锅苗苗鱼火锅、戴大肉、李大肉、周仔鹅、水乡冰粉、白家肥肠粉等特色餐饮店；二期定位为五省会馆区，主要有香道馆、功夫茶馆、三道宴特色餐饮、重庆老灶火锅、德国现酿啤酒馆等；三期则有“印象·水街”“乡情·水街”和“市景·水街”三部分，以旅游购物、特色小吃为主，代表性小吃有甜水面、酥糖、重庆小面、纸包鱼、麻饼、卤鹅等。整个街区美食种类丰富，味道多样，注重带给游客难忘的美食体验之旅。

Sandaoyan, as an AAAA tourist attraction, derived its name from three nearby weirs which were made of bamboo baskets with pebbles in to intercept the water and irrigate the field. With one thousand-year history, it has ever been renowned ferry terminal and business areas. It is reputed as "Waterland of Sichuan" and awarded with National Town with Beautiful Environment, Provincial Demonstration Town of Comprehensive Environmental Harness and Beautiful Environment, Provincial Garden Town, Top 10 Charming Town in Chengdu and Top-ten Leisure Tourism Town. Covering an area of 50,000 square meters, the 2,000-meter-long Huili Gourmet Street is the core scenic area of Sandaoyan. Relying on the high-quality ecological resources, it is themed by wharf and water-oriented culture in the late Qing Dynasty and the early Republic of China. It takes great efforts to reshape the precious historical sites and enriched cultural connotation to revivify the prosperity of Sandaoyan as an important town of western Sichuan through integrating the architectural style of traditionally ancient western Sichuan buildings with that of Anhui buildings and focusing on old street and old alley, time-honored stores, wharf landscape and courtyard. The gourmet street can be divided into three phases. The first phase majors in traditional delicacy of Sandaoyan, where there are specialty stores like Yellow-head Catfish Hotpot in Stone Pot, Meat of the Dai's, Meat of the Li's, Goose of the Zhou's, Ice Jelly of Waterland and Sweet Potato Noodles with Pig Colon of Baijia Village. The second phase is defined as areas of five provincial assembly halls, where there are incense lore stores, kung fu tea houses, specialty restaurants of Sandaoyan, Chongqing hot pot restaurants and German brewed beer stores. The third phase is divided into three parts, namely "Impression of Water Street", "Nostalgia of Water Street" and "Townscape of Water Street", which consist of tourism shopping and special snacks. The representatives include Sweet and Spicy Noodles, Crunchy Candy, Chongqing Spicy Noodles, Paper-Wrapped Fish, Sesame Cake and Marinated Goose. The whole block boasts a variety of gourmet food with numerous tastes, bringing an unforgettable gourmet experience for tourists.

三道堰惠里特色美食街

郫都区战旗村乡村十八坊（郫都区融媒体中心/供）

八、战旗村旅游美食集聚区

战旗村旅游美食集聚区位于郫都区唐昌镇战旗村，是集非遗产品生产、开发、科普教育与乡村美食为一体的文化旅游地。

VIII. Tourism and Gourmet Food Cluster of Zhanqi Village

Located in Tangchang Town, Pidu District, the Tourism and Gourmet Food Cluster of Zhanqi Village is a cultural tourist destination integrating production and development of intangible cultural heritage, science popularization education and country food.

战旗村地处横山脚下、柏条河畔，国家AAAA级旅游景区，先后荣获“全国文明村”“全国乡村旅游重点村”“全国乡村振兴示范村”和“四川省集体经济十强村”等称号，是战旗村重点打造的多功能产业区，主要包括乡村美食产品的生产、销售，以及饮食文化博览、乡村文明传承等，主要景点有乡村十八坊、郫县豆瓣博物馆、天府农耕文化博物馆、战旗村村史馆等。其中，乡村十八坊是战旗村旅游美食核心区，由非遗展示区、文化大院、美食餐饮区三部分组成，有郫县豆瓣、豆腐乳、酱园、榨油、粽子、唐昌布鞋、蜀绣、酿酒、三编等30余个作坊。这里传统特色美食比比皆是，如先锋萝卜干、界农酱菜、手工牛肉酱、糖油果子、蜀脆麻花等，现做现卖，香气扑鼻，令人垂涎。此外，这里还有四川浪大爷食品有限公司，着力点为旅游美食产品研发，该公司生产的唐昌红油豆腐乳、浪大爷豆豉等旅游食品深受游客喜爱，成为乡村十八坊中的热销产品。

Zhanqi Village is at the foot of Hengshan Mountain and near the Baitiaohe River. As an AAAA tourist attraction, it is honored with National Civilized Village, National Key Village for Rural Tourism, National Demonstration Village for Rural Revitalization and Top-ten Village of Collective Economy in Sichuan Province. The Tourism and Gourmet Food Cluster of Zhanqi Village is the industrial area which is developed in priority in the village. It has multiple industrial functions, mainly including production and sale of rural gourmet products, food culture exposition and rural civilization heritage. The main tourist attractions include Eighteen Rural Workshops, Pixian Chili Bean Paste Museum, Farming Culture Museum of Land of Abundance and Zhanqi Village History Hall. Among them, the Eighteen Rural Workshops is the core area of the Tourism and Gourmet Food Cluster of Zhanqi Village, consisting of three parts, namely intangible heritage exhibition area, cultural courtyard and gourmet food area, as well as over 30 workshops including that of Pixian chili bean paste, fermented bean curd, soy sauce, oil manufacture, Zongzi, Tangchang cloth shoes, Sichuan embroidery, beer brewing and products woven by bamboo, grass and palm fiber. The traditional specialty is everywhere, Xianfeng Dried Radish, Jienong Vegetables Pickled in Soy Sauce, handmade beef paste, sweet deep-fried dough sticks and crispy fried dough twist. They are made at the scene with fragrance striking the nose, which make people's mouth water. At the same time, Sichuan Langdaye Food Limited Company is established to speed up the research and development of tourism food products. The Tangchang Spicy Fermented Bean Curd and Langdaye Fermented Soybeans produced by this company are hot-sale products in the Eighteen Rural Workshops.

九、唐昌大椿巷特色小吃街区

大椿巷特色小吃街区位于郫都区唐昌古镇核心区，是历史底蕴深厚、最具当地传统风味的特色美食街区。

IX. Dachun Alley Snack Block at Tangchang Town

The Dachun Alley Snack Block is the core area of Tangchang Ancient Town, which is a special gourmet food block with historical profundity and most traditional flavor of local characteristics.

唐昌古镇旧称崇宁县，始建于唐代仪凤二年（公元677年），至今已有1 300余年历史，曾是川西坝子最繁华、最富裕的乡镇，现已成为“四川省历史文化名镇”“四川省实施乡村振兴战略工作先进乡镇”。大椿巷是因巷内有一参天大椿树而得名，曾为明朝进士姚激宅院，清中叶开始相继修建公馆。清末民初，崇宁著名文人易象乾、同盟会川西领导人杨靖中先后寄居于此。大椿巷的建筑风格以川西民居和中西合璧为主，灰墙黑瓦、青石铺路、巷道幽静、庭院深深，巷内主要景点有易象乾故居、赵公馆、杨靖中故居、西南马道等，历史文化气息浓厚，是唐昌古镇文化旅游的核心景区。大椿巷及其附近街道的美食主要以特色小吃为主，巷内外有杨抄手、朱记牛肉馆、老号风吹吹、满堂红串串香、崇宁茶社等特色餐饮店；小吃品种特色突出，葱葱卷、施鸭子、豆腐撸冰、紫山药豆饼、马凉面等，均为当地传统小吃，物美价廉，深受游客喜爱。

Tangchang Ancient Town, historically called Chongning County, was firstly established in the second year of Yifeng Period in the Tang Dynasty (677 A. D.). With 1,300-year history, it has ever been the most prosperous and wealthy town in the western Sichuan. Nowadays, it becomes the famous historical and cultural town in Sichuan and the advanced town for rural revitalization in Sichuan. The Dachun Alley derived its name from a towering cedrela in the alley. It was the house of Yao Ji, a Jinshi (a successful candidate in the highest imperial examinations) in the Ming Dynasty. In the middle of the Qing Dynasty, the mansions were established here one after another. In the late Qing Dynasty and the early Republic of China, Yi Xiangqian, a well-known scholar in Chongning, and Yang Jingzhong, a leader of Chinese Revolutionary League in western Sichuan, lived here successively. The architectural style of Dachun Alley mainly includes residence of western Sichuan and that of a combination of Chinese and Western elements. The wall is gray, the tile is black, the road is made of flagstone, the alley is quiet and the courtyard is deep. The scenic spots within the alley include Former Residence

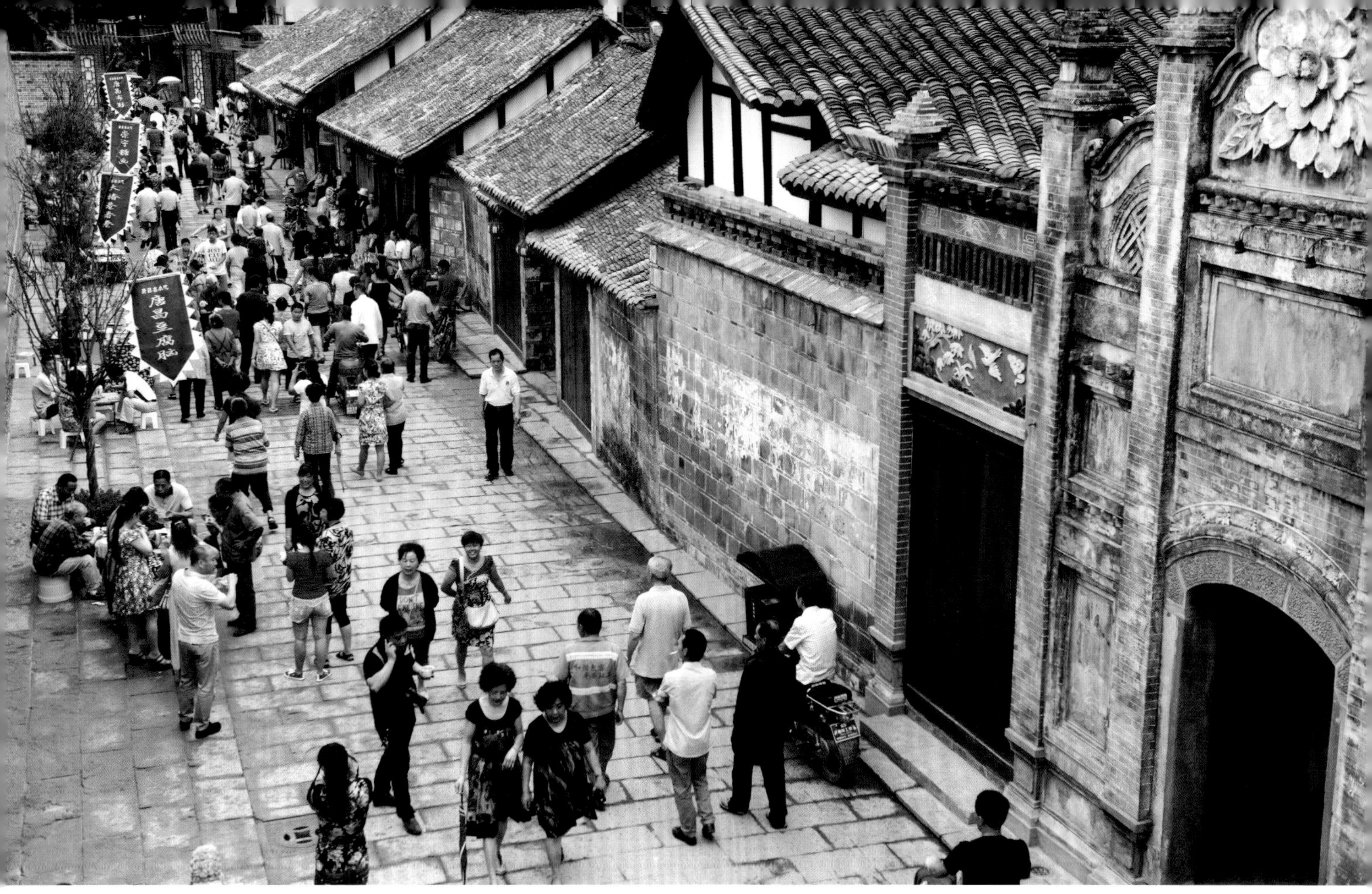

唐昌特色小吃街区（唐昌镇/供）

of Yixiangqian, Zhao's Mansion, Former Residence of Yang Jingzhong and Horse Road in the Southwest. It is the core attraction of the cultural tourism in Tangchang, which has strong historical and cultural atmosphere. There are special snacks in Dachun Alley and its neighboring streets. The alley gathers lots of specialty restaurants, like Yang's Wonton, Restaurant of Zhu's Beef, Time-honored Fengchuichui Spicy Hotpot, Mantanghong Skewers and Chongning Teahouse. The snacks are various and distinctive. The traditional local food, such as thin wrapper with shredded radish, Spiced Duck of the Shi's, ice jelly, purple yam bean cake and Ma's cold noodles in sauce, are attractive in price and quality and popular among tourists.

第三节 川菜文化深度游

Section Three In-Depth Cultural Tour of Sichuan Cuisine

郫都区川菜文化底蕴深厚，这里建有成都川菜博物馆、中国·川菜文化体验馆、中国川菜博览馆、绿城川菜小镇和郫县豆瓣工业旅游基地等川菜文化旅游景点，既各具特色，又彼此融合，都以各自不同的方式讲述着川菜的悠久历史，珍藏着川菜文化的宝贵遗产，展示着川菜产业发展的最新成就，讲述着川菜文化古今相通、薪火相传、守正创新的可贵精神。畅游其中，既可以观赏展品，感受川菜历史厚重的文化气息，又可亲自参与川菜制作、互动体验，还可享受川菜之美，购买川菜的相关产品，深度感受川菜文化的博大精深和无穷魅力。

As the cultural deposits are rich in Sichuan cuisine of Pidu District, there are Sichuan cuisine cultural tourist attractions such as Sichuan Cuisine Museum of Chengdu, China • Sichuan Cuisine Cultural Experience Museum, Sichuan Cuisine Exhibition Hall of China, Lvcheng Sichuan Cuisine Town and Pixian Chili Bean Paste Industrial Tourism Base.

They have distinctive features and add radiance to each other. Telling the long history of Sichuan cuisine in different ways, they enshrine its precious cultural heritage, showcase the achievement made in the industry development and demonstrate its spirit of keeping good tradition and striving for innovation throughout the ancient and modern times which is passed down from generation to generation. When the tourists visit there, they can view the exhibits to feel the historical and cultural atmosphere, or cook the Sichuan cuisine by themselves and have an interactive experience, or buy related products, taste Sichuan cuisine, enjoy the cuisine and deeply understand its broad, profound and glamorous culture.

一、成都川菜博物馆

成都川菜博物馆位于郫都区三道堰，毗邻郫都区古城遗址，建于2005年，占地面积约2.7公顷，藏品6 000余件，是国家AAA级旅游景区、国家三级博物馆，也是一座可以“吃”的博物馆。

I. Sichuan Cuisine Museum of Chengdu

Located at the Sandaoyan Subdistrict, Pidu District, Sichuan Cuisine Museum of Chengdu was established in 2005 and occupies an area of about 2.7 hectares. Owning 6,000 collections, it is an AAA tourist attraction and national third-level museum. It is also a museum where you can enjoy food.

成都川菜博物馆将典型的川西民居与新派古典园林相融合，烘托出一派古蜀文明的余韵流风。该馆布局完整，主要有典藏馆、灶王祠、川菜原料展示区、川菜原料加工工具展示区、互动演示馆、品茗休闲

馆、老川菜馆一条街等。典藏馆中的藏品有用青铜、牙骨、陶、瓷、铁、木、竹等不同材质做成的煮食器、盛食器、酒器、用餐器、茶具等，其收藏的川菜饮食器皿就达到3 000余件。灶王祠是成都供奉灶王最大的祠堂，每逢成都国际美食旅游节和农历腊月二十三，都有祭祀灶王的活动，包括傩戏、舞狮、锣鼓及古装祭祀等。川菜原料展示区集中展示了川菜烹饪所使用的常用原料，诸如蔬菜、家禽等，游客还可在现场体验郫县豆瓣的传统制作工艺。川菜原料加工工具展示区，不仅展示了一系列川菜厨师曾经使用过的加工工具，游客还可参与亲自体验。互动演示馆重在现场演示川菜的刀功、火候及成菜过程，游客可通过“玩做菜”参与互动体验，通过品尝来感受川菜的独特魅力。品茗休闲馆则体现了川菜文化“茶饭相随、饮食相依”的特点，游客可在此感受四川的传统茶馆文化。老川菜馆一条街展现了民国时期著名川菜餐馆的名称，游客不仅能从中领略到民国时期老四川的街景、民风，还可在包间中体验川菜制作工艺。总之，成都川菜博物馆借助“活态博物馆”的理念，不仅能让游客眼观川菜的历史与文化，还能亲自动手做川菜、品川菜，全面、立体了解和体验川菜文化。

Sichuan Cuisine Museum of Chengdu integrates the typical Western Sichuan folk houses with the new classical gardens, setting off the lingering charm of ancient Shu civilization. The museum is well-arranged, including Collection Museum, Kitchen King Temple, Sichuan Cuisine Raw Materials Display Area, Display Area of Processing Tools of Sichuan Cuisine Raw Material, Interactive Demonstration Hall, Tea Lounge and Old Sichuan Restaurant Street. The Collection Museum owns cooking ware, food container, drinking vessel, table ware and tea set made of bronze, dentale, pottery, porcelain, iron, wood and bamboo, with more than 3000 food utensils of Sichuan cuisine. Kitchen King Temple is the largest one to worship kitchen king in Chengdu. In Chengdu International Gourmet and Tourism Festival and the 23rd of the 12th lunar month, there are activities to worship Kitchen King, including Nuo Opera, lion dance, percussion performance and worship in ancient costume. Sichuan Cuisine Raw Materials Display Area demonstrates the raw materials including vegetables and poultry. The tourists can experience the traditional manufacturing process of Pixian chili bean paste at the scene. In Display Area of Processing Tools of Sichuan Cuisine Raw Material, a series of processing tools which were used by chefs are demonstrated for tourists to have a try. In the Interactive Demonstration Hall, the cutting skills of Sichuan cuisine, duration and degree of heating and cooking process are demonstrated. The tourists can join in interactive activities like “cooking the dishes” and feel the charm of Sichuan cuisine by tasting the dishes. Tea Lounge embodies the characteristics of “the food going with the tea” where tourists can feel the culture of Sichuan teahouse. Old Sichuan Restaurant Street demonstrates the architecture style of residence of western Sichuan and the names of well-known Sichuan cuisine restaurants. In this area, the tourists can not only appreciate the old street view and folk custom in Sichuan during Republic of China, but also make Sichuan cuisine personally in the private room. Generally, in virtue of “vivid museum” concept, Sichuan Cuisine Museum of Chengdu can not only make tourists see the history and culture of Sichuan cuisine, but also cook and taste the cuisine by themselves. In this way, they can comprehensively understand and experience this culture.

二、中国·川菜文化体验馆

中国·川菜文化体验馆位于郫都区安德街道川菜汇美食体验街，2013年落成，建筑面积4 000余平方米，收藏了许多川菜文化史料、陶瓷器具及出土文物，是郫都区宣传和展示川菜文化的核心景区。

II. China • Sichuan Cuisine Cultural Experience Museum

Founded in 2013, China • Sichuan Cuisine Cultural Experience Museum is located at the Huimeishi Gourmet Experience Street, Ande Subdistrict, Pidu District. As the key tourist attraction in Pidu to publicize and show the culture of Sichuan cuisine, it occupies an area of over 4,000 square meters and collects historical material, ceramic utensil and unearthed relics about Sichuan cuisine.

中国·川菜文化体验馆

中国·川菜文化体验馆共分为川菜印象、川菜历史、川菜魅力、郫县豆瓣、川菜产业城、现代川菜形成与繁荣六个展示区及蜀都特产中心，“以川菜为题，以豆瓣为魂”，高度浓缩了川菜的文化与历史，生动再现了郫县豆瓣的风雨历程。其中，第一部分以“川菜印象——体验川菜文化，品味巴蜀魅力”为主题，通过高清影像加以呈现；第二部分以“川菜历史——辛香之源，滋味绵长”为主题，梳理、展现了川菜的历史发展脉络；第三部分以“川菜魅力——清鲜醇浓，麻辣多姿”为主题，全面介绍了川菜的调味料、味型、烹饪技法、风味流派、特性、筵宴等，充分展示了川菜“一菜一格、百菜百味”的魅力；第四部分以“郫县豆瓣——食都味典，川菜之魂”为主题，深度梳理了郫县豆瓣的起源、传承和发展；第五部分以“川菜产业城——秉承传统，开启未来”为主题，重点陈述了川菜产业城的建设成就、荣誉及发展前景；第六部分以“近现代川菜的形成与繁荣——融合创新，誉满世界”为主题，详细介绍了近现代川菜的发展情况，展示了不同时期的名店、名菜、名师及川菜文化产品，最后以“世界美食之都——成都”收篇。在本馆游览，可让游客一览历史文化名城、美食之都的民情风貌。

China • Sichuan Cuisine Cultural Experience Museum includes six exhibition areas, namely Sichuan Cuisine Impression, Sichuan Cuisine History, Sichuan Cuisine Glamor, Pixian Chili Bean Paste, Sichuan Cuisine Industrial City, Formation and Prosperity of Modern Sichuan Cuisine as well as the Chengdu Specialty Center. “Its theme of Sichuan cuisine with the bean paste as the soul” condenses the history of Sichuan cuisine culture and represents ups and downs of Pixian chili bean paste. Among these areas, the first part is themed by “Sichuan Cuisine Impression — Experiencing Sichuan Cuisine History and Feeling the Ba and Shu Charm” shown via HD video. With the theme of “Sichuan Cuisine History — Origin of Spicy Flavor with a Lingering Aftertaste”, the second part reviews and shows the historical development of Sichuan cuisine. Themed by “Sichuan Cuisine Glamor — Fresh and Strong in Taste with Gradations of Spice”, the third part introduces the seasonings, different styles and tastes, cooking methods, characteristics and feasts and shows the unique style and flavor of each dish in detail. The fourth part is themed by “Pixian Chili Bean Paste — The Typical Taste and Soul of Sichuan Cuisine”, revealing the origin, inheritance and current development of Pixian chili bean paste as well as its seasoning legend from multiple levels. The fifth part is themed by “Sichuan Cuisine — Following the Traditions and Embarking on Future”, and introduces the building, accomplishment, reputation and future development of Ande Sichuan Cuisine Industrial City. With theme of “Formation and Prosperity of Sichuan Cuisine during Modern and Contemporary Times — Famous around World with Innovation Integrated”, the sixth part introduces the development of Sichuan cuisine during modern and contemporary times, and shows the famous restaurants, dishes, masters and the cultural products of Sichuan cuisine at different times. "Gourmet Capital of the World — Chengdu" is the final part, showing the tourists the popular sentiment and customs of the famous historical and cultural city and the gourmet capital of the world.

川菜文化体验馆在呈现方式上立足科技，突出"与游客互动"的特色。在这里，游客不仅可通过高清影片、电子地图、场景还原、幻影成像等高科技手段了解川菜文化，还可通过虚拟厨房、味觉博士、鸣堂叫菜等互动方式，从视觉、听觉、味觉、触觉上全面感受川菜的独特魅力。川菜文化体验馆已成为追溯川菜历史、传承川菜文化、展示川菜产业发展的重要窗口，已成为成都文化旅游的一张靓丽名片。

Sichuan Cuisine Cultural Experience Museum presents these contents based on scientific technology, highlighting the feature of "interaction with the tourists". Here, the tourists can not only understand Sichuan cuisine culture through technologies like HD video, electronic map, scene representing and holographic image, but also completely feel the unique charm of Sichuan cuisine from the sense of sight, hearing, taste, touch through interaction activities like virtual kitchen, taste doctor and shouting out the ordered dishes to the chef. Sichuan Cuisine Cultural Experience Museum has become an important window for reviewing the history of Sichuan cuisine, carrying forward the Sichuan cuisine culture and showing the development of Sichuan cuisine industry and the calling card of Chengdu cultural tourism.

三、中国川菜博览馆

中国川菜博览馆位于郫都区安德街道川菜产业核心区，始建于2021年，建筑面积6 000多平方米。该馆以川菜产业为核心，是集川菜产业展示、文化体验、行业交流、会议与会务功能为一体的川菜产业博览中心、川菜行业交流中心、川菜产业城规划展示中心。

III. Sichuan Cuisine Exhibition Hall of China

Located at core area of Sichuan cuisine industry at Ande Subdistrict, Pidu District, Sichuan Cuisine Exhibition Hall of China was founded in 2021 and occupies an floor area of over 6,000 square meters. With Sichuan cuisine industry as the core, the hall creates Sichuan Cuisine Industry Expo Center, Sichuan Cuisine Sector Exchange Center and Display Center of Sichuan Cuisine Industrial City Planning with international exposure and integrating the functions of Sichuan cuisine industry exhibition, cultural experience, profession exchange and conference and general affairs.

中国川菜博览馆共分为三层，一层以"植根巴蜀、川香世界——川菜历史演进之路"为主题，以历史的自然推进为时间轴，全面梳理了从先秦时期到21世纪20年代，川菜从无到有、从弱到强的发展历程，客观追溯了川菜不断创新、生生不息、绵延至今的动力源泉；二层以"融合创新、繁荣四方——川菜产业繁荣之景"为主题，以地理空间为坐标，详细梳理了川菜从内陆巴蜀不断走向世界、成为中华代表性风味流派的发展历程，充分展示了改革开放以来川菜产业的发展成就；三层为多功能会议中心，是川菜文化研学、交流场所，主要用于临时展览、新闻发布、会议举办及学术报告等活动。

Sichuan Cuisine Exhibition Hall of China is divided into three floors. The first floor is themed by "Historical Evolution of Sichuan Cuisine — Rooting in Ba and Shu and Spreading throughout the World". It shows chronologically the development history from pre-Qin period to 1920s of Sichuan cuisine which grew out of nothing and finally thrived. Meanwhile, it discovers the dynamic source for Sichuan cuisine's inclusive innovation, continuous and endless development. The second floor is themed by "Prosperity of Sichuan Cuisine Industry — Integrating Innovation and Flourishing Everywhere", which reveals the geographical development where Sichuan cuisine entering the world from inland Ba and Shu thus becoming the representative flavor style. It fully shows the accomplishment of Sichuan cuisine industry since the reform and opening-up. The third floor is the multifunction conference center, the place for study and exchange of Sichuan cuisine. It is mainly used for activities like temporary exhibition, press release, meeting and academic report.

中国川菜博览馆

中国川菜博览馆在表现方式上具有更强、更具时代特征的科技感，采用数字化控制系统，利用声、光、电等现代科技手段，结合触控交互、纱幕投影、红外感应、感官体验、艺术装置、大型实体灯光沙盘造景、多媒体影片、场景营造、5G适时连线等多种表现手法，互动感更强，能够让游客身临其境，全方位、沉浸式了解川菜的发展历程及产业建设成就，感受川菜文化的博大精深和无限魅力。

Sichuan Cuisine Exhibition Hall of China applies high technologies with the feature of the times in the presentation modes. It adopts the digital control system and the modern technological means like sound, light and electric equipment, combining with various presentation modes including touch control interaction, curtain projection, infrared sensor, sensual experience, artistic installation, large lighting sand table models, multimedia films, scene creating, timely 5G connection, etc. Such a strong sense of interaction gives the tourists the feeling of immersion, who can wholly understand the development process of Sichuan cuisine and the accomplishment of Sichuan cuisine industry, also feel the extensive and profound Sichuan cuisine culture with endless glamour.

四、绿城川菜小镇

绿城川菜小镇位于郫都区安德街道，紧邻成灌公路、成灌高速与第二绕城高速，属成都市“半小时经济圈”范围。绿城川菜小镇占地面积670多公顷，是紧紧围绕“川菜文化”与“现代都市田园”两大IP而打造的集文旅、田园、休闲、康养、教育、宜业、宜居于一体的生活小镇。

IV. Lvcheng Sichuan Cuisine Town

With Chengdu-Dujiangyan Highway, Chengdu-Dujiangyan Expressway and the second beltway expressway around, Lvcheng Sichuan Cuisine Town is located at Ande Subdistrict, Pidu District, belonging to “half-hour economic circle” in Chengdu. The town occupies an area of over 670 hectares and creates a town full of vitality integrating tourism, garden, recreation, rest, education and business operation based on two representatives “Sichuan Cuisine Culture” and “Modern Urban Garden”. It is also of livability.

在绿城川菜小镇的川味生活美学馆，游客能以沉浸式、全方位、多角度的不同方式来体验川菜的魅

力。未来，绿城川菜小镇还将建设占地面积约15公顷的“川菜·科创里”高品质科创空间，并导入研发设计、创新转化、场景营造、社区服务等功能，整合关联科技资源，搭建川菜产业研究院、川菜学院等科研创新平台，使这里成为安逸田园生活的缩影、川菜文化的集成展示地和郫都川菜产业链上的重要环节。

The tourists will be immersed in the charm of Sichuan cuisine from all-round and multiangle perspectives in Sichuan-style Life Aesthetics Museum at Lvcheng Sichuan Cuisine Town. In the future, Lvcheng Sichuan Cuisine Town plans to build the high-quality “Scientific Innovation Space of Sichuan Cuisine” of about 15 hectares with functions like R&D, designing, innovation transforming and community services. It will integrate relevant scientific resources and

郫县豆瓣工业旅游基地——阳光晒场

create the scientific innovation platforms like Sichuan Cuisine Industry Institute and Sichuan Cuisine College, making the town an epitome of comfortable countryside life and the place for displaying Sichuan cuisine culture, thus becoming the important one within the Sichuan cuisine industrial chain in Pidu.

五、郫县豆瓣工业旅游基地

郫县豆瓣工业旅游基地位于郫都区安德镇四川省郫县豆瓣股份有限公司内，该基地将工业生产与旅游有机融合，是中国川菜产业城中工业旅游的典型代表。

V. Pixian Chili Bean Paste Industrial Tourism Base

Located inside the Sichuan Pixian Chili Bean Paste Incorporated Company, Ande Town, Pidu District, Pixian Chili Bean Paste Industrial Tourism Base combines the industrial production and tourism together and becomes a typical example of industrial tourism in China's Sichuan Cuisine Industrial City.

该基地规模庞大、布局科学、功能齐全，共分为企业文化展示厅、阳光晒场、自动化包装车间、郫县豆瓣非物质文化遗产体验基地四个部分，游客可近距离、全方位参观郫县豆瓣的加工、制作、包装等全部生产流程，亲身参与郫县豆瓣的非遗体验活动，还可选购心仪的郫县豆瓣产品。该基地每年接待游客上万名，有效带动了中国川菜产业城工业旅游的发展势头，同时还为提升郫县豆瓣的工业旅游价值、传承郫县豆瓣文化、开拓郫县豆瓣的产品市场发挥了重要作用。

With large scale, scientific layout and multiple functions, the base area is divided into the display area of enterprise culture, the sunning ground, automatic packing workshop as well as the intangible cultural heritage experience area of Pixian chili bean paste. The tourists could closely and thoroughly watch the processing, making and packing of Pixian chili bean paste. Also, they can participate in the intangible cultural heritage experience activities by themselves. Finally, people could choose to buy their favored Pixian chili bean paste products. Every year, the Industrial Tourism Base Area of Pixian Chili Bean Paste receives over 10 thousand tourists, effectively driving the development of industrial tourism in China's Sichuan Cuisine Industrial City. Meanwhile, it plays an important role in improving the tourism value of Pixian chili bean paste, carrying forward the Pixian chili bean paste culture, and extending the market for Pixian chili bean paste products.

温馨小贴士：川菜文化寻根体验之旅线路推荐

Warm Tips: Recommended Routes for Root-Seeking Experience Tour of Sichuan Cuisine Culture

为了全方位展示川菜文化旅游的无穷魅力，打造川菜文化旅游品牌，郫都区特制定了川菜文化寻根之旅一日游、二日游线路，为游客提供川菜文化寻根之旅指南，带给游客川菜文化寻根之旅非凡体验。

In order to display the infinite charm of Sichuan cuisine cultural tourism in an all-round way and create brand for Sichuan cuisine cultural tourism, Pidu District has specially formulated one-day and two-day routes of root-seeking tour of Sichuan cuisine culture so as to provide a guide to tourists and bring them an extraordinary experience.

一日游线路规划：享安逸郫都·品百菜百味

One-Day Tour Route: Enjoying Comfortable Pidu • Tasting Various Sichuan Cuisine

China • Sichuan Cuisine Cultural Experience Museum

中国·川菜文化体验馆

活动内容：深度参观
博物馆探秘
博物馆摄影等

Intangible Cultural Heritage Experience Base of Pixian Chili Bean Paste

郫县豆瓣传统技艺非遗体验

活动内容：微生物研究
温光水研究
豆瓣集市
小食神体验行等

Nongke Village
Qinggangshu Village
Sichuan Embroidery Park

农科村/青杠树/蜀绣公园

活动内容：学做一道正宗川菜
赛龙舟活动
望丛祠赛歌会等郫都民俗活动相结合。

二日游线路规划：追寻川菜文化之根，探秘熊猫成长之谜

Two-Day Tour Route: Seeking Root of Sichuan Cuisine Culture • Exploring Mystery of Giant Panda

China • Sichuan Cuisine Cultural Experience Museum

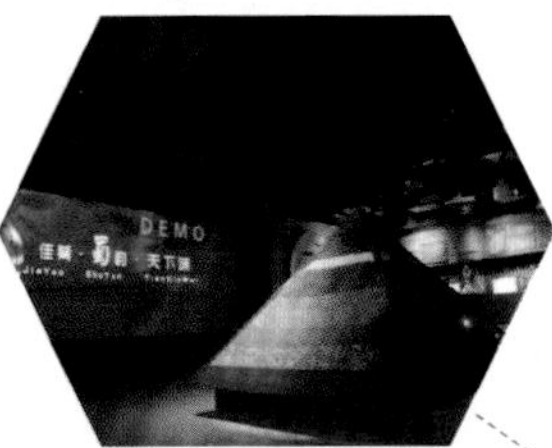

中国·川菜文化体验馆

活动内容：深度参观
博物馆探秘
文化交流
课题研讨
博物馆摄影等

Sichuan Pixian Chili Bean Paste Incorporated Company

活动内容：微生物研究
地理标志性产品研究
非遗技艺体验
豆瓣剧场等

郫县豆瓣传统技艺非遗体验

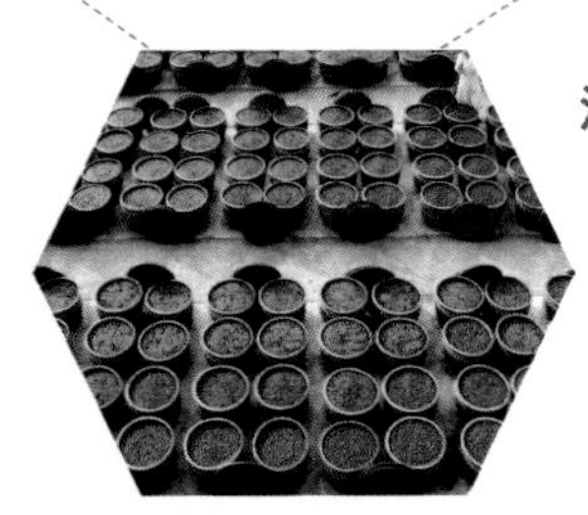

Dujiangyan Irrigationy Project - Landmark Sichuan Cuisine

拜水都江堰

活动内容：追寻水利文化
探究水利过程
品鉴地域美食等

第六章 CHAPTER SIX

饮食名人 群贤毕集

Celebrities Related with Delicacy

古往今来，郫都汇聚了众多与饮食文化相关的杰出人物，有的生于郫都、长于郫都；有的则是出于各种原因来到郫都居住、生活，他们都与郫都美食产生了千丝万缕的联系。在郫都这片土地上，历朝历代的原住民、移民与访客共同创造了引人注目的郫都饮食文化。

Since ancient times, many celebrities related to food culture have emerged in Pidu. Some were born and grew up in Pidu, some came to settled down in Pidu for various reasons. They are all inextricably linked with Pidu delicacy. On the land of Pidu, aborigines of dynasties, immigrants and visitors have jointly created a brilliant Pidu food culture.

第一节 古代郫都名人与饮食典故

Section One Ancient Celebrities in Pidu and Food Allusions

在古代郫都，有通过诗词歌赋记录郫都及蜀地美食的文学家、诗人，如扬雄通过《蜀都赋》描绘了一幅汉代蜀中的饮食画卷；杜甫、陆游、苏轼等名家则用诗词歌颂了郫都的美酒、美食；还有酿造郫筒酒的晋代郫县县令山涛，以及推动中国瓷器制造工艺的官员何稠、杨景宗等。他们都从不同角度对郫都饮食及中国饮食文化发展做出了重要贡献。

In ancient Pidu, there were writers and poets who recorded Pidu and Sichuan cuisine through poetry and songs. *Story of Chengdu* by Yang Xiong described a picture of Sichuan cuisine in Han Dynasty. Poets like Du Fu, Lu You and Shu shi praised food and wine in Pidu. There is Shan Tao, the magistrate of Pixian County in the Jin Dynasty who brewed Pitong Wine, and He Chou and Yang Jingzong, officials who promoted the improvement of Chinese porcelain manufacturing technology. They all contributed to the development of food culture in Pidu and China from different aspects.

一、西汉大儒扬雄与《蜀都赋》

扬雄，字子云，生于公元前53年，是我国汉代著名的文学家、哲学家和语言学家，他对饮食也有自己的见解，并首次以赋的形式生动描绘了四川的饮食风貌，具有开创之功。

I. Yang Xiong — Great Confucianist in the Western Han Dynasty and His *Story of Chengdu*

Yang Xiong, with the courtesy name of Ziyun, was born in 53 B. C. He was a famous writer, philosopher and linguist in Han Dynasty, and had his own views on delicacy and recorded the dishes and cooking in Sichuan in the form of compose for the first time.

扬雄像

扬雄的高祖扬季官至庐江太守，汉武帝时举家迁到今郫都友爱镇境内。由于蜀中才子司马相如以辞赋名闻天下，使得蜀中辞赋之风极盛，扬雄也深受其影响。扬雄四十岁后到长安，因“文似相如”而待诏宫廷，任黄门郎等职。他在为汉成帝连续作了许多辞赋之后，又潜心研究学术，并写下了《法言》《太玄》《方言》等著述。

Yang Ji, great-great-grandfather of Yang Xiong, came to the governor of Lujiang. During the reign of Emperor Wu of the Han Dynasty, the whole family moved to Youai Town, Pidu. As Sima Xiangru in Sichuan got his fame all over the country by writing compose, which we called Fu nowadays. The style of Fu in Sichuan is very prosperous, and Yang Xiong is deeply influenced by it. When Yang Xiong went to Chang'an at the age of 40, he got the reputation as court official (Huang Men Lang) because his Fu was as good as Sima Xiangru. After writing a lot of Fu for Emperor Cheng of Han, he focused on academic research and wrote *Fayan, Taixuan, Fangyan,* etc.

扬雄所处的时代，正是巴蜀经济蓬勃发展的时期。由于土地肥沃、物产丰富、百姓富庶，人们便“以富相尚”，因重视饮食，而致佳肴众多、筵宴不断。面对一派繁荣富饶的景象，扬雄极为自豪，便用自己最喜爱的辞赋形式，描绘和赞美了家乡的物产、美肴和筵宴。伴随着扬雄《蜀都赋》的横空出世，使更多的世人知晓了蜀地丰饶的山川物产、特殊的风土人情、繁荣的商业经济，以及奢华的豪门盛宴……

Yang Xiong is in a vigorous economic development period in Bashu. Because the land was fertile, the products were abundant, and the people were rich, local people pay attention to diet, enjoy feast & banquets frequently. Yang Xiong was very proud of the prosperous hometown. He described and praised the products, delicacies and banquets of in the form of his favorite Fu, and created *Story of Chengdu,* which was popular among people at that time.

扬雄在《蜀都赋》中共记录了70余种食材。首先记叙了蜀中东南西北四方的著名物产，如川东石鳙，川南菌芝，川西井盐，川北野兽，接着又详细描绘了水陆所产，还有江河溪流中的各种水生植物及龟鳖鱼类：“其浅湿则生苍葭蒋蒲，藿芧青苹，草叶莲藕，茱华菱根……其深则有猵獭沉鳝，水豹蛟虵，鼋蟺鳖龟，众鲜鳎鰗。”“蒋”即茭白，“蒲”即香蒲，“藿”即豆叶，“藕”即莲藕，“鳝”为鳝鱼，“鳎”俗称娃娃鱼，这些都是当时蜀中的常用食材。陆地上的食材也极为丰富，园圃田野里果实累累，“黄甘诸柘，柿桃杏李枇杷，杜樼栗柰，棠黎离支，杂以梴橙，被以樱梅……五谷冯戎，瓜瓠饶多……往往姜栀附子巨蒜，木艾椒蓠，蒟酱酴清，众献储斯。盛冬育笋，旧菜增伽，百华投春，隐隐芬芳”。“木艾”“椒蓠”即指食茱萸籽和花椒，它们与生姜、大蒜、蒟酱等共同组成蜀中独特的调味料。此外，《蜀都赋》中还展现了当时豪门筵宴的情景：“若其吉日嘉会……置酒乎荥川之闲宅，设座于华都之高堂，延惟扬幕，接帐连冈。众器雕琢，早刻将皇。……厥女作歌，舞曲转节”。可以说，扬雄是最早对四川的食材、烹饪技术、菜肴及宴会进行系统描述的大家，他所作的《蜀都赋》已成为了解汉代成都饮食的重要文献资料和郫都饮食文化的重要组成部分。

Yang Xiong recorded more than 70 kinds of ingredients in his *Story of Chengdu*. Firstly, it recorded famous products in all directions, like fish in eastern Sichuan, mushroom in southern Sichuan, well salt in western Sichuan and animals in northern Sichuan. Secondly, he described in details the various aquatic plants and turtles and fish produced by land and water, rivers and streams. "In the shallow water, it produces Cangjia, Jiang, Pu, Huo, Qing Ping, grass leaves, lotus roots, Zhu Hualing roots; in the deep water area, there are Marmot, Xi, Turtles, Ta." Jiang is water oat. Pu is Typha orientalis Presl. Huo is bean leaf. Xi is Monopterus albus. Ta is Andrias davidianus. All of them are normal food ingredients in Sichuan at that time. The food ingredients on the land are also very rich, and the gardens and fields are full of fruits. "Huang Gan, Zhu Zhe, Persimmon, Peach, Apricot, Plump, Loquat, Apple,etc., mixed with Oranges, Cherry; Cereals & melon is abundant; Ginger, Aconitum carmichaelii Debx, big Garlic, Muai & Jiaoli, etc. The paste is nice and clean. The bamboo shoots are cultivated in winter, the vegetables on table are added, and hundred flowers cast into spring, which is faint and fragrant." Muai & Jiaoli is Cornus and pepper, which forms a unique seasoning together with ginger, garlic and catsup in Sichuan. Moreover, *Story of Chengdu* had revealed the scene of great feast at that time, "when people gather together in a good day, they would prepare good wine in a leisure house near the river or a proper hall in the city. The guests are so many that the host has to keep adding tables, some of which even reach to the mountains nearby. And the carved decorations are beautiful." In the meantime, "there are girls singing" and "accompaniment along with the beat of dancing". We can say that Yang Xiong is the first master to systematically describe food ingredients, cooking methods, dishes and feasts in Sichuan. His *Story of Chengdu* has become a very important reference to understand the delicay in ancient Chengdu, and become an important part of Pidu food culture.

时至今日，作为扬雄故里的郫都，依然在不断继承和发扬扬雄文化，不仅修缮了扬雄墓，还规划建设了扬雄纪念馆、扬雄文化苑、扬雄绿道等。2021年4月，郫都区举办了“西道孔子・扬雄春祭大典”活动，以古今交融之风范，让先贤高风在当代大放光彩。

Today, Pidu, the hometown of Yang Xiong, continues to inherit and carry forward his culture. It not only repaired tomb of Yang Xiong, but also planned to build memorial hall of him, Yang Xiong cultural park, Yang Xiong greenway, etc. In April, 2021, Pidu District hosted a spring memorial ceremony for Yang Xiong, the "Confucianist • Yang Xiong Spring Festival Ceremony". With the style of blending ancient and modern, this activity carried forward the spirit of ancient sage in modern time.

二、郫县县令山涛与郫筒酒

在历史上，四川很早便有酿酒的记载，《华阳国志》中载古蜀开明王时有“以酒曰醴”之说。西汉时，由于蜀地富豪们经常痛饮宿醉，故有“成都有累月之醉”的说法。古代郫都农业发达、物产丰富，为酿酒业的发展提供了良好的物质基础，于是，郫筒酒应运而生。

II. Shan Tao — County Magistrate and Pitong Wine

There was the record of brewing wine long ago in Sichuan. *The Records of Huayang* recorded that wine was been invented when Kaiming was the king of ancient Sichuan. In the Western Han Dynasty, Chengdu was recognized as drunken city, because the rich people there drank over night frequently at that time. The developed agriculture and abundant products in ancient Pidu built a good foundation for the development of wine industry. Pitong Wine was born at the right moments.

在郫都，一直流传着西晋名士山涛到郫县做县令时酿制郫筒酒的故事。山涛，字巨源，西晋散文家，河南怀县（今河南武陟县）人，与嵇康、阮籍等交游，为“竹林七贤”之一。司马师执政时，山涛出仕为官，历任吏部尚书、右仆射等职。“竹林七贤”都喜饮酒，并且常常通过聚会饮酒的方式来表现超脱世俗的气度。其中，山涛的酒量很大，但极为理智，适量而止，从不过量。《晋书・山涛传》记载说：“涛饮酒至八斗方醉，帝欲试之，乃以酒八斗饮涛，而密益其酒，涛极本量而止。”正是出于山涛对酒的喜爱，他在郫县任上发现并酿制郫筒酒也就不足为怪了。

山涛像

In Pidu, the story of Shan Tao, a proser of the Western Jin Dynasty, one of the "seven sages of the bamboo grove" in the Jin Dynasty, acting as the county leader of Pixian and brewing the Pitong Wine there, is always circulating. He was born in Huaixian County (current Wuzhi County), Henan province, with courtesy name Juyuan and made friends with Ji Kang, Ruan Ji, etc. When Sima Shi was in power, Shan Tao became an official and successively served as the Minister of the Ministry of officials, the Right Servant and so on. The seven sages all loved drinking. The frequently gathered to drink to show their unworldly grace. Shan Tao was very good at drinking, but he was very reason and never drank too much. *Jin Shu • Biography of Shan Tao* recorded, "Shan Tao started to get drunk after drinking 8 Dou of wine. The emperor wanted to test his drinking capacity and sent him 8 Dou of wine to drink. In fact, more wine had been sent to Shan Tao secretly. Shan Tao just stopped at his own drinking capacity." It is not surprising that Shan Tao found and brewed Pitong Wine during his tenure in Pixian County because of his love for wine.

据清代同治年间《郫县志》载：山涛"晋初为郫令，常刳巨竹节酿酴醾，蔽以蕉叶，缠以藕经，宿熟，世目为郫筒酒"，即将酴醾酒放置在竹节中，盖上芭蕉叶，用藕带缠绕，酿制一夜后成熟即可。用此方法酿制的郫筒酒味道醇厚，有竹的清香，并因其浓郁的地方风味而逐渐称誉全川，并走出四川，凡品尝过此酒的历代文人墨客无不赞赏有加。杜甫、陆游、苏轼等都为郫筒酒留下了千古名句，范成大、仇兆鳌、袁枚等还记录了郫筒酒的酿造方式及风味特色，袁枚就在其所著的《随园食单》中写道："郫筒酒，清冽彻底，饮之如梨汁蔗浆，不知其为酒也，但从四川万里而来，鲜有不味变者。余七饮郫筒，惟杨笠湖刺史木簰上所带为佳。"郫筒酒清凉爽透，清澈见底，喝起来像梨汁、甘蔗浆，并且已经传播到了四川以外的地区。

Chronicles of Pixian in Tongzhi period of the Qing Dynasty recorded, "as the county leader in Pixian in the early Jin Dynasty, Shan Tao used to cut out large bamboo tubes to brew the sweet rice wine, with the steamed rice including the starter covered by banana leaf and tied up by lotus stem, thus making the wine after just one night, which called Pitong Wine by people". Pitong Wine brewed in this way is mellow with bamboo fragrance. Due to its strong local flavor, Pitong Wine has gradually become famous throughout Sichuan and other provinces for a long time. All literati and writers of all dynasties praised the wine after drinking. Du Fu, Lu You, Su Shi wrote well known poems for it. Fan Chengda, Qiu Zhaoao, Yuan Mei recorded the way of brewing Pitong Wine and its flavor characteristics. In *Food List*, Yuan Mei wrote that "Pitong Wine was clean and clear, had a flavor of pear juice and sugar cane juice. People even don't know it is wine in fact. However, it's difficult to keep original taste as transported from so far away from Sichuan. I tried Pitong Wine 7 times, only the Pitong Wine brought by the provincial governor Yang Lihu was good -tasting."

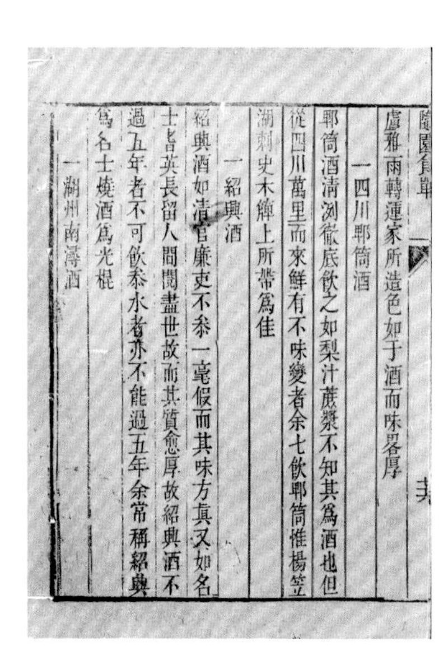

隨園食單

盧雅雨轉運家所造色如于酒而味略厚

一四川郫筒酒

郫筒酒清冽徹底飲之如梨汁蔗漿不知其爲酒也但從四川萬里而來鮮有不味變者余七飲郫筒惟楊笠湖刺史木簰上所帶爲佳

一紹興酒

紹興酒如清官廉吏不參一毫假而其味方真又如名士耆英長留人間閱盡世故而其質愈厚故紹興酒不過五年者不可飲參水者亦不能過五年余常稱紹興爲名士燒酒爲光棍

一湖州南潯酒

袁枚《随园食单》有关郫筒酒的记载

民国时期，郫都的益丰和、元丰源、绍丰和三家酱园都生产过郫筒酒。新中国成立后，朱德到郫都视察时也曾询问过郫筒酒的生产情况。2008年5月，郫筒酒传统酿造技艺被列为郫都区第二批非物遗保护名录。郫筒酒的传统酿造技艺保护单位为了保护郫筒酒这一千年历史品牌，逐渐进行了较大批量的生产。2020年，郫筒酒传统酿造技艺被列入第七批市级非遗代表性项目名录。在郫都区各方的共同努力下，使得郫筒酒的传统制作技艺得到了持续、有效地保护与传承。

During the Republic of China, Yifenghe paste shop, Yuanfengyuan paste shop and Shaofenghe paste shop had produced Pitong Wine. After the foundation of PRC, marshal Zhu De had asked about the production of Pitong Wine when he visited Pidu. In May, 2008, the traditional brewing technique of Pitong Wine has been added into the second batch of Intangible Cultural Heritage Protection list of Pidu District. In order to protect the thousand-year old brand of Pitong Wine, the protection unit of traditional brewing technology of Pitong Wine gradually produces mass production of it. In 2020, the traditional brewing technique of Pitong Wine has been added into the seventh batch of city municipal Intangible Cultural Heritage Representative Projects. With the efforts of all parties, the traditional brewing technique has been protected and inherited persistently.

三、隋朝发明家何稠与瓷器烧造技术的改进

中国的瓷器烧造技术在隋代取得了较大进步，在其发展过程中，郫都人何稠为此做出了突出贡献。何稠，字桂林，生于公元543年，是中国历史上著名的建筑师、发明家。何稠祖父名细胡，因通商入蜀，定居今郫都，号为西州大贾。其父何通，善斫玉。何稠幼年时即学会制造工艺品，思考问题细致周到。父亲去世后，他跟随叔父何妥生活、学习，十余岁即跟随何妥到长安，后历任御府监、太府丞、太府少卿、太府卿等，为皇室制造车辆、服装、仪仗、兵器和其他器物，并负责营建宫室和陵庙。隋文帝时，波斯（今伊朗）国王献金绵锦袍一件，隋文帝命何稠仿制，他制成后竟比原件更精美。在瓷器制作方面，何稠对当时的瓷器烧造技术进行了大量改进，对中国的瓷器生产历史产生了深远影响。何稠在研究前人琉璃制造技术的基础上，采用新平镇（江西景德镇）的瓷土、制瓷技术，提高烧窑温度而烧制出绿瓷。据《隋书·何稠传》载："时中国久绝玻璃之作，匠人无敢措意，稠以绿瓷为之，与真不异。"何稠制造琉璃，不仅对当时的琉璃生产有直接影响，而且还对提高景德镇的制瓷技术，起到了重要的促进作用，推动了景德镇制瓷业的发展，使瓷器在中国社会各阶层的饮食生活中得到更为广泛的应用，大大丰富了中国瓷质餐饮器具的品类。

何稠像

III. He Chou — Inventor of the Sui Dynasty and the Improvement of Porcelain Firing Technology

The porcelain firing technique has been improved much in the Sui Dynasty, which was inseparable from the contribution made by He Chou. He Chou, with courtesy name Guilin, was born in about 543 A. C. He was a famous architect, inventor, and an outstanding technology figure in Chinese history. He Chou's grandfather was He Xihu, the great merchant in west of Central Plains. He did business in Sichuan and then settled down in current Pidu. He Chou's father was He Tong who was good at dealing with jade. He Chou learned to make handicrafts and to think carefully when he was young. When his father passed away, He Chou lived with his uncle He Tuo and visited Chang'an with him as a teenager. He Chou worked for the emperor as Yu Fu Jian, Tai Fu Cheng, Tai Fu Shao Qing, Tai Fu Qing, etc. He made vehicles, clothing, ceremonial equipment, weaponry and other equipment. He also established palace and temple of emperors. During Emperor Wen of the Sui Dynasty, the king of Persia (now Iran) presented golden cotton robes. Emperor Wen ordered He Chou to copy one. It was more exquisite than the original one when he made it. In terms of porcelain making, He Chou improved the porcelain firing technology at that time, which had a far-reaching impact. On the basis of studying the colored glaze manufacturing of predecessors, He Chou adopted the porcelain clay and porcelain making technology of Xinping town (Jingdezhen, Jiangxi Province), increased the kiln temperature and fired Lvci porcelain with functional characteristics such as anti-bacterial, easy to clean, etc. According to *Sui Shu • Biography of He Chou*, "it wasn't masterpiece of glass in China for a long time. Craftsmen didn't dare to try. Lvci porcelain by He Chou is the one." He Chou 's colored glaze not only had a direct impact on the production of colored glaze at that time, but also played an important role in promoting the improvement of porcelain making technology in Jingdezhen, driving the development of Jingdezhen porcelain making industry, making porcelain more widely used in Chinese society, and enriching the categories of Chinese porcelain tableware.

四、圆悟克勤与茶禅一味

佛教自汉代传入巴蜀，至唐宋而兴盛。特别是宋代，成都地区的禅宗（汉传佛教宗派之一）得到空前发展，苏轼曾赞叹道："成都，西南大都会也，佛事最胜。"曾两度住持成都昭觉寺的圆悟克勤，就是宋代成都地区涌现出的一位禅宗著名人物。

圆悟克勤像

IV. Yuan Wu Ke Qin and Culture of Tea and Dhyana

Buddhism was introduced to Sichuan in the Han Dynasty and reached its peak in the Tang and Song Dynasties. Dhyana (one of the sects of Buddhism) in Chengdu region has developed greatly in the Song Dynasty. Even Shu Shi once said, "Chengdu is the metropolis of Southwestern China, where the Buddhism is so prosperous." Yuan Wu Ke Qin, who had been the chief of the Zhaojue Temple in Chengdu for twice, was the famous figure of Dhyana in the Song Dynasty.

圆悟克勤，俗姓骆，字无著，法名克勤，崇宁县（今郫都区唐昌镇附近）人。曾参谒宋代禅宗主流临济宗杨岐派传人五祖法演，后弘法于各地，遍历楚水吴山，成为名冠南北的一代佛教宗师。皇帝曾多次召其问法，并赐"佛果禅师"之号，后又赐号"圆悟"。他不仅撰写了禅学奇书《碧岩录》，而且手书"茶禅一味"四字真诀，对禅宗和茶文化的发展都产生了深远影响。

Yuan Wu Ke Qin was born in Chongning Couty (near current Tangchang Town in Pidu). His family name was Luo, courtesy name was Wuzhu, and religion name was Ke Qin. He once paid a formal visit to Master Fa Yan, who was the Fifth Master of Yang Qi Division, Lin Ji Sect (the main stream of Dhyana) in the Song Dynasty. Then he travelled and spread Buddhism all over the country. Finally, he became a Buddhism master in China. The emperor asked him for Buddhism questions several times, and granted him the religion name of Fo Guo Master and Yuan Wu. He wrote a master piece *Bi Yan Lu* and wrote the four-word admonition "Cha Chan Yi Wei" (tea culture and Dhyana culture are in common), which has influenced Dhyana and food culture deeply.

作为一种饮品，茶与佛教有千丝万缕的联系。唐代陆羽所著《茶经》，就记载了当时僧人饮茶的情况。在日常茶事中，茶道与佛教之间在思想内涵方面产生了越来越多的共通之处，被人们所共识，从而形成了“禅茶”这一具有浓郁佛教特色的茶道风格。佛教认为茶有三德：一是坐禅时可以通宵不眠；二是满腹时帮助消化；三是茶为不发之药。于是，喝茶就成为佛教僧人日常生活不可或缺之事，以茶参道、以茶悟禅、以茶对禅机的故事，历史上也是层出不穷。

As a drink, tea is inextricably linked with Buddhism. *The Book of Tea* written by Lu Yu in the Tang Dynasty recorded monks drank tea at that time. There are more and more commonalities in the ideological connotation between teaism and Buddhism, which are recognized by people. Then Zen tea, a teaism style with strong Buddhism characteristic has been formed. In Buddhism, tea has three virtues. First, tea can keep people sleepless overnight during sitting in meditation. Second, tea can help digestion. Third, tea can control lust. Therefore, drinking tea has become a necessary part of monk’s daily life. There were lots of stories about seeking Taoism from tea, understanding Dhyana from tea, etc.

僧人与寺院也促进了茶叶生产的发展和制茶技术的进步，名刹出名茶，自古有之。名刹多位于名山，名山多在深山云雾之中，既有野生之茶树，也宜于茶树的种植。“扬子江中水，蒙山顶上茶”，四川雅安蒙顶山出产的蒙顶茶，相传是汉代甘露寺普贤禅师吴理真亲手所植，晋朝开始作贡茶。福建武夷山出产的武夷岩茶是乌龙茶的始祖，它以当年佛寺天心观生产的大红袍最为著名，僧人还根据不同时节将采回的茶叶分别制成“寿星眉”“莲子心”和“凤尾龙须”三种名茶。

Monks and temples have promoted the development of tea production and technique. In ancient time, famous tea was produced in famous temple. Famous temple was usually located in famous mountain with deep mountain clouds, where there is wild tea plants and is suitable for tea planting. There is a saying “Water from Yangtze river and tea from top of Mengding mountain”. It is said that Mengding tea produced in Mengding Mountain, Ya'an, Sichuan was personally planted by Wu Lizhen, Puxian Zen master of Ganlu Temple in the Han Dynasty, and became the royal tea in the Jin Dynasty. Mt. Wuyi’s cliff-grown tea from Wuyi Mountain in Fujian province is the origin of oolong tea. Da Hong Pao produced by Tianxingguan is the most famous one. Monks have made tea leaves picked in different seasons into three famous tea — Shouxingmei, Lianzixin and Fengwei Longxu.

“茶禅一味”四字真诀传入日本后，深刻地影响了日本茶道，成为日本代代相传的国宝，如村田珠光（1423–1502年）创设“四叠半”茶室，将禅的精神与饮茶方式密切结合在一起，他为茶道所确立的禅的指南——“和敬清寂”成了日本茶道思想的核心，一直延续至今。

When the four-word admonition “Cha Chan Yi Wei” (tea culture and Dhyana culture are in common) was introduced to Japan, it deeply influenced the teaism and became a national treasure by generations in Japan. Murata Shukō (1423-1502) established the Yojouhan Tea Room to combine the Dhyana with the way of drinking tea. The Dhyana he established for the teaism — “harmony, respectation, purity, tranquility” has become the principle of Japanese teaism till now.

第二节 近现代郫都名人与饮食典故

Section Two Modern Celebrities in Pidu and Food Allusions

近现代郫都饮食名人更是人才济济、群星璀璨。其中，有创制郫县豆瓣，开办顺天、益丰和酱园，促进郫县豆瓣发展的福建移民陈氏家族；有曾经寓居郫都的国画大师张大千；也有郫都本地的糖画艺人高邦金、名厨钟泽先，等等。通过梳理他们与饮食的精彩故事，更能深刻地感受郫都饮食文化的厚重与丰富多彩。因陈氏家族与郫县豆瓣在第三章中已有详讲，这里不再赘述。

There are many celebrities related with delicacy in modern Pidu. The Chen family, who immigrated from Fujian, invented Pixian chili bean paste, established Shuntian and Yifenghe paste shops to promote the development of Pixian chili bean paste. The great master of traditional Chinese painting, Zhang Daqian, was born in Neijiang but lived in Pidu once. There are local sugar painting artist Gao Bangjin and famous chef Zhong Zexian, etc. By sorting out stories about those celebrities, we can feel the richness of Pidu food culture. Chen family and Pixian chili bean paste have been introduced in Chapter Three, hence we won't repeat here.

一、张大千在郫都的饮食生活

张大千（1899～1983年），四川内江人，是我国近现代著名的书法家、绘画大师。生长于崇尚饮食的四川，张大千从小耳濡目染，也钟情于饮食，他的一生，主要从事中国传统书画艺术的学习、创作和研究，因周游世界各国，画风熔中西风格于一炉，赋予了传统中国山水画别开生面的现代气息，并取得重大成就。与此同时，他还抽出许多时间和精力从事烹饮实践。著名画家徐悲鸿在《张大千画集》序中说："大千蜀人也，能治蜀味，性酣高谈，往往入厨作羹飨客。"张大千自己认为："以艺术而论，我善烹饪更在画艺之上。"张大千不仅喜吃善烹，形成了以川味为基础、融会其他美味的"大千风味"系列菜肴，精心设计出了"大风堂酒席"，而且将烹饪艺术与绘画艺术视为同等重要的艺术形式。他的烹饪理论与实践，对中国饮食文化的发展和走向世界都起到了积极的助推作用。

I. Zhang Daqian and Pidu Delicacy

Zhang Daqian (1899-1983) was born in Neijiang, Sichuan province. He is a famous master of calligraphy and painting in modern China. Sichuan Province attaches great importance to delicacy. Zhang Daqian was influenced by this culture and was passionate for delicacy. In his life, he was mainly engaged in the study, creation and research of Chinese traditional calligraphy and painting. As a result of traveling around the world, his painting style melts China and the west, which endows Chinese landscape painting with a modern flavor, and made great achievements. Xu Beihong, a famous painter, said in the preface in *Zhang Daqian's Collection of Paintings*, "Zhang Daqian can book Sichuan dishes as he is local people. He is talkable and usually cooks for guests in person." Zhang Daqian thought that his "cooking was even better than painting". Zhang Daqian not only liked eating and cooking, but also created "Daqian flavor" dishes based on Sichuan flavor and other delicacies. He has designed "Dafengtang feasts", and regarded cooking as equally important painting art forms. Zhang Daqian's cooking theory and practice have contributed much to the development of Chinese cooking and food culture, which also helped the culture to go forward to the world.

张大千与郫都饮食之间的故事发生在抗战期间。当时，张大千举家迁来成都，最初居住在青城山，后迁至郫都犀浦太和场（今郫都区团结镇永定村）的钟家大院，这是一座川西风格的百年老宅，青砖黛瓦、花木遍植、环境清幽，院外不远处就是热闹的大街，茶馆、客栈、饭馆、商铺……应有尽有。赏心悦目、

郎静山拍摄的张大千

交通便利的环境，不仅激发了他的创作热情，更是方便了他与好友宴饮聚会。张大千在作画之余，喜欢去街边的茶馆坐一坐，也喜爱上了闻名遐迩的犀浦酱油和香糟豆腐乳。他以犀浦酱油为主要调料，亲手独创了有名的“大千鱼”，他说：“犀浦酱油色、香、味俱佳，真是名不虚传；香糟豆腐乳质软可口，柔糯带肥，三餐宜备也。”大千先生待人热情大方，非常好客，逢年过节，如端午包粽子、中秋吃月饼，以及从青城山避暑归来，买回核桃、板栗等，都会请邻居一同享用或给每户送一份，共享口福。某年秋，太和场小学教师李英兰请他画两把檀香折扇，他不但应允，而且留她吃午饭，并亲自上灶做了红烧狮子头来款待她。大千先生在寓居太和场4年间，不仅创作了众多名作，而且留下了很多饮食故事，至今还在当地民间广为流传。

The story between Zhang Daqian and Pidu delicacy happened during anti Japanese invasion war. Zhang Daqian and his family moved to Chengdu at that time. Firstly they settled down in Mount Qingcheng, and then moved to the Yard of Zhong Family in Tai Hechang, Xipu, Pixian (currently Yongding village, Tuanjie Town, Pidu District). Yard of Zhong Family was a 100 years house with green brick and wooden structure, which was standard western Sichuan style. The environment was quiet and beautiful, with flowers and fruit trees in the courtyard. Not far away from the courtyard, there were busy streets with tea houses, inns, shops, etc. That wonderful environment not only inspired Zhang Daqian's painting, but also facilitated his party with friends. After painting, Zhang Daqian loved to sit in the tea house nearby and was in found of the famous Xipu soy sauce and fermented tofu at that time. He created “Daqian Fish” which made Xipu soy sauce as the main seasoning. He also noted, “Xipu sauce is well-known for its nice color, smell and taste; fermented tofu tofu is soft and delicious, tasty and a little fatty. It is suitable for three meals a day.” Mr. Daqian is generous and hospitable. At every festival like Dragon Boat Festival, Mid-Autumn Day, he would share with neighbors Zongzi and moon cake. When he returned from Mount Qingcheng, he purchased walnuts and Chinese chestnut and shared with others. One autumn, Li Yinglan, teacher of Taihechang Primary School invited Zhang Daqian to paint two sandalwood folding fans. He not only painted for her but also invited her for lunch, supplied with Braised Lion Head Meatballs cooked by himself. Lived in Taihechang for four years, he not only created many famous paintings, but also left stories relate to delicacy, which have been widely circulated.

二、糖画艺人高邦金

成都的糖画制作在清朝咸丰、同治年间就颇为盛行，当时成都有个老艺人叫廖洪江，徒弟众多，每逢青羊宫花会，他们师徒众人都会齐集二仙庵，彼此交流、相互竞赛，久而久之，糖画之艺日益兴旺起来，而来自郫都安德镇的高邦金就是廖洪江众多徒弟中的佼佼者。

II. Sugar Painting Artist Gao Bangjin

Sugar painting was very popular in Chengdu in Xianfeng and Tongzhi Periods of the Qing Dynasty. There was an old artist named Liao Hongjiang with lots of apprentices. During flower fair in Qingyang Palace, Liao and his apprentices gathered in Erxian'an Taoists Temple to compete. Sugar painting has become prosperous since then. Gao Bangjin was one of the best apprentices of Liao.

“糖画”又称“倒糖饼”，是四川乃至西南地区人们喜爱的民间工艺食品之一，因所“倒”之物包罗万象而得名。糖画艺人一挑担子，所有家什尽在其中，两张矮方桌，面前一张放上一块光洁的石板，右边放一个熬糖的小火炉儿，左边设一转盘，摊顶张一把大油布伞，摊边立个草把儿，上插制作好的各式作品招徕顾客。转盘上画有龙、凤、鸟、鱼等动物图案，盘中一指针，用手轻轻一拨即快速旋转，待指针停下，指龙“倒”龙，指鱼“倒”鱼，孩子们尤喜得“彩”，而大人们更在意的是欣赏艺人的精湛技艺。

Sugar painting, also called as “drawing sugar cake”, is one of the favorite folk snacking in Sichuan and the whole

Southwestern China. It is famous for painting everything. Sugar painting artists put all issues into a load. There are two low square tables. A smooth slate is placed in front of one, a small stove for boiling sugar is placed on the right, a turnplate is set on the left, a large oilcloth umbrella is placed on the top of the booth, a straw handle is set at the edge of the booth, and all kinds of works are on it to attract customers. On the surface of turnplate, there are pictures of dragon, Phoenix, bird, fish, etc. There is a pointer in the turnplate for people to rotate. When it stops and points to the dragon, you can get the dragon sugar painting. When it points to the fish, you get the fish sugar painting. Children pay more attention to what they can rorate, while adults are more concerned about appreciating the exquisite skills of artists.

高邦金十分热爱糖画艺术，师从廖洪江从事糖画制作。改革开放后，他辞去粮店工作，开始上街摆起了糖画摊。随着技艺的不断提高，特别是随着人们文化水平的提升和审美情趣的变化，高邦金对糖画的传统造型越来越不满足，决心开拓新的题材，改革传统工艺。他购买了大量的素描、速写、国画等画册，潜心研究，把传统国画的用笔技法巧妙地运用于糖画艺术中。1983年，中国民间美术博物馆馆长曹振峰一行来到郫都采风，在看到高邦金的精湛技艺后，将其事迹在《新观察》发表，引起了国内外众多糖画爱好者的追捧，郫都区政府也因此在望丛祠内为其专设了一个摊位。时至今日，高邦金糖画技艺作为郫都民俗文化的代表，不断传承，为弘扬中华优秀传统文化做出了重要贡献。

Gao Bangjin loves sugar painting very much and learned from Liao Hongjiang. After reform and opening up, he quit the job in grain store and opened a sugar painting booth on street. With the development of skill, especially with the improvement of people's cultural level and the change of aesthetic taste, Gao Bangjin was more and more dissatisfied with the traditional painting style, and is determined to explore new themes and reform the traditional technology. He purchased a large number of sketch, traditional Chinese painting and other albums, devoted himself to research, and applied the skills of traditional Chinese painting to the art of sugar painting. In 1983, Cao Zhenfeng, curator of Chinese folk art museum, collected folk arts in Pidu. After seeing Gao Bangjin's exquisite skills, he published an article about this in *New Observation*, which attracted many sugar painting lovers at home and abroad. Government of Pidu District provided a booth for him inside Wangcong Temple then. Till now, as the representative of Pidu's folk culture, Gao Bangjin's sugar painting has been continuously inherited and made important contributions to carrying forward Chinese traditional culture.

三、艺传三代的名厨钟泽先

钟泽先，1937年出生于郫都安德镇的一个餐饮世家，20世纪80年代后在全国烹饪大赛中多次获奖，被评为特三级烹调师，是郫都名厨的优秀代表。

III. Zhong Zexian — Three Generations of Famous Chef

Zhong Zexian was born in a catering family in Ande Town, Pidu District in 1937. In 1980s, he won many awards in the national cooking competition and was awarded as a special third-class cook. He is a typical representative of Pidu famous chefs.

钟泽先的父亲钟光龙曾学艺于成都“隆盛园”，精于红、白两案，1931年在安德镇开了一家饭馆，生意很好。抗战期间，社会名流邵从恩经常光顾该店，对他的菜肴极为称赞。20世纪30年代，钟光龙被誉为五县烹饪界四条金龙之一。钟泽先幼时从父学艺，十多岁时便已能立灶烹饪、独立操作。钟泽先不善言辞，但头脑灵活，做事认真，也善于向他人学习。在钟光龙的潜心培养下，钟泽先的烹饪技艺日渐精进，红、白两案均有很高造诣。

Zhong Guanglong, father of Zhong Zexian, learned cooking in Longshengyuan Restaurant in Chengdu. He was good at cooking meat dishes and pastry, and opened a restaurant in Ande Town. During anti-Japanese invasion war, celebrity Shao Cong'en often visited his restaurant and praised his dishes. In 1930s, Zhong Guanglong was awarded one of the 4 masters in cooking industry of five counties. Zhong Zexian learned to cook from his father when he was young. He could cook dishes independently as a teenager. Although not good at talking, he was clever, careful and like to learn from others. With the training of Zhong Guanglong, Zhong Zexian's skill had been improved greatly and got new achievement in making meat dishes and pastry.

从1956年至20世纪70年代，钟泽先在成都军区司令部管理局、成都军区招待所事厨并任厨师长，其精湛的厨艺得到首长们的交口称赞。进入20世纪80年代后，各种烹饪赛事不断兴起，在1988年2月川菜第一届“旭水杯”烹饪技术比赛中，钟泽先获得冷菜金牌；1988年，在全国第二届烹饪大赛中，解放军组团参加大赛，钟泽先一人就获得五枚奖牌。鉴于钟泽先在烹饪大赛中为解放军代表团取得的优异成绩和获得的诸多荣誉，成都军区司令部管理局为钟泽先荣记了二等功。

From 1956 to 1970s, Zhong Zexian worked as head chef of Administration Bureau of the Headquarters and Guest House of Chengdu Military Region. His exquisite skill have been praised by high level military officers. In 1980s, different kinds of Sichuan cuisine cooking contests were organized. Zhong Zexian had won lots of rewards. In the first “Xu Shui Cup” Cooking Contest held in February, 1988, Zhong Zexian got the golden medal for cold dishes. In the Second National Cooking Contest, Zhong Zexian won 5 medals as the representative of Chinese People's Liberation Army (PLA). In consideration of Zhong Zexian's outstanding achievements and important contributions to the PLA delegation in cooking competition, the Administration Bureau of the Headquarters of Chengdu Military Region recorded second-class merit for Zhong Zexian.

四、川菜厨艺有传人

中华人民共和国成立后，既让郫都餐饮业进入到一个新的发展时期，也同时造就了一支优秀的烹饪名厨队伍，并且一直薪火相传。

IV. Successors of Sichuan Cuisine

After the foundation of the People's Republic of China, the catering industry of Pidu has entered new development stage. A team of famous chefs has been formed and cooking skills been inherited .

在20世纪70年代前后，除了钟泽先外，郫都还涌现了黄云福、任福奎、吴观远、吴观元、刘绪麟、李德明、廖洪述、廖明春、张育弼等一大批烹饪大师或名师。他们不仅练就了精湛的烹饪厨艺，积累了丰富的烹饪专业知识，而且还将这些宝贵财富传授给众多徒弟，培养了一大批优秀的川菜烹饪人才。如任福奎拜黄云福为师，而他又带出了吴观元、刘绪麟等多位优秀的徒弟。李德明曾担任成都市餐饮服务业技术委员会委员和成都市西城区饮食技术委员会副主任，多次参与行业技术培训，担任职称考核评委，并为编写《中国名菜谱（七）》提供宝贵资料。李德明的多位徒弟都成了特级厨师。刘绪麟曾先后在郫都早期比较著名的四季春饭店、迎宾饭店、嘉宾饭店担任主厨，长期培训和指导餐饮工作人员，他们通过多种形式培养的大批烹饪人才，为改革开放后郫都餐饮业的快速发展，打下了良好的人才储备基础。

Around 1970s, in addition to Zhong Zexian, a large number of famous chefs such as Huang Yunfu, Ren Fukui, Wu Guanyuan, Wu Guanyuan, Liu Xulin, Li Deming, Liao Hongshu, Liao Mingchun and Zhang Yubi also emerged in Pidu. They have got exquisite cooking skills, accumulated rich cooking sills, imparted those knowledge to numerous apprentices, and trained large quantity of excellent talents for Sichuan cuisine. For example, Ren Fukui learned from Huang Yunfu. He also taught lots of good students like Wu Guanyuan, Liu Xulin. Li Deming was member of Chengdu

Catering Service Industry Technical Committee and deputy director of Chengdu Xicheng District Catering Technical Committee. He has participated in industrial training for many times, worked as a judge for professional title assessment, and provided valuable materials for the *Chinese Famous Recipes (VII)*. Many apprentices of Li Deming have become famous chefs. Liu Xulin once worked as head chef in Sijichun Restaurant, Yingbin Restaurant, Jiabin Restaurant. He has trained and guided staffs in culinary industry for a long time. Those talents trained by them have established the foundation for the rapid development of culinary industry in Pidu after reform and opening up.

改革开放后，尤其是进入21世纪以来，郫都餐饮业的繁荣发展，与一大批活跃在郫都餐饮企业的烹饪名厨密切相关，如官燎、康龙虎、胡小勇、黄其华、顾云彬、陈洪光、谭军、童逊、杨挺等，他们在继承传统的基础上勇于创新，既为郫都川菜产业的发展做出了重要贡献，同时又为郫都培养了大批后备烹饪人才。

After reform and opening up, especially in 21st century, the prosperity and development of catering industry in Pidu are closely related with numerous famous chefs active in catering enterprise in Pidu, such as Guan Liao, Kang Longhu, Hu Xiaoyong, Huang Qihua, Gu Yunbin, Chen Hongguang, Tan Jun, Tong Xun, Yang Ting, etc. While by inheriting traditional dishes and keeping innovating, they not only made contributions to the development of Pidu's Sichuan cuisine industry, but also trained a large number of reserve cooking talents for Pidu.

庄生晓梦迷蝴蝶

望帝春心托杜鹃